Discretization in Differential Equations and Enclosures

Mathematical Research · Mathematische Forschung

Wissenschaftliche Beiträge
herausgegeben von der
Akademie der Wissenschaften der DDR
Karl-Weierstraß-Institut für Mathematik

Band 36

Discretization in Differential
Equations and Enclosures

Discretization in Differential Equations and Enclosures

edited by

Ernst Adams
Rainer Ansorge
Christian Großmann
Hans-Görg Roos

Akademie-Verlag Berlin 1987

Herausgeber:
Prof. Dr.-Ing. Ernst Adams,
Universität Karlsruhe, Institut für Angewandte Mathematik

Prof. Dr. Rainer Ansorge
Universität Hamburg, Institut für Angewandte Mathematik

Prof. Dr. Christian Großmann
Doz. Dr. Hans-Görg Roos
Technische Universität Dresden, Sektion Mathematik

Die Titel dieser Schriftenreihe werden vom Originalmanuskript der Autoren reproduziert

ISBN 3-05-500362-4
ISSN 0138-3019

Erschienen im Akademie-Verlag Berlin, DDR-1086 Berlin, Leipziger Str. 3-4

Lizenznummer: 202 · 100/409/87
Printed in the German Democratic Republic
Gesamtherstellung: VEB Kongreß- und Werbedruck, 9273 Oberlungwitz
Lektor: Dr. Reinhard Höppner
LSV 1065
Bestellnummer: 763 734 0 (2182/36)

03200

P R E F A C E

Engineering analysis rests on mathematical modelling and numerical simulation. In this context, differential equations are of particular interest. In simulation, procedural and rounding errors are unavoidable. Additionally, there may be the problem of data imprecision. The availability of more and more powerful computers leads to the employment of well known numerical methods in the case of very large and very complex systems, which may produce a significant uncontrolled and unacceptable error propagation. This is a challenge to analyse the propagation of various kinds of errors and, if possible, to generate sharp error bounds or at least realistic quantitative and safe error estimations. In contrast to the determination of an approximation followed by the one of an a posteriori error estimate, a more advantageous approach is the construction of algorithms yielding directly solution enclosures. The practical determination of enclosures of solutions of differential equations requires not only a reduction to a finite-dimensional setting but also a suitable computer arithmetic. This is provided by recent developments of computer languages such as PASCAL-SC.

The present volume is concerned with a collection of original papers investigating the problem area outlined above and related fields. Prior to this publication, the majority of the papers have been presented and discussed at the ISAM-86 workshop on "Numerical solution of differential equations with emphasis to the generation of two-sided bounds" held in Weißig, March 10th - 14th 1986.

The subjects of the various papers included here range from the presentation and the analysis of new methods generating bounds for the solution of differential equations to effective techniques for solving finite-dimensional subproblems arising in discretization. The authors of the following articles treat the numerical analysis of the generation of bounds for solutions of ordinary differential equations: ADAMS/CORDES/LOHNER, BAUCH/WIEBIGKE, GROSSMANN, TOBISKA and SHOKIN. Furthermore, ROOS extends some investigations to 1D-parabolic problems. In the paper of ADAMS/CORDES/LOHNER, also computer realizations using PASCAL-SC are presented and applications of the developed methods to a problem in mechanical engineering. The problem of adapted error estimates is considered by FRÜHNER, LEHMANN, SCHOLZ and SENS. Especially the numerical solution of stiff systems of ordinary differential equations requires specific techniques. Such tasks are the subject of the papers of SCHOLZ, SPIJKER, STREHMEL/WEINER. Monotonicity principles play an essential role in getting bounds. In the paper of WINDISCH, the author investigates several discrete maximum principles.

Appropriate approximations of practically important problems in electronics and fluid dynamics are considered by ČERMAK and HEMKER, respectively. In this context, HEMKER and MAREK use multigrid- and aggregation techniques, respectively. ANSELONE/ANSORGE investigate general convergence behavior of discretization methods. BURMEISTER/DIETZE/HESS/SCHMIDT consider effective methods for solving finite-dimensional problems which also arise in discretization.

The authors would like to thank their colleagues for the stimulating discussions and the fruitful exchange of ideas at the ISAM-86 workshop. The editors are grateful to the Akademie-Verlag, especially Dr.Höppner, for the permanent interest in our work and the kind support.

October 1986 The editors

C O N T E N T S

ENCLOSURE OF SOLUTIONS OF ORDINARY INITIAL VALUE PROBLEMS AND APPLICATIONS

E. Adams, D. Cordes, R. Lohner [1)]

This Paper is dedicated to Professor W.F. Ames
on the occasion of his 60th birthday.

Summary

Linear or nonlinear ordinary initial value problems with explicit differential equations $y' = f(t,y,c)$ are considered without or with intervals for the initial vector $y(t_o)$ or the parameter vector c. Componentwise bounds $\underline{y}$ and $\overline{y}$ of the solution(s) y^* are determined by use of the mean value theorem, which provides a representation of y^* in terms of parameters standing for local input or procedural errors. Several examples are given. If f is independent of t or depends periodically on t, the uniform boundedness of y^* can be verified, making use of $\underline{y},\overline{y}$, if an enclosure condition is satisfied. Applications are concerned with Mathieu's equation and vibrations of gear drives.

1. Introduction

In this paper linear or nonlinear systems of explicit ODEs (= ordinary differential equations) $y' = f(t,y,c)$ with initial condition $y(t_o) = \eta$ are considered. The input vectors c or η are either fixed or they are arbitrary within given intervals. Vectors $\underline{y},\overline{y}$ will be determined whose components $\underline{y}_i$, $\overline{y}_i$ are bounds for the components y_i^* of the solution y^* or every solution $y^*(c,\eta)$. The bounding property holds with respect to the influences of all local errors, i.e., procedural, rounding, and input errors. Since $\overline{y}-\underline{y}$ is positive in general, a set of values $S_v \subset \mathbb{R}^n$ is simultaneously enclosed by $\underline{y}$ and $\overline{y}$. This is true even in the case of fixed vectors c and η. Consequently, there is a "blow-up" influence in the construction of $\underline{y},\overline{y}$ favoring an exponential growth of $\overline{y}-\underline{y}$ as t increases. If $y'=f(t,y,c)$ is a system in $\mathbb{R}^n$ with $n>1$, the blow-up influence is enhanced by the fact that $[\underline{y}(t),\overline{y}(t)] \subset \mathbb{R}^n$ is an interval, whereas ∂S_v may possess any shape. Consequently, enclosure methods must be sophisticated in order to achieve sufficiently small overestimates. Since the property of inverse monotonicity or any other monotonicity are exceedingly rare in mathematical simulation, assumptions on the presence of this or any other such property will not be made subsequently. The actual construction of $[\underline{y},\overline{y}]$ is obviously restricted to finite intervals of time t. The uniform boundedness for all $t>0$ of $\underline{y},\overline{y}$ and

---

1) Institut für Angewandte Mathematik, Universität Karlsruhe,
D-7500 Karlsruhe, Kaiserstraße 12.

thus y^*, can be verified if an enclosure condition is satisfied. This will be carried out in the case of linear ODEs with periodic coefficients, with applications in the mathematical simulation of gear drive vibrations.

Due to a lack of space, the following aspects are not discussed subsequently: (i) the relationship with enclosure methods for IVPs in papers by other authors or (ii) the construction of enclosures for distribution functions over $[\underline{y}_i(t),\overline{y}_i(t)]$ in the case of input intervals for c or η, see [25],[26].

This paper presupposes a working knowledge of interval arithmetics, see for example [3],[20],[21]. An expression in brackets will denote an interval, e.g., $[a]:=[\underline{a},\overline{a}]$ with $a,\underline{a},\overline{a}\in\mathbb{R}$ or $\mathbb{R}^n$. Correspondingly, $[A]:=[\underline{A},\overline{A}]$ with $A,\underline{A},\overline{A}\in L(\mathbb{R}^n)$ where $L(\mathbb{R}^n)$ is the set of real n×n-matrices. For any function $f:\mathbb{R}^n\to\mathbb{R}^m$, the symbol $f([a])$ denotes an interval containing the set of values $\{f(a)\,|\,a\in[a]\}$.

2. Construction of Enclosures for Solutions of Ordinary IVPs

2.1 The Admitted IVPs

The following class of vectorial autonomous systems of ODEs is considered:

(2.1) $\quad u' = f(u)$ for $t\in\mathbb{R}_0^+$, $f:D\to\mathbb{R}^n$, $D\subset\mathbb{R}^n$ is open, $f\in C^p(D)$.

In view of applications, f is assumed to be a composition of rational and certain standard functions. Here, $C^p(D)$ denotes the set of functions defined on D whose derivatives up to and including order p exist and are continuous. All initial vectors

(2.2) $\quad u(0)\in[u_0]:=\tilde{u}_0+[z_0]$, $[u_0]\subset D$, $\tilde{u}_0\in[u_0]$ is fixed,

are admitted where $[z_0]$ and $\tilde{u}_0$ are given.
In the case of non-autonomous systems, the variable t is replaced systematically by use of the following extension of (2.1):

(2.3) $\quad u_{n+1}:=t,\ u'_{n+1}=f_{n+1}(u):=1,\ u_{n+1}(0)=0.$

If f as given in Section 1 depends on a parameter vector $c\in[c]\subset\mathbb{R}^m$, (2.1), (2.2) may be extended as follows:

$$(2.4)\quad \left\{\begin{array}{l} v:=\begin{bmatrix}u\\c\end{bmatrix}\in\mathbb{R}^{n+m},\ g(v):=g\left(\begin{bmatrix}u\\c\end{bmatrix}\right):=\begin{bmatrix}f(u,c)\\0\end{bmatrix}\in\mathbb{R}^{n+m},\\[2ex] v(0)\in\begin{bmatrix}[u_0]\\ [c]\end{bmatrix},\ v'=g(v).\end{array}\right.$$

These extensions (2.3) or (2.4) generate a convenient fixed system of ODEs (2.1); intervals (if there are any) appear only in the initial conditions. Subsequently, it is assumed that the system has been extended, if that was necessary.

According to Section 1, functions $\underline{u},\overline{u}$ are to be determined with the following bounding properties:

(2.5) $u^*(t)\in[u(t)]:=[\underline{u}(t),\overline{u}(t)]\subset D$ for all $t\in[0,T]$ and all $u(0)\in[u_o]$,

where T is given or the time of a truncation of the algorithm, e.g., because of exponent overflow. Beginning here, the superscript * will be omitted.

2.2 The Enclosure Method Using Exact Interval Arithmetic

In order to simplify discussions, an exact interval arithmetic is assumed in the present and the following subsection. The employment of rounded interval arithmetic will be discussed in Subsection 2.4. A not necessarily equidistant time grid is adopted, with grid points $t_o=0<t_1<\ldots<t_j<$ $<\ldots$ and the step sizes $h_j:=t_{j+1}-t_j$. For convenience of the discussion, from now on, the grid is assumed to be equidistant with constant step size h. An explicit one-step method of order p is adopted, with the local discretization error z_{j+1}/h :

(2.6) $u(t_{j+1})=u(t_j)+h\Phi(u(t_j))+z_{j+1},\ j+1\in\mathbb{N},\quad u_o=u(0)\in[u_o],\ z_o=0,$

where Φ should be (i) continuously differentiable for all $u\in D$ and (ii) chosen such that z_{j+1} can be expressed easily in terms of u and/or derivatives of u. If Φ is generated by means of a Taylor-expansion of $u_{j+1}:=u(t_{j+1})$ in a neighborhood of the point of expansion t_j, the following basis of the construction to be presented is obtained:

$$(2.7)\quad \begin{cases} z_{j+1}=\dfrac{h^p}{p!}u^{(p)}(\tau_{j+1})=\dfrac{h^p}{p!}f^{(p-1)}(u(\tau_{j+1})),\ \tau_{j+1}\in(t_j,t_{j+1}),\\ u^{(p)}:=d^pu/dt^p, \qquad p\in\mathbb{N}\text{ is fixed},\\ f^{(o)}(u):=f(u),\ f^{(1)}(u):=u''=f'(u)f(u),\ldots,\ f^{(p-1)}(u):=u^{(p)}. \end{cases}$$

Due to discussions following (2.26), there is no need for a formal derivation of the sequence of functions $f^{(v)}$. According to [1],[2],[16], the basic idea of the following construction is as follows: the sequences $\{u_1,u_2,\ldots\}$ with $u_k:=u(t_k)$ and $\{z_1,z_2,\ldots\}$ in (2.6) are embedded in families of such sequences. For the vectors in these families, enclosing intervals will be determined such that the special vectors u_k and z_k in (2.6) are contained in these intervals. The new family of vectors $u_1,u_2,\ldots$ is defined as follows by use of the new sequences of vectors:

(2.8) $u_{j+1}=u_{j+1}(z_o,\ldots,z_{j+1}):=u_j+h\Phi(u_j)+z_{j+1}$ for $j+1\in\mathbb{N},\ u_o\in[u_o]$.

With respect to the component $u_{i,j+1}$ of u_{j+1}, the classical mean value theorem yields, for $i=1(1)n$,

$$(2.9)\quad \left\{ u_{i,j+1}=\tilde{u}_{i,j+1}+\sum_{k=0}^{j+1}\frac{\partial u_{i,j+1}(\hat{z}_o^{(i,j+1)},\ldots,\hat{z}_{j+1}^{(i,j+1)})}{\partial z_k}(z_k-s_k)\right.$$

where (a) the unknown intermediate vectors $\hat{z}_\ell^{(i,j+1)}$ depend on the component index i and on the time index j+1, (b) the fixed vectors $s_o,\ldots,$

s_{j+1} will be chosen in (2.16),(2.27) and _(c)_ $\tilde{u}_{j+1}$ is defined as follows:

(2.10) $\tilde{u}_{j+1} = \tilde{u}_{j+1}(s_0,\ldots,s_{j+1}) := \tilde{u}_j + h\Phi(\tilde{u}_j) + s_{j+1}$ for $j+1\in\mathbb{N}$, $s_0=0$,

by use of $\tilde{u}_0$ from (2.2). Because of (2.8), the Fréchet derivatives in (2.9) can be expressed as follows:

$$(2.11)\quad \frac{\partial u_{j+1}}{\partial z_k} = \begin{cases} (I+h\Phi'(u_j))\dfrac{\partial u_j}{\partial z_k} & \text{for } k=0(1)j, \\ I & \text{for } k=j+1, \end{cases}$$

where $I\in L(\mathbb{R}^n)$ denotes the identity matrix. By use of

$$(2.12)\quad A_k := I+h\Phi'(u_k) \text{ and } A_{j+1,k} := \frac{\partial u_{j+1}(\ldots)}{\partial z_k} \quad \text{for } k=0(1)j+1$$

(where the intermediate argument has been omitted),

$$(2.13)\quad A_{j+1,k} = A_j\cdot A_{j,k} = \ldots = \begin{cases} A_j\cdot A_{j-1}\cdots A_k & \text{if } k\le j, \\ I & \text{if } k=j+1. \end{cases}$$

In the following, interval matrices $[A_k]$ will be introduced which contain _(i)_ the set of all matrices A_k for the subsequently admitted vectors z_0, $z_1,\ldots,z_k$, see (2.2),(2.26) and _(ii)_ the special A_k depending on the unknown intermediate vectors $\hat{z}^{(a,b)}$ in (2.9). Due to (2.8)-(2.13),

$$(2.14)\quad u_{j+1} = \tilde{u}_j + h\Phi(\tilde{u}_j) + \sum_{k=0}^{j} A_{j+1,k}(z_k - s_k) + z_{j+1} \quad \text{for } j+1\in\mathbb{N},$$

where s_{j+1} has dropped out.

In the spirit of an induction, it is assumed that the following intervals or vectors are known:

(2.15) $[u(t)]$ with the property (2.5) for all $t\in[0,t_j]$,

(2.16) $[z_k]\subset\mathbb{R}^n$ and $s_k = m([z_k])$ for $k=0(1)j$, and

(2.17) $[A_k]\subset L(\mathbb{R}^n)$ for $k=0(1)j-1$,

where $m([a])$ denotes the midpoint of an interval $[a]\subset\mathbb{R}^n$.

The step from j to $j+1$ of the induction is started with

$$(2.18)\quad [A_j] := I+h\Phi'([u_j]).$$

An algorithm for the computation of the Fréchet derivative Φ' and for its interval evaluation is presented at the end of this subsection. Due to (2.2), the intervals $[u_0]$ and $[z_0]$ are known at time $t=0$. In order to continue the construction of $[u(t)]$ for $t\in[t_j,t_{j+1}]$, a constant "a priori interval" $[u^0_{j+1}]\subset\mathbb{R}^n$ is chosen such that $[u_j]\subset[u^0_{j+1}]$. Since every IVP admitted in (2.1),(2.2) is equivalent with a Volterra integral equation, there holds

$$(2.19)\quad u(t) = u_j + \int_{t_j}^{t} f(u(\sigma))\,d\sigma$$

for all solutions $u_j = u(t_j)$ of (2.1),(2.2). Since f consists of a composition of rational and certain standard functions, the following interval can be determined by use of interval arithmetic:

(2.20) $$[u^1_{j+1}] := [u_j] + [0,h]f([u^0_{j+1}])$$

If the "enclosure condition"

(2.21) $$[u^1_{j+1}] \subset [u^o_{j+1}]$$

is satisfied, the Banach fixed point theorem asserts that there holds for the solution u of (2.1),(2.2), [28,p.80]

(2.22) $u(t,u(0)) \in [u^1_{j+1}]$ for all $u(0) \in [u_o]$ and all $t \in [t_j, t_{j+1}]$,

due to the assumption of the induction at time $t=t_j$. If (2.21) is not true, either a smaller $h=t_{j+1}-t_j$ is used or an "ε-inflation" is employed which consists in the new choice of

(2.23) $[u^o_{j+1}] := (1+\varepsilon)[u^1_{j+1}] - \varepsilon[u^1_{j+1}] \Rightarrow m([u^o_{j+1}]) = m([u^1_{j+1}])$,

by use of a suitably small $\varepsilon \in \mathbb{R}^+$.

The interval enclosure $[u(t)]=[u^1_{j+1}]$ obtained so far for $t \in [t_j, t_{j+1}]$ can be refined as follows:

(A) An interval $[u_{j+1}]$ is determined

(2.24) $$[u_{j+1}] = \tilde{u}_j + h\Phi(\tilde{u}_j) + [\, A_j \sum_{k=0}^{j} A_{j,k}(z_k - s_k) + z_{j+1}\,],$$

where the evaluation of the right hand side will be discussed in Subsection 2.3.

(B) A "forward facing cone" is defined by use of (2.19):

(2.25a) $[c_{j+1}(t)] := [u_j] + (t-t_j)\, f([u^1_{j+1}])$ for $t \in [t_j, t_{j+1}]$.

(C) A "backward facing cone" is defined analogously:

(2.25b) $[d_{j+1}(t)] := [u_{j+1}] - (t_{j+1} - t)\, f([u^1_{j+1}])$ for $t \in [t_j, t_{j+1}]$.

(D) A new interval $[u^o_{j+1}]$ is chosen as the smallest interval containing the intersection $[c_{j+1}(t)] \cap [d_{j+1}(t)] \cap [u^1_{j+1}]$ for all $t \in [t_j, t_{j+1}]$.

If the new interval $[u^o_{j+1}]$ is strictly contained in the interval $[u^1_{j+1}]$ an iteration can be started with (2.20).

This new interval $[u^o_{j+1}]$ will now be employed to generate the interval evaluation of the right hand side of (2.7) to obtain $[z_{j+1}]$. For a computation of $[z_{j+1}]$, there is no need to determine formally the functions $f^{(1)}, \dots, f^{(p-1)}$ in (2.7). In fact, an algorithm in [21,p.24-29] yields interval enclosures of the values of the derivatives of any order of a sufficiently smooth function $\varphi: \mathbb{R} \to \mathbb{R}$ for either an input $x \in \mathbb{R}$ or $[x] \subset \mathbb{R}$. The new interval $[u^o_{j+1}]$ generated in step (D) may thus be used in order to determine recursively interval enclosures for the components of the functions $f^{(1)}, \dots, f^{(p-1)}$ defined in (2.7). As a result of this recursion, $f^{(p-1)}([u^o_{j+1}])$ has been enclosed. By construction, the interval

(2.26) $$[z_{j+1}] := (h^p/p!)\, f^{(p-1)}([u^o_{j+1}])$$

contains the product of h and the unknown local discretization error, see (2.6). According to (2.16), s_{j+1} is chosen as follows:

$$(2.27) \qquad s_{j+1} := m([z_{j+1}]).$$

The induction starting with (2.15)-(2.18) has thus been completed.

The Taylor-expansion of the solution u in a neighborhood of t_j with a remainder term of order p gives rise to the following final Spline-like interval enclosure of the solution(s) u of (2.1),(2.2)

$$(2.28) \begin{cases} u(t) \in [u_j] + (t-t_j)[u'_j] + \ldots + \dfrac{(t-t_j)^{p-1}}{(p-1)!}[u_j^{(p-1)}] + \\ \qquad + \left[\dfrac{t-t_j}{h}\right]^p [z_{j+1}] \quad \text{for } t\in[t_j,t_{j+1}]. \end{cases}$$

Here, $[u_j^v]$ is an enclosure of $d^v u(t_j)/dt^v$.

An expression for Φ' in (2.18) may be obtained by means of a Taylor-expansion of the solution u of (2.1),(2.2),

$$(2.29) \qquad u_{j+1} = u_j + hf^{(o)} + \ldots + (h^{p-1}/(p-1)!)f^{(p-2)} + z_{j+1},$$

making use of (2.7) and an evaluation of $f^{(o)},\ldots,f^{(p-2)}$ at u_j.

Differentiation of $u'=f(u)$ with respect to u(0) yields the matrix-ODE

$$(2.30) \qquad U' = \frac{\partial f(u)}{\partial u} U, \quad U := (\frac{\partial u_i}{\partial u_j(0)}).$$

Analogously to (2.29),

$$(2.31) \quad U_{j+1} \approx U_j + hU'_j + \ldots + (h^{p-1}/(p-1)!)U_j^{(p-1)}, \quad U_j := U(t_j).$$

Due to an induction, and with an evaluation of F,F',... at u_j,

$$(2.32) \begin{cases} U^{(v)} = F^{(v-1)}U, \quad F^{(o)} := \dfrac{\partial f(u)}{\partial u}, \quad F^{(v)} := \dfrac{\partial f^{(v)}}{\partial u} = \dfrac{\partial F^{(v-1)}}{\partial u} f^{(o)} + F^{(v-1)}F^{(o)} \\ \quad \text{for } v=1(1)p-1 . \end{cases}$$

A differentiation of (2.29) with respect to z_k, for $k \le j$, yields

$$(2.33) \quad \frac{\partial u_{j+1}}{\partial z_k} = B_j \frac{\partial u_j}{\partial z_k}, \qquad B_j := I + hF^{(o)} + \ldots + (h^{p-1}/(p-1)!)F^{(p-2)}.$$

A comparison of (2.11) with (2.33) shows that $A_j = B_j$. The interval $[A_j]$ is obtained via the determination of interval enclosures of $F^{(o)},\ldots,F^{(p-2)}$ at $t=t_j$ making use of <u>(i)</u> the final interval enclosure $[u_j]$ which is due to (2.24) (as applied to the preceding time step) and <u>(ii)</u> the algorithm in [21,p.24-29] which is also used in the generation of $[z_{j+1}]$ in (2.27).

<u>2.3 Propagation of the Influence of the Local Errors</u>

According to (2.9) and (2.12),

$$(2.34) \qquad r_{j+1} := u_{j+1} - \tilde{u}_{j+1} = \sum_{k=0}^{j+1} A_{j+1,k}(z_k - s_k),$$

where r_{j+1} expresses the influence of the preceding local error vectors $z_o,\ldots,z_j$, in addition to the influence of the present error vector z_{j+1}. The bounds $\underline{u}_{j+1}, \overline{u}_{j+1}$ are sensitive with respect to the construction of the required enclosure of r_{j+1}. Here, four alternative methods (I)-(IV) are discussed, which are mainly concerned with the evaluation of the right hand side of (2.24).

<u>Enclosure Method (I), Via an "Interval Vector"</u>

According to (2.2) or (2.9), (2.13), respectively,

(2.35) $\quad [r_0] := [z_0]-s_0, \; [r_{j+1}] := [A_j][r_j]+[z_{j+1}]-s_{j+1}$ for $j \geq 0$.

Whereas the storage and computational requirements are small, an enclosure $[u_{j+1}]$ due to (2.24),(2.35) is too crude in many cases. This is particularly true in the presence of oscillatory solutions, see Section 3. According to numerical experience, the enclosure method (II) is advantageous in the case of oscillatory solutions with little damping since the sets of vectors r_{j+1} to be generated are suitable parallelepipeds. For the class of solutions under consideration, these sets generally admit a closer outer approximation of the set of solutions than intervals with respect to a time-independent basis of $\mathbb{R}^n$. The idea of a parallelepiped as an enclosure has been used before, see in particular [8]. This was preceded by the related idea of the employment of a suitably rotated basis of $\mathbb{R}^n$, see [11] and [20].

Methods (II) and (III) are defined as follows:

$$(2.36) \quad \begin{cases} r_0 \in B_0[\hat{r}_0], \; B_0 := I, \quad [\hat{r}_0] := [z_0]-s_0, \\ r_{j+1} \in B_{j+1}[\hat{r}_{j+1}], \\ [\hat{r}_{j+1}] := (B_{j+1}^{-1}[A_j]B_j)[\hat{r}_j] + B_{j+1}^{-1}([z_{j+1}]-s_{j+1}) \text{ for } j \geq 0. \end{cases}$$

where different choices of regular matrices B_{j+1} will be made in the subsequent presentations of methods (II) and (III).

As compared with (II) and according to numerical experience, the enclosure method (III) is advantageous under more general conditions, e.g., in the case of oscillatory solutions with considerable damping.

<u>Enclosure Method (II), Via a "Parallelepiped"</u>

Any fixed regular choice of $B_{j+1} \in [A_j]B_j$ is admitted, e.g.,

(2.37) $$B_{j+1} := m([A_j]B_j)$$

or a machine approximation of the right hand side. This choice of a fixed matrix B_{j+1} is motivated as follows:

<u>(i)</u> For any $j \in \mathbb{N}$, (2.36) generates an interval matrix $B_{j+1}^{-1}[A_j]B_j$ which is almost diagonal and contains the identity matrix provided B_{j+1} is sufficiently well-conditioned and $\overline{A}_j-\underline{A}_j$ is sufficiently small. Then, there is an almost negligible overestimate in the interval-arithmetic evaluation of the underlined part of the following expression:

(2.38) $$[\hat{r}_{j+1}] = \underline{(B_{j+1}^{-1}[A_j]B_j)[\hat{r}_j]} + B_{j+1}^{-1}([z_{j+1}]-s_{j+1}).$$

(ii) The affine mapping of $\hat{r}_{j+1}\in[\hat{r}_{j+1}]$ by B_{j+1} generates the set

(2.39) $$S_{j+1}:=\{r_{j+1}=B_{j+1}\hat{r}_{j+1} \mid \hat{r}_{j+1}\in[\hat{r}_{j+1}]\}$$

of "remainder vectors" r_{j+1} which is a parallelepiped in $\mathbb{R}^n$ whose basis vectors are parallel to the column vectors of B_{j+1}.

(iii) In the step from t_j to t_{j+1}, $[\hat{r}_o],\ldots,[\hat{r}_j]$ are lumped into $[\hat{r}_{j+1}]$. Provided conditions corresponding to (i) are satisfied for $k=0(1)j$, the premultiplication of $[\hat{r}_{j+1}]$ by B_{j+1} generates a relatively favorable enclosure of the set of vectors r_{j+1}, according to numerical experience. In the case of the interval $B_{j+1}^{-1}([z_{j+1}]-s_{j+1})$, this is not necessarily true. Since $([z_{j+1}]-s_{j+1})$ is relatively small, this is irrelevant, provided B_{j+1} is not ill-conditioned.

Any fixed choice $B_{j+1}\in[A_j]B_j$ does not necessarily yield a regular matrix or an acceptably well conditioned matrix. This is immediately seen by means of a linear system (2.1). Then, B_{j+1} is an approximation of the fundamental matrix of this system, at time t_{j+1}. If any one of the chosen matrices $B_{j+1}\in[A_j]B_j$ is ill-conditioned, an application of the following enclosure method (III) is advantageous. Whereas a presentation of (III) will appear in [19], here only the basic idea is stated in (2.40) and a numerical application is given in Example 4.2.

Enclosure Method (III), Via a "QR-Factorization"

Here, the fixed choice of $B_{j+1}\in[A_j]B_j$ is replaced as follows by use of an orthogonal matrix Q_{j+1} and an upper triangular matrix R_{j+1}:

(2.40) $$B_{j+1}:=Q_{j+1},\ \check{A}_{j+1}=:Q_{j+1}R_{j+1},\ \check{A}_{j+1}\in[A_j]B_j \text{ is fixed.}$$

The enclosure methods (I), (II), and (III) yield necessarily overestimates at each time t_k with $k\in\mathbb{N}$, which are mainly due to

(A) the relatively crude enclosure of z_k according to (2.27), making use of an interval enclosure of the value of $f^{(p-1)}$,

(B) the outer approximation in $\mathbb{R}^n$ of the true set of values at t_k,

(C) the lumping of the influence of $z_o,\ldots,z_j$ into one interval, $[r_{j+1}]$, and

(D) the independent treatment of $z_o,z_1,\ldots,z_{j+1}$ even though z_{j+1} depends on $z_o,\ldots,z_j$.

If $u'=f(u)$ in (2.1) is a linear ODE, the linear dependency of f on u causes relatively small overestimates in the computation of interval enclosures for $u',\ldots,u^{(p)}$.

Enclosure Method (IV), Via a "Non-Lumped" Treatment of $z_o,\ldots,z_{j+1}$

Sharper enclosures $[u(t)]$ can possibly be determined via a replacement of the lumping of $M_j:=\{z_o,\ldots,z_j\}$ by a lumped treatment only of each subset in a suitable partition of M_j. If $\overline{z}_o-\underline{z}_o$ is considerably larger than $\overline{z}_k-\underline{z}_k$ for $k>0$, M_j may be partitioned into the subsets $\{z_o\}$ and $\{z_1,\ldots,z_j\}$. In this case, (2.34) is replaced as follows:

$$(2.41)\quad r_{j+1}=u_{j+1}-\tilde{u}_{j+1}=A_{j+1,o}(z_o-s_o)+\check{r}_{j+1},\quad \check{r}_{j+1}:=\sum_{k=1}^{j+1}A_{j+1,k}(z_k-s_k)$$

and, using a fixed and not necessarily regular choice of $C_{j+1}\in[A_{j+1,o}]$,

$$(2.42)\qquad [r_{j+1}]:=C_{j+1}([z_o]-s_o)+[\check{\check{r}}_{j+1}]$$

where $[\check{\check{r}}_{j+1}]$ accounts for $[A_{j+1,o}]-C_{j+1}$.

2.4 Modifications Due to Rounded Interval Arithmetic

When the method is programmed on a computer, then all operations have to be carried out in rounded interval arithmetic. This arithmetic (see [14]) is automatically performed in the programming language PASCAL-SC which was used for programming the method. Because of rounding, there arise problems with (i) the computation of the fixed quantities $\tilde{u}_{j+1}$ and (ii) the determination of the inverse matrices B_{j+1}^{-1}. Concerning (i), first an interval

$$(2.43)\qquad [\tilde{\tilde{u}}_{j+1}]:=\tilde{u}_j+h[\Phi(\tilde{u}_j)]+[z_{j+1}]$$

is computed (in rounded interval arithmetic). Here $[\Phi(\tilde{u}_j)]$ accounts for the rounding errors in the evaluation of Φ. Now a machine number $\tilde{\tilde{u}}_{j+1}\in[\tilde{\tilde{u}}_{j+1}]$ is chosen (e.g. the rounded midpoint) and an interval $[z^R_{j+1}]$ is defined by the rounded result of

$$(2.44)\qquad [z^R_{j+1}]:=[\tilde{\tilde{u}}_{j+1}]-\tilde{u}_{j+1}.$$

This interval $[z^R_{j+1}]$ then replaces the interval $[z_{j+1}]-s_{j+1}$ in the method. Now $\tilde{u}_{j+1}$ allows the required representation (2.10) with $s_{j+1}=\tilde{\tilde{u}}_{j+1}-\tilde{u}_j-h\Phi(\tilde{u}_j)$ (which is in general not a machine number). Due to $[z_{j+1}]-s_{j+1}=(\tilde{u}_j+h\Phi(\tilde{u}_j)+[z_{j+1}])-\tilde{u}_{j+1}$, it is obvious from (2.43), (2.44) that $[z^R_{j+1}]$ contains all rounding errors in addition to the procedural errors contained in $[z_{j+1}]-s_{j+1}$. Concerning (ii), a standard PASCAL-SC routine has been used.

3. A Comparison of Enclosure Methods in $\mathbb{R}^2$

As a sample problem, the following set of IVPs is considered:

$$(3.1)\quad \left\{ \begin{pmatrix}u_1'\\u_2'\end{pmatrix}=\begin{pmatrix}u_2\\-u_1\end{pmatrix}\text{ for } t>0,\ \begin{pmatrix}u_1(0)\\u_2(0)\end{pmatrix}=\begin{pmatrix}1+\varepsilon_1\\\varepsilon_2\end{pmatrix}\in I_u(0):=\begin{pmatrix}1+[-\varepsilon,\varepsilon]\\ [-\varepsilon,\varepsilon]\end{pmatrix}. \right.$$

$\varepsilon\in(0,1)$ is fixed.

The solutions can be represented as follows by use of the fundamental matrix U_F:

$$(3.2)\quad \begin{bmatrix}u_1^*(t,\varepsilon_1,\varepsilon_2)\\u_2^*(t,\varepsilon_1,\varepsilon_2)\end{bmatrix}=U_F(t)\begin{bmatrix}1+\varepsilon_1\\\varepsilon_2\end{bmatrix},\quad U_F(t):=\begin{bmatrix}\cos t & \sin t\\-\sin t&\cos t\end{bmatrix}.$$

This yields the set of solutions, S_u, and the smallest interval, I_u, containing S_u:

$$(3.3)\quad S_u(t):=\left\{\begin{bmatrix}u_1^*\\u_2^*\end{bmatrix}\;\middle|\;\begin{bmatrix}\varepsilon_1\\\varepsilon_2\end{bmatrix}\in\begin{bmatrix}[-\varepsilon,\varepsilon]\\ [-\varepsilon,\varepsilon]\end{bmatrix}\right\}\subseteq I_u(t)=U_F(t)\cdot\begin{bmatrix}1+[-\varepsilon,\varepsilon]\\ [-\varepsilon,\varepsilon]\end{bmatrix},$$

where the dot denotes the interval-arithmetic evaluation of this product. Obviously, the width of I_u is a periodic function of t with period $\pi/2$.

$(\hat{IA})$ In order to simulate the numerical execution of an enclosure method in the case of method (I) of Section 2.3 as applied to (3.1), an equidistant time grid with step size $h:=t_{j+1}-t_j$ is chosen. Starting from $\hat{I}_u(0)$, U_F is used to generate the parallelepiped $\hat{S}_u(h):=S_u(h)$. Starting at $t_1=h$ from the smallest interval, $\hat{I}_u(h)$, containing $\hat{S}_u(h)$, U_F is used to generate the parallelepiped $\hat{S}_u(2h)$, etc. Then, $\hat{S}_u(t_{j+1})$ is enclosed by the smallest interval, $\hat{I}_u(t_{j+1})$, containing $\hat{S}_u(t_{j+1})$. Here, $\hat{S}_u(t_k) = \hat{I}_u(t_k)$ if and only if $h=k\pi/2$ with any $k\in\mathbb{N}$. With the exception of this special case, $\hat{I}_u(t_k)$ contains points outside of $\hat{S}_u(t_k)$. It is easily confirmed that the sequence of intervals $\hat{I}_u(t_k)$ possesses an exponentially increasing width in each component unless $h=k\pi/2$.

$(\hat{IB})$ An exponential increase of an interval enclosure is also obtained by means of employing the theory of differential inequalities, e.g. [29, p.93]. Due to the theorem of Müller in this theory, the following auxiliary system is assigned to (3.1):

$$(3.4)\quad \begin{bmatrix} u_1'(t) \\ u_2'(t) \end{bmatrix} \in \begin{bmatrix} [\underset{[u_2]}{\mathrm{Min}}\, u_2,\ \underset{[u_2]}{\mathrm{Max}}\, u_2] \\ [\underset{[u_1]}{\mathrm{Min}}(-u_1),\ \underset{[u_1]}{\mathrm{Max}}(-u_1)] \end{bmatrix} = \begin{bmatrix} [\underline{u}_2(t),\overline{u}_2(t)] \\ [-\overline{u}_1(t),-\underline{u}_1(t)] \end{bmatrix} \quad \text{for } t>0,$$

where $[\underline{u}_1(0),\overline{u}_1(0)]$ and $[\underline{u}_2(0),\overline{u}_2(0)]$ are given.

An inspection of the corresponding system of Volterra integral equations shows that there holds

$$(3.5)\quad (\overline{u}_1',\overline{u}_2',\underline{u}_1',\underline{u}_2')^T = (\overline{u}_2,-\underline{u}_1,\underline{u}_2,-\overline{u}_1)^T.$$

A fundamental matrix of (3.5) is

$$(3.6)\qquad \hat{U}_F(t) := \begin{bmatrix} e^t & e^{-t} & \cos t & \sin t \\ e^t & -e^{-t} & -\sin t & \cos t \\ -e^t & -e^{-t} & \cos t & \sin t \\ -e^t & e^{-t} & -\sin t & \cos t \end{bmatrix}.$$

$(\hat{II})$ For an application of the mean value theorem according to method (II) of Section 2, the definition $v_{ik} := \partial u_i/\partial\varepsilon_k$ is introduced. A differentiation of (3.1) with respect to ε_k yields the auxiliary system

$$(3.7)\qquad \begin{bmatrix} v_{1k}' \\ v_{2k}' \end{bmatrix} = \begin{bmatrix} v_{2k} \\ -v_{1k} \end{bmatrix}, \quad \begin{bmatrix} v_{1k}(0) \\ v_{2k}(0) \end{bmatrix} = \begin{bmatrix} \delta_{1k} \\ \delta_{2k} \end{bmatrix},$$

where δ_{ij} is the Kronecker symbol. Consequently, the following exact representation of the solutions of (3.1) is obtained:

$$(3.8)\quad \begin{cases} \begin{bmatrix} u_1^*(t,\varepsilon_1,\varepsilon_2) \\ u_2^*(t,\varepsilon_1,\varepsilon_2) \end{bmatrix} = \begin{bmatrix} u_1^*(t,0,0) \\ u_2^*(t,0,0) \end{bmatrix} + \begin{bmatrix} v_{11}(t) & v_{12}(t) \\ v_{21}(t) & v_{22}(t) \end{bmatrix} \begin{bmatrix} \varepsilon_1 \\ \varepsilon_2 \end{bmatrix} = \\ = U_F(t) \begin{bmatrix} 1+\varepsilon_1 \\ \varepsilon_2 \end{bmatrix} \text{ for all } \begin{bmatrix} \varepsilon_1 \\ \varepsilon_2 \end{bmatrix} \in \begin{bmatrix} [-\varepsilon,\varepsilon] \\ [-\varepsilon,\varepsilon] \end{bmatrix}. \end{cases}$$

This discussion exhibits the advantage of method (II) as compared with method (I), both of which are defined in Section 2.3.

4. Numerical Examples

For the subsequent results, a microcomputer with a 68000 CPU, running at 8 MHz, has been used which supports the language PASCAL-SC. This language (i) has a 13-digit decimal arithmetic, (ii) executes an exact scalar product, and (iii) admits the definition of general operators, see [9], [13], [14] for details.

Example 4.1: The enclosure method (II) of Subsection 2.3 has been applied in the case of the IVP (3.1), with ε_1 and ε_2 equated to zero. Table 4.1 presents numerical results which were obtained by use of the choices p=10 and h=0.32, i.e., for a total of 31 250 time steps.

t	2000	4000	6000	8000	10000
$\overline{u}_1-\underline{u}_1$	1.1E-8	2.2E-8	3.0E-8	3.7E-8	4.6E-8
$\overline{u}_2-\underline{u}_2$	9.5E-9	2.3E-8	3.3E-8	3.0E-8	5.3E-8

Table 4.1

Example 4.2: The restricted three body problem is considered in the case of the motion of a small satellite in a plane defined by these three heavenly bodies [27]. The initial conditions are chosen so as to approximate a periodic orbit of the satellite with the period T≈6.192169331396.

$$
(4.1)\quad \begin{cases}
u_1' = u_3, \qquad u_2' = u_4, \\
u_3' = u_1+2u_4-\dfrac{\lambda(u_1+\mu)}{((u_1+\mu)^2+u_2^2)^{1.5}}-\dfrac{\mu(u_1-\lambda)}{((u_1-\lambda)^2+u_2^2)^{1.5}}, \\
u_4' = u_2-2u_3-\dfrac{\lambda u_2}{((u_1+\mu)^2+u_2^2)^{1.5}}-\dfrac{\mu u_2}{((u_1-\lambda)^2+u_2^2)^{1.5}}, \\
\mu = 1/82.45, \qquad \lambda = 1-\mu, \\
u_1(0)=1.2, \quad u_2(0)=0, \quad u_3(0)=0, \quad u_4(0)=-1.04935750983.
\end{cases}
$$

For selected t∈[0,T], Table 4.2 presents upper bounds for the relative errors of u_1, u_2, u_3 and u_4, due to method (III) of Subsection 2.3 and the listed manual controls of h and p. The entries --- in the last line of the table indicate that the computed intervals contain the number zero; the absolute error in these cases was less than 2.8E-5. The enclosure methods (I) and (II) produced overflow after (I) t=4.68 and (II) t=1.462.

t	h	p	u_1	u_2	u_3	u_4
1.3	0.025	14	1.1E-9	9.5E-10	2.0E-9	1.8E-9
1.4	0.0025	12	2.1E-8	3.1E-8	1.2E-8	1.3E-8
1.5	0.0005	12	6.1E-8	2.1E-7	1.1E-7	4.4E-8
1.8	0.0025	12	6.1E-8	4.6E-8	2.3E-8	3.5E-8
2.0	0.025	14	4.5E-8	4.3E-8	1.7E-8	1.1E-7
4.0	0.05	14	1.5E-7	2.6E-7	1.3E-7	1.4E-6
4.7	0.005	12	5.0E-5	1.5E-5	9.3E-6	3.4E-6
4.725	0.0005	12	4.2E-5	6.3E-5	3.0E-5	2.9E-5
4.735	0.00025	12	1.6E-4	4.8E-5	2.9E-5	7.4E-5
4.75	0.0005	12	2.3E-5	1.1E-4	2.0E-5	6.9E-6
4.9	0.0025	12	1.6E-5	2.0E-5	8.1E-6	6.7E-6
5.2	0.025	14	1.4E-5	2.2E-5	1.2E-5	3.6E-5
6.15	0.05	14	1.1E-5	6.8E-4	3.2E-4	2.6E-5
T	T-6.15	14	1.2E-5	---	---	2.8E-5

Table 4.2

5. Test for Uniform Boundedness for Solutions of Linear ODEs With Periodic Coefficients

5.1 On Stability for Linear ODEs With Periodic Coefficients

Within the class of linear IVPs, the subclass with periodic coefficients is of particular interest since (I) the pertinent Floquet theory yields relatively simple theoretical conditions for the stability of all solutions pertaining to a given parameter vector, (II) the quantitative application of these conditions leads to difficult numerical problems, and (III) linear ODEs with periodic coefficients come up in mathematical simulation in various applied fields, in particular, in mechanical engineering. Subsequently in this paper, numerical results of applications concerning gear drive vibrations will be discussed briefly. Rotor-dynamics is another domain of important applications of these ODEs in mechanical engineering. In view of these applications, it is of particular interest that special cases are known in literature where there is no stable point in the parameter space. According to [4,p.62], this is true in the case of the ODEs

(5.1) $$u''+a^2u+e\cdot(\sin\omega t)v=0,\quad v''+b^2v+e\cdot(\cos\omega t)u=0.$$

A supplemented version to be presented of the enclosure method (II) of Section 2 has been applied to yield safe quantitative results in numerical applications of the Floquet theory underlying the subclass of linear ODEs under discussion here:

(5.2) $$\begin{cases} y'=M(t)y,\quad M:\mathbb{R}\to L(\mathbb{R}^n),\quad M(t+T)=M(t) \text{ for all } t\in\mathbb{R}, \text{ with} \\ \text{a fixed } T\in\mathbb{R}^+,\quad M \text{ is sufficiently smooth.} \end{cases}$$

The number T is fixed and represents the smallest common period of the periodic coefficient functions in the expression for M.
According to the Floquet theory [4,p.55-59],

(5.3)
- (I) every fundamental matrix $Y:\mathbb{R}\to L(\mathbb{R}^n)$ of (5.2) satisfies $Y(t+T)=Y(t)\cdot C$ for all $t\in\mathbb{R}$ with a matrix $C\in L(\mathbb{R}^n)$ which depends on Y and whose eigenvalues are independent of Y; if $Y(0)=I$, then $Y(T)=C$;
- (II) if and only if $|\sigma|<1$ for all eigenvalues σ of $Y(T)$, all solutions y^* of (5.2) are asymptotically stable as $t\to\infty$ (in the sense of Lyapunov).

Provided an additional condition is satisfied, ordinary stability in this sense holds on the boundaries of stability, where $|\sigma|=1$. A precise determination of these boundaries by means of usual methods is not possible practically since (i) approximations for C and the spectral radius of C are unreliable and (ii) safe criteria for a verification of $|\sigma|\leq1$ are usually over-conservative. Since a safe verification of $|\sigma|<1$ is desired almost up to the true boundary of stability, enclosure methods will be used. A perhaps conservative character of the results for the spectral radius of C will have to be investigated by means of comparisons of results with literature.

<u>Remark</u>: In the engineering literature on the verification of stability, usually truncated asymptotic expansions in powers of e are used for this purpose, where e is a coefficient of the periodic functions in the ODEs.

A special case of (5.2) is the system of ODEs which is equivalent with the (damped) Mathieu equation

(5.4) $y'' + cy' + (d + e\cos(\omega t))y = 0$ for $t \in \mathbb{R}$, with $c, d, e, \omega \in \mathbb{R}$.

The general solution of (5.4) may be represented as follows:

(5.5) $y^*(t) = Ap_1(t)\exp(\mu t) + Bp_2(t)\exp(-\mu t)$ for $t \in \mathbb{R}$,

where $A, B \in \mathbb{C}$ are free constants. The $(2\pi/\omega)$-periodic functions p_1 and p_2, and the number $\mu \in \mathbb{C}$ depend on c, d, e, ω. The solution y^* is bounded if $\mu = ib$ is purely imaginary such that $b \neq N/2$ for all $N \in \mathbb{N}$.
In the case of (5.4), there are several special methods in literature (e.g.[7]) which admit a reliable numerical verification of $|\sigma| < 1$, at least for sufficiently small $|e|$ and m, where H_m with $m \in \mathbb{N}$ is a domain of instability of (5.4). According to [7, p.143-147], the verification of stability <u>or</u> instability for (5.4) with c, d, e, ω fixed can be reduced to an inequality for $|\cosh(2\pi\mu)|$ where $\cosh(2\pi\mu)$ follows from a solution y^* of (5.4) with fixed initial data. Since this solution y^* can be enclosed by means of Section 2, the safe evaluation of this inequality offers the possibility of a rigorously valid comparison with results for $|\sigma| < 1$ due to the methods derived in this paper.

5.2 The Sufficient Test for Uniform Boundedness

In applications, (5.2) usually has the following form:

(5.6) $\begin{cases} y' = (A + B(t))y \text{ for } t \in \mathbb{R}, \quad A \in L(\mathbb{R}^n), \quad B: \mathbb{R} \to L(\mathbb{R}^n), \; B(t+T) = B(t) \\ \text{for all } t \in \mathbb{R} \text{ and a fixed } T \in \mathbb{R}^+, \; B \text{ is sufficiently smooth.} \end{cases}$

For the test to be derived, the following set of initial values will be admitted:

(5.7) $y(0) \in S_o := [-\delta, \delta]d$, any $\delta \in \mathbb{R}^+$, $d := (d_i) \in \mathbb{R}^n$, $d_i := 1$ for $i = 1(1)n$.

The solutions can be represented as follows:

(5.8) $y^*(t, y(0)) = Y(t) \cdot y(0) \in S_t := \{y(t) \mid y' = (A + B(t))y, y(0) \in S_o\} \subset \mathbb{R}^n$,

where the set of solutions, S_t, is a parallelepiped at any fixed t. By construction, S_t is due to the affine mapping of S_o by $Y(t)$. Because of (5.3(I)), the uniform boundedness of S_t for all $t > 0$ and fixed δ depends on $C = Y(T)$.

The column vectors of $Y(t)$ can be enclosed for $t \in [0, T]$, making use of Section 2. This yields the intervals $[c_{ij}] := [\underline{c}_{ij}, \overline{c}_{ij}] \subset \mathbb{R}$ containing the elements c_{ij} of $C =: (c_{ij})$. There holds

$$(5.9)\quad \begin{cases} y^*(T)=C\cdot y(0) \in S_T \subseteq [C]S_o = ([-\delta,\delta][C])d = [-\delta,\delta]\cdot a^{(1)}, \\ \text{where } [C]=([c_{ij}]),\ a^{(1)}:=(a_i^{(1)})\in\mathbb{R}^n, \\ a_i^{(1)}:=\sum_{j=1}^{n}|[c_{ij}]|, \text{ and } |[c_{ij}]|:=\mathrm{Max}\{|\underline{c}_{ij}|,|\overline{c}_{ij}|\}. \end{cases}$$

The row sum criterion

$$(5.10)\qquad \underset{i=1(1)n}{\mathrm{Max}}\ \left(\sum_{j=1}^{n}|[c_{ij}]|\right) < 1$$

implies that $0\leq a_i^{(1)}<1=d_i$ and, therefore, that

$$(5.11)\qquad S_T \subset S_o.$$

If (5.11) is true, the uniform boundedness of S_t for all $t>0$ can be shown directly, making use of the analysis following (5.14). Due to the oscillatory character of the solutions, however, S_T is in a rotated position in $\mathbb{R}^n$, as compared with S_o. Consequently, if (5.6) is a mathematical model of a real world problem, the model must possess a sufficiently strong dissipation (damping) property if (5.11) is to be true.

If (5.11), and thus (5.10), are not true, the combination of dissipation and rotation of S_t may cause the existence of a time $t^*=kT$ with a $k\in\mathbb{N}$ such that $S_{t*}\subset S_o$. This can be verified as follows: Due to (5.3), $Y(2T)=C^2$ and, generally, $Y(kT)=C^k$. Analogously to (5.9) and by use of the notation $[C^k]$ for an enclosure of $C^k=(Y(T))^k$, see (5.16), there holds

$$(5.12)\quad \begin{cases} y^*(kT) = C^k y(0) \in S_{t*} \subseteq [C^k]S_o = ([-\delta,\delta][C^k])d = [-\delta,\delta]a^{(k)} \\ \text{where } C^k\in[C^k]=:([c_{ij}^{(k)}])\subset L(\mathbb{R}^n),\ a^{(k)}:=(a_i^{(k)})\in\mathbb{R}^n, \\ \text{and } a_i^{(k)}:=\sum_{j=1}^{n}|[c_{ij}^{(k)}]|. \end{cases}$$

The "enclosure condition"

$$(5.13)\qquad \underset{i=1(1)n}{\mathrm{Max}}\ \left(\sum_{j=1}^{n}|[c_{ij}^{(k)}]|\right) < 1$$

implies the validity of

$$(5.14)\qquad S_{t*} \subset S_o.$$

If S_{t*} is replaced by its superset S_o, the admission of all vectors $y(t^*)$ $\in S_o$ yields a set of solutions at $t=2t^*$ which is contained in S_o and which, in turn, contains S_{2t*} because of $t^*=kT$ and $B(t+T)=B(t)$. Generally, there holds

$$(5.15)\qquad S_{mt*} \subset S_o \text{ for all } m\in\mathbb{N}.$$

Since S_t is bounded for $t\in[0,t^*]$, (5.13) implies the validity of (5.15) and, thus, of the uniform boundedness of the set of solutions for all $t>0$. Due to linearity, this result is valid for any (and even an arbitrarily large) choice of the number $\delta\in\mathbb{R}^+$ in the definition (5.7) of the initial set S_o.

Remarks: 1.) For practical applications, the property of uniform boundedness is at least as useful as the one of asymptotic stability.

2.) The verification of the enclosure condition (5.13) is a practical implementation of the stability condition $|\sigma|<1$ in (5.3) since $[C]$ contains the unknown true $C=Y(T)$; (5.13) implies the validity of $\varrho(C^k)<1$ for all $C\in[C]$, and thus also $\varrho(C)<1$ for $C=Y(T)$.

The symbol $[C^k]$ introduced in (5.12) denotes a safe enclosure of the set of matrices generated by means of the following algorithm [5],[6] which rests on method (II) of Subsection 2.3:

$$(5.16)\quad \left\{\begin{array}{l} \text{choose } B_o=I,\ [E_o]=I,\ B_i\in[C]B_{i-1}, \\ \text{determine } [E_i]:=(B_i^{-1}[C]B_{i-1})[E_{i-1}], \\ \text{determine } [C^i]:=B_i[E_i], \end{array}\right\} \text{ for } i=1(1)k\ .$$

The execution of this algorithm requires, in particular, the determination of an interval enclosure $[B_i^{-1}]$ for the inverse of $B_i\in L(\mathbb{R}^n)$ in <u>every</u> step of this algorithm. In view of the relatively large involved computation time, (5.16) has been replaced by a simplified version for practical applications of the algorithm. In this version, a new matrix B_i is chosen only in selected steps of the iteration and retained inbetween. Via this simplified version of the algorithm, the sequence of interval matrices $[C^k]$ has been determined by use of the FORTRAN subroutine library ACRITH and an IBM 4331 where, in particular, a routine is available for the guaranteed enclosure $[B_i^{-1}]$ of the inverse of B_i.

<u>Remarks:</u> <u>1.)</u> The question whether or not (5.13) is suitable for the practical verification of $\varrho(C)<1$ depends on the condition of the matrices C and B_i in (5.16).

<u>2.)</u> In the absence of numerical examples, a correspondingly over-conservative character of the sufficient test (5.13) for $\varrho(C)<1$ is in particular possible if $C=Y(T)$ and if the parameter vector in (5.6) yields a point inside a domain of stability but close to its boundary. This difficulty is obviously aggravated by the fact that $\varrho(C)<1$ has to be verified for all $C\in[Y(T)]$. Consequently, Subsections 5.3 and 5.5 are devoted to a comparison of results for boundaries of stability for a given system (5.6) which <u>(i)</u> have been reported in literature or <u>(ii)</u> determined by use of (5.13). For all practical purposes, these comparisons reveal that (5.13) as applied to $C\in[Y(T)]$ does not underestimate the extent of domains of stability, at least as far as pertinent results are available in literature in the case of the investigated examples.

<u>3.)</u> A qualitative explanation of this success of the method employed here is as follows: In these examples, the width of the enclosure $[C]$ is sufficiently small and all $C\in[C]$ are sufficiently well-conditioned such that $\varrho(C)<1$ can be recognized via (5.13) for all $C\in[Y(T)]$. If this is true, then $\{\varrho(C^k)\}$ is a decreasing sequence for all $C\in[C]$. Consequently, it may be possible to verify $\varrho(C)<1$ by means of the row sum criterion for a sufficiently large $k\in\mathbb{N}$ in cases where this is not possible for $k=1$, i.e. (5.10).

5.3. Application of the Sufficient Test in the Case of the Mathieu-Equation

According to [7,p.143-147], the question of stability or instability of the solutions y^* of (5.4) with c,d,e,ω fixed can be reduced to the safe evaluation of an inequality for $\cosh(2\pi\mu)$ (see end of Subsection 5.1). This provides the possibility of a rigorously valid test for the perhaps conservative character of (5.13).

Figure 5.1 exhibits the following results for the boundaries ∂H_0, ∂H_1, and ∂H_2, [5],[6]:

(A) the solid curves according to one of the conventional approximations in literature, see [10,p.369], (within graphical accuracy, selected points of these curves coincide with results from the inequality mentioned above).

(B) the dots "•" indicating that (5.13) could be satisfied for a $k<3000$,

(C) the crosses "×" indicating that either a $k<3000$ could not be found or that the sequence of intervals $[c^k]$ diverged.

Numerical experience concerning (B) shows in particular that (B1) $k\approx 20$ suffices for "stable points" far from one of the boundaries ∂H_0, or ∂H_1, or ∂H_2, (B2) As k increases, the width of $[c^k]s_0$ either increases initially, followed by a gradual decrease (due to dissipation), or it "blows up".

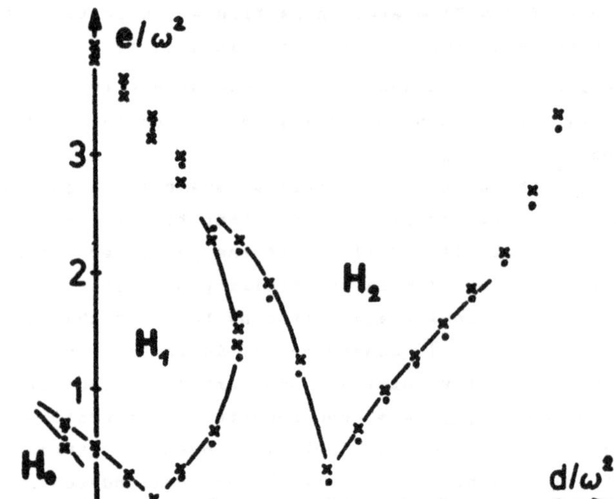

Figure 5.1: Stability Chart for the Mathieu-Equation (5.4), $c/\omega = 0.1$

5.4 On Vibrations of Gear Drives

In the operation of a gear drive, there is a periodic change in the number of teeth in contact in any pair of mated gears. This induces vibrations whose dynamic loads may reach values of up to three times the ones of the static external loads, [23, p.80]. Points in a pertinent parameter space with such relatively large dynamic loads are called "unstable". In "domains of stability", vibrations can be simulated by means of suitable solutions of a modelling system (5.6). In engineering literature, stability is "tested" either (i) directly by use of truncated expansions of the solutions of a simulating system (5.6) or by means of simplified models from mechanics replacing the ODEs employed in (5.6) or (ii) indirectly via the (approximate) determination of the dynamic loads for $t \in [0,qT]$ with q as small as, e.g., 20; see Subsection 5.6 for an enclosure of the dynamic loads in domains of stability. In several engineering doctoral dissertations, results due to "tests" of type (i) have been compared with (approximately valid) applications of the Floquet theory, e.g., [12] and [22]. In the dissertation [22], sufficient tests for stability due to methods in functional analysis have been applied. These tests require a considerable preparatory effort to recast the given system of ODEs such that the new system (i) is admissible for one of these tests and (ii) does not yield overpessimistic results in the verification of stability.

5.5 Numerical Results for a Gear Drive Model With n=8

In the doctoral theses [12] and [22], vibrations of two mated spur-type gears are simulated by means of a system (5.6) of order n=8. In the ori-

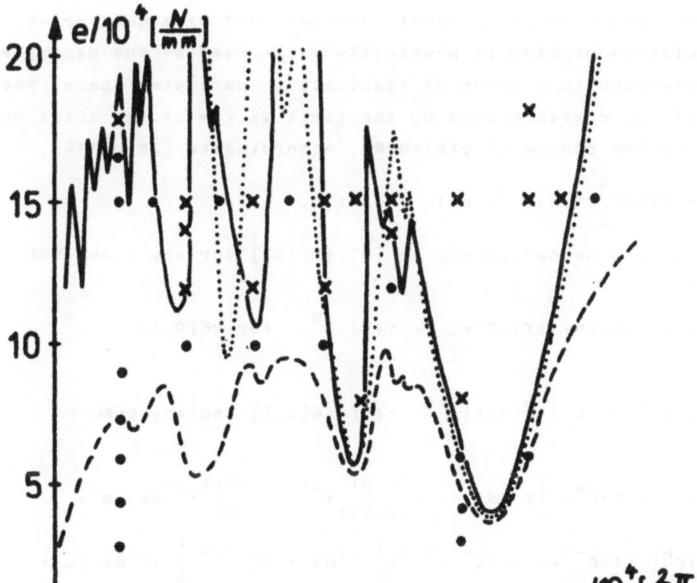

Figure 5.2: Stability Chart for Gear Drive Vibrations [12],[22] with e denoting the coefficient of the periodic terms in the ODEs.

ginal model, there are four ODEs of the second order each, corresponding to the admitted two rotational and two translational degrees of freedom. Figure 5.2 exhibits the following results, [5],[6]:

(A) the solid curves according to an approximately valid execution of the Floquet theory, with numerical results reported in [12],

(B) the dots "•" indicating that (5.13) could be satisfied for a $k<200$,

(C) the crosses "×" indicating that either a $k<200$ could not be found or that the sequence of intervals $[C^k]$ diverged,

(D) the dashed curves represent the upper envelope of results due to the sufficient tests employed in [22], see Subsection 5.4,

(E) the dotted curves are due to the usual "engineering test" of stability consisting of a two-term asymptotic expansion.

Obviously, (D) is over-conservative and (E) fails to distinguish between stability and instability.

Remark: For the results (B) and (C) in Figure 5.2, the enclosure method (II) of Section 2 has been used for $t\in[0,T]$ with $h=T/40$ and $p=16$. Then, at least 8 decimals of the mantissa agree in the computed enclosure $[Y(T)]=[C]$.

5.6. Enclosure of Dynamic Loads in a Gear Drive

Here, (5.6) is replaced by

(5.17) $\quad y'-(A+B(t))y = b \quad$ for $t>0$, $b\in\mathbb{R}^n$, etc. as in (5.6),

where b represents the time-independent external load of a gear drive. This linear simulation problem is physically well-posed if the parameters entering A+B correspond to a point of stability in parameter space. The dynamic loads will be characterized by the limit as $t\to\infty$ of the solution y^* which is due to the choice of $y(0)=0\in\mathbb{R}^n$. According to [28,p.116],

(5.18) $\quad y^*(t) = Y(t)\left(\int_0^t (Y(s))^{-1} b\, ds\right) \quad$ for $t>0$.

This solution will now be considered on $[NT,(N+1)T]$ for any fixed $N\in\mathbb{N}$. Because of (5.3),

(5.19) $\quad Y(t+NT) = Y(t+(N-1)T)\cdot C = \ldots = Y(t)\cdot C^N \quad$ for $t\in[0,T]$.

Consequently,

(5.20) $\quad (Y(t+pT))^{-1} = (C^{-1})^p (Y(t))^{-1} \quad$ for $t\in[0,T]$ and any $p\in\mathbb{N}$.

Therefore,

$$(5.21)\quad \begin{cases} y^*(t+NT) = Y(t)C^N\left(\int_0^T Y^{-1}ds + \ldots + \int_{(N-1)T}^{NT} Y^{-1}ds + \int_{NT}^{NT+t} Y^{-1}ds\right)b = \\ \quad = Y(t)C^N\left((I+C^{-1}+\ldots+(C^{-1})^{N-1})\int_0^T Y^{-1}ds + (C^{-1})^N \int_0^t Y^{-1}ds\right)b = \\ \quad = Y(t)\left(\sum_{k=0}^{N-1} C^k\right)y^*(T) + y^*(t) \quad \text{for } t\in[0,T]. \end{cases}$$

If the spectral radius $\varrho(C)$ of C satisfies

(5.22) $\quad \varrho(C) < 1,$

there follows for $t \in [0,T]$

(5.23) $$\lim_{N\to\infty} y^*(t+NT) = Y(t)(I-C)^{-1}y^*(T) + y^*(t) \quad \text{for } t\in[0,T],$$

making use of (5.18). An enclosure of the right hand side of (5.23) involves (i) the enclosure $[Y(t)]$ for $t\in[0,T]$ which has already been determined in the analysis leading to $[C]=[Y(T)]$ and (ii) an enclosure of $y^*(t)$ for $t\in[0,T]$ which can be determined by means of Section 2.

Remarks: 1.) The final result in (5.23) is still true if b is a continuous periodic function of t with period T.

2.) A result analogous to (5.23) holds if b is continuous and periodic with period $\hat{T}$ such that $pT=q\hat{T}$ with fixed $p,q\in\mathbb{N}$. Then, y^* is bounded. This choice of periodic functions b comes up in the analysis of gear drive vibrations with a simulation of the wear of teeth.

References

[1] Adams,E., Lohner,R., Error Bounds and Sensitivity Analysis, in: Scientific Computing, Editor: R.S. Stepleman, North-Holland Publ. Co., Amsterdam, 1981, p.213-222.

[2] Adams,E., Invers-Monotonie, Direkte und Indirekte Intervallmethoden, Berichte der Math.-Statist. Sektion im Forschungszentrum Graz, Graz, 1982, Number 185.

[3] Alefeld,G., Herzberger,J., Introduction to Interval Computations, Academic Press, New York, 1983.

[4] Cesari,L., Asymptotic Behavior and Stability Problems in Ordinary Differential Equations, Springer-Verlag, Berlin, 1963, 2nd edition.

[5] Cordes,D., Adams,E., Test for Uniform Boundedness for Problems Admitting Parameter Resonance, Proc. 11th IMACS World Congress, Oslo, Norway, Aug.5-9,1985, vol.1, p.103-105.

[6] Cordes,D., Adams E., Test for Uniform Boundedness for Dynamic Problems Admitting Parameter and Combination Resonance, will appear in: Computer Arithmetic, Scientific Computation and Programming Languages, Editors: E. Kaucher, U. Kulisch, Ch. Ullrich, G. Teubner, Stuttgart, J. Wiley, New York.

[7] Eicher,N., Einführung in die Berechnung parametererregter Schwingungen, TUB-Dokumentation Weiterbildung, Technische Universität Berlin, 1981, Heft 1.

[8] Eijgenraam, The Solution of Initial Value Problems Using Interval Arithmetic, Stichting Mathematisch Centrum, Amsterdam, Mathematical Centre Tracts 144, 1981.

[9] Kaucher,E., Klatte,R., Ullrich,Ch., Programmiersprachen im Griff, vol.2, PASCAL, Bibliographisches Institut, Mannheim, Hochschultaschenbücher, vol.796, 1981.

[10] Klotter,K., Technische Schwingungslehre, Berlin, 1951, vol.1, 2nd edition.

[11] Krückeberg,F., Ordinary Differential Equations, in: Topics in Interval Analysis, Editor: E. Hansen, Clarendon Press, Oxford, 1969.

[12] Kücükay,F., Über das dynamische Verhalten von einstufigen Zahnradgetrieben, Fortschrittsberichte der VDI-Zeitschriften, Reihe Schwingungstechnik-Lärmbekämpfung, VDI-Verlag, Düsseldorf, 1981, Number 43.

[13] Kulisch,U., Miranker,W., Computer Arithmetic in Theory and Practice, Academic Press, New York, 1981.

[14] Kulisch,U., Miranker,W.L., A New Approach to Scientific Computation, Academic Press, New York, 1983.

[15] Lang,R., Adams,E., Mathematical Analysis of Dependency of Dynamic Perturbations on Constitutive Equations, Proc.11th IMACS World Congress, Oslo, Norway, Aug.5-9.1985, vol.1, p.63-66.

[16] Lohner,R., Adams,E., Einschließung der Lösung gewöhnlicher Anfangs- und Randwertaufgaben, Zeitschrift für Angew. Math. und Mechanik, vol.64, 1984, T295-T297.

[17] Lohner,R., Enclosing the Solutions of Ordinary Initial- and Boundary-Value Problems, Proc. 11th IMACS World Congress, Oslo, Norway, Aug.5-9, 1985, vol.1, p.99-102.

[18] Lohner,R., Enclosing the Solutions of Ordinary Initial and Boundary Value Problems, will appear in: Computer Arithmetic, Scientific Computation and Programming Languages, Editors: E.Kaucher, U.Kulisch, Ch.Ullrich, G.Teubner, Stuttgart, J.Wiley, New York.

[19] Lohner,R., Über die Fehlerfortpflanzung bei Einschließungsverfahren für gewöhnliche Anfangswertaufgaben, will be published in ZAMM.

[20] Moore,R.E., Interval Analysis, Prentice-Hall, Englewood Cliffs, N.J., 1966.

[21] Moore,R.E., Methods and Applications of Interval Analysis, SIAM, Philadelphia, 1979.

[22] Naab,K., Stabilitätsuntersuchungen an linearen Systemen mit periodisch zeitveränderlichen Parametern, Fortschrittsberichte der VDI-Zeitschriften, Reihe Schwingungstechnik-Lärmbekämpfung, VDI-Verlag, Düsseldorf, 1981, Number 44.

[23] Niemann,G., Winter,H., Maschinenelemente, Springer-Verlag, Berlin, 1983, 2nd edition, vol.2.

[24] Rump,S.M., Kleine Fehlerschranken bei Matrixproblemen, Dissertation, Institut für Angewandte Mathematik, Universität Karlsruhe, 1980

[25] Spreuer,H., Adams,E., Holzmüller,A., Probability Distributions Over Output Intervals for (Non)linear ODEs With Given Distributions Over Input Intervals, Proc.11th IMACS World Congress, Oslo, Norway, Aug. 5-9,1985, vol.1, p.107-109.

[26] Spreuer,H., Adams,E., Holzmüller,A., Enclosure of Output Distributions for Ordinary Differential Equations With Stochastic Input, will appear in: Computer Arithmetic, Scientific Computation and Programming Languages, Editors: E.Kaucher, U.Kulisch, Ch.Ullrich, G. Teubner, Stuttgart, J. Wiley, New York.

[27] Stroud,A.H., Numerical Quadrature and Solution of Ordinary Differential Equations, Springer-Verlag, Berlin, 1974.

[28] Walter,W., Gewöhnliche Differentialgleichungen, Springer-Verlag, Berlin, 1972. 1st edition

[29] Walter,W., Differential and Integral Inequalities, Springer-Verlag, Berlin, 1970.

DISCRETE CLOSURE AND ASYMPTOTIC (QUASI-) REGULARITY IN DISCRETIZATION ALGORITHMS.

P. M. Anselone [1) 2)] and R. Ansorge [3)]

1. Introduction

This paper is concerned with general principles that underlie approximation procedures in numerical analysis, particularly difference methods for differential equations. Central roles are played by convergence, compactness, regularity, continuity, and closure properties. The aim is to determine minimal conditions on the operators and their approximations which imply the convergence of approximate solutions, independently of existence and uniqueness. This paper extends the theory we presented in [2].

A typical situation in Numerical Analysis is the following one: There are given sets $\mathring{X}$, $\mathring{Y}$ which become metric spaces X, Y by certain metrics d_X and d_Y, respectively:

$$X = (\mathring{X}, d_X) \quad , \quad Y = (\mathring{Y}, d_Y). \tag{1}$$

Moreover, given an operator A with domain $D(A) \subset \mathring{X}$ and range $R(A) \subset \mathring{Y}$. For given $y \in \mathring{Y}$, solutions $x \in \mathring{X}$ of the problem

$$Ax = y \tag{2}$$

are sought. The numerical procedure is defined by a given sequence $\{A_n\}$ of operators each of which maps its domain $D_n \subset \mathring{X}$ into $R(A_n) \subset \mathring{Y}$, and by the equations

$$A_n x_n = y_n \quad n \in N = \{1,2,\ldots\} \tag{3}$$

where $y_n \in R(A_n)$ and

$$y_n \to y \quad \text{in} \quad Y. \tag{4}$$

The solutions x_n of (3) are considered as numerical approximations of a solution x of (2). Let

$$S := \{x \in D(A) \mid Ax = y\},$$

---

1) Mathematics Department, Oregon State University, Corvallis, OR 97331, USA

2) The research was initiated while this author was a Senior Humboldt Fellow at the University of Hamburg

3) Institut für Angewandte Mathematik, Universität Hamburg, Bundesstr.55, D-2000 Hamburg 13, Bundesrepublik Deutschland

$$S_n := \{x_n \in D_n \mid A_n x_n = y_n\} \quad (n=1,2,\ldots).$$ [1)]

It is part of the definition of set convergence that if $S_n \to S$ and S_n is nonvoid for infinitely many n, then S is nonvoid. This is a standard technique for proving the existence of solutions of equations.

If we ask for minimal properties of the pair $[\{A_n\}, A]$ which ensure

$$S_n \to S \tag{5}$$

for arbitrary admissible right hand sides y and $y_n \to y$, we must take into account that property $S_n \to S$ can be considered as a combination of two basic properties of the operators A_n, A:

1. Because of (3) and (4), $\{A_n x_n\}$ is compact in Y, and it is expected that every subsequence of $\{x_n\}$ has a convergent subsequence, i. e. to be compact in X, too. This behaviour of a sequence $\{A_n\}$ is called "asymptotic regularity" (cf. [2]).
   The concept of asymptotic regularity was developed by Grigorieff [7] and Stummel [11]. Special cases were given earlier by Petryshin. See also [9], p. 299.

2. If asymptotic regularity of $\{A_n\}$ is shown and if - therefore - $\{x_n\}$ is compact in X, if - moreover - x is the limit of a convergent subsequence $\{x_n \mid n \in \mathbb{N}' \subset \mathbb{N}\}$, it must be ensured that $x \in D(A)$ and that Ax = y. Thus we need - besides asymptotic regularity - the following property of $[\{A_n\}, A]$:

$$x_n \to x \text{ (in X)} \quad (n \in \mathbb{N}' \subset \mathbb{N}), \quad A_n x_n \to y \text{ (in Y)} \tag{6}$$
$$\Rightarrow x \in D(A) \text{ and } Ax = y.$$

Property (6) of a pair $[\{A_n\}, A]$ was first described by Grigorieff [6] in 1969 (see also [7]); he called such a pair "discretely closed".

Although a proof of $S_n \to S$ can - therefore - be reduced to a proof of asymptotic regularity of $\{A_n\}$ and to a proof of discrete closure of $[\{A_n\}, A]$, the term "discretely closed" did not become well known. In convergence proofs of numerical methods, one normally tried to show - instead of (6) - the property

$$x_n \to x \text{ (in X)} \Rightarrow A_n x_n \to Ax \text{ (in Y)}. \tag{7}$$

---

1) If the solution x of (2) exists uniquely and if $S_n \to S$, then the full sequence $\{x_n\}$ converges to x, i. e. set convergence then leads to convergence of the numerical method.

In the case of $D_1 = D_2 = D_3 = \ldots$, property (7) is called "continuous convergence" of $\{A_n\}$ to A, and we write

$$A_n \overset{c}{\to} A.$$

Continuous convergence was defined by du Bois-Reymond [4] and by Courant [5], was introduced into Numerical Analysis - particularly in connection with the numerical treatment of nonlinear partial initial value problems - in [3] and was then used by Stummel [11] within his general theory of discretization.

Of course, (7) is stronger than (6), but leads in the case of continuous operators A_n, A to equicontinuity of the approximate operators A_n (cf. [10], p. 78), and this equicontinuity is also sufficient for continuous convergence of individually continuous and pointwise convergent operator sequences.

On the other hand, if $D_1 = D_2 = \ldots$ and if the operators A_n are asymptotically compact, then discrete closure and continuous convergence are equivalent (cf. Theorem 2.3). Asymptotic compactness is weaker than collective compactness which is presented in [1].

It is the purpose of this paper to make these connections clear in detail, both theoretically and by examples. These examples are chosen in such a way that they are easy to survey. They concern the numerical treatment of differential equations. In particular, Example 3.4 shows that much weaker conditions than formulated in [2] are sufficient to establish convergence, provided that discrete closure is used instead of continuous convergence and that the conception of asymptotic regularity is replaced by a new and a little weaker term which we call "asymptotic quasi-regularity".

2. Basic Definitions and Theorems

We complete the definition of discrete closure given in the first section as follows:

<u>Definition 2.1</u>: If (6) holds, the pair $[\{A_n\}, A]$ is called "discretely closed with respect to the triple $(X, D(A), \{D_n\})$". 1)

---

1) Because we write $(X, D(A), \{D_n\})$ instead of $(\overset{\circ}{X}, D(A), \{D_n\})$, also the metric used on $\overset{\circ}{X}$ is especially mentioned in the definition.

Remark 2.1: Let $\overset{\square}{X} \subset \overset{\circ}{X}$, $\overset{\square}{D}(A) = \overset{\square}{X} \cap D(A)$, $\overset{\square}{D}_n = \overset{\square}{X} \cap D_n$.
Let $\tilde{X} := (\overset{\square}{X}, d_X)$ be closed. Then from simple arguments: $[\{A_n\}, A]$ discretely closed with respect to $(X, D(A), \{D_n\})$ $\Rightarrow$ $[\{A_n\}, A]$ discretely closed with respect to $(\tilde{X}, \overset{\square}{D}(A), \{\overset{\square}{D}_n\})$.

Theorem 2.1: Let y, $\{y_n\}$ be given and $y_n \to y$.
Let there be at least one solution x_n of equation (3) for every $n \in \mathbb{N}$. Assume $[\{A_n\}, A]$ to be discretely closed with respect to $(X, D(A), \{D_n\})$ and let $\{S_n\}$ be d-compact (cf. [2]).
Then $S \neq \emptyset$ and $S_n \to S$.

Proof: $\{S_n\}$ d-compact $\Rightarrow \exists \{x_n \mid x_n \in S_n, n \in \mathbb{N}' \subset \mathbb{N}\}$,
$\exists$ x, such that $x_n \to x$.
Hence - from (3) - $A_n x_n \to y$.
Because $[\{A_n\}, A]$ is discretely closed, we find $x \in D(A)$ and $Ax = y$, i. e. $x \in S$.

Remark 2.2: Obviously, the assumption of d-compactness of $\{S_n\}$ is fullfilled if $\{A_n\}$ has the property of asymptotic regularity mentioned in Section 1 of this paper. We repeat the definition of this property here: $\{A_n\}$ is called asymptotically regular if the following implication holds:

$$\{x_n\} \text{ bounded (in X)}^{1)}, \ \{A_n x_n\} \text{ compact in Y} \quad \Rightarrow \{x_n\} \text{ compact in X.} \tag{8}$$

Remark 2.3: Often the situation is such that $\overset{\circ}{X} \subset \overset{\circ}{Y}$ but the metric d_X is not induced by d_Y. In such cases, normally d_X pays better regard to the particular properties of the elements of the subset $\overset{\circ}{X}$ than d_Y does, e. g. $X = C^1[0,1]$ with C^1-norm, $Y = C^0[0,1]$ with C^0-norm. Nevertheless - from a numerical point of view - it often seems to be satisfactory if - instead of $x_n \to x$ in X - at least

$$x_n \to x \quad \text{in } \hat{X} := (\overset{\circ}{X}, d_W) \tag{9}$$

can be ensured where d_W is a metric on a set $\overset{\circ}{W}$ with $\overset{\circ}{X} \subset \overset{\circ}{W} \subset \overset{\circ}{Y}$, and $W := (\overset{\circ}{W}, d_W)$.

Remark 2.3 gives rise to a certain restriction of the concept of asymptotic regularity defined by (8):

---

1) $\{x_n\}$ is called "bounded" (in X) if there is a constant M and an element $\hat{x} \in \overset{\circ}{X}$ such that $d_X(x_n, \hat{x}) \leq M$ for all $n \in \mathbb{N}$.

<u>Definition 2.2</u>: In the case of $\overset{\circ}{X} \subset \overset{\circ}{W} \subset \overset{\circ}{Y}$ with $\overset{\circ}{W} \neq \overset{\circ}{X}$ and $d_W \neq d_X$, $\{A_n\}$ is called "asymptotically quasi-regular" if the following implication holds:

$$\left.\begin{array}{r}\{x_n\} \text{ bounded (in X)} \\ \{A_n\, x_n\} \text{ compact in Y}\end{array}\right\} \Rightarrow \{x_n\} \text{ compact in } \hat{X}. \qquad (10)$$

If one is contented with quasi-regularity, then this weaker type of regularity is often already a consequence of discrete closure such that the two steps of a proof of convergence mentioned in Section 1 reduce to only one:

<u>Theorem 2.2</u>: Let

(i) $\overset{\circ}{X} \subset \overset{\circ}{W} \subset \overset{\circ}{Y}$, $\overset{\circ}{W} \neq \overset{\circ}{X}$ and $d_X \neq d_W$

(ii) $\{x_n\}$ bounded in X $\Rightarrow$ $\{x_n\}$ compact in W

(iii) $[\{A_n\}, A]$ discretely closed with respect to $(W, D(A), \{D_n\})$.

Then $\{A_n\}$ is asymptotically quasi-regular.

Proof: Assume $\{x_n\}$ to be a sequence bounded in X and $\{A_n\, x_n\}$ to be compact in Y. Hence,

$\exists\, y \in \overset{\circ}{Y}$, $\exists\, \mathbb{N}' \subset \mathbb{N}$ such that

$A_n\, x_n \to y \quad (n \in \mathbb{N}')$ in Y.

Because of (ii), $\exists\, x \in \overset{\circ}{W}$ and $\mathbb{N}'' \subset \mathbb{N}'$ such that

$x_n \to x$ in W $(n \in \mathbb{N}'')$.

From (iii), $x \in D(A) \subset \overset{\circ}{X}$ [1], i. e. $\{x_n\}$ compact in $\hat{X}$.

This simple theorem can lead to important consequences as will be shown in Section 3 by examples.

The next theorems of this section concern the connections between discrete closure and continuous convergence, also already mentioned in Section 1. $\overset{\circ}{X} \subset \overset{\circ}{Y}$ is no longer necessary.

<u>Theorem 2.3</u>: Assume

(i) $D_1 = D_2 = \ldots$

(ii) $[\{A_n\}, A]$ discretely closed with respect to $(X, D(A), \{D_n\})$

(iii) $\{A_n\}$ asymptotically compact, i.e.:

---

1) Thus, $[\{A_n\}, A]$ discretely closed with respect to $(\hat{X}, D(A), \{D_n\})$

$$\{x_n\} \subset D_1 \text{ bounded (in } X) \Rightarrow \{A_n x_n\} \text{ compact in } Y.$$

Then:

$$A_n \overset{c}{\to} A \text{ on } D_1.$$

Proof:

Let $x_n \to x$ (in X). Then $\{x_n\}$ is bounded, hence $\{A_n x_n\}$ compact in Y (cf. (iii)). Thus

$$\exists N' \subset N, \ \exists z \in \mathring{Y}: \ A_n x_n \to z \qquad (n \in N').$$

From (ii) we get $z = Ax$. Therefore z is unique, and $A_n x_n \to Ax$ $(n \in N')$. The uniqueness of z then implies $A_n x_n \to Ax$ $(n \in N)$, i. e. $A_n \overset{c}{\to} A$.

Remark 2.4: Vice versa, in the case of $D_1 = D_2 = \ldots$, the continuous convergence $A_n \overset{c}{\to} A$ leads immediately to discrete closure of $[\{A_n\}, A]$ - as already mentioned in Section 1 - :

$$x_n \to x \text{ and } A_n x_n \to y \text{ and } A_n \overset{c}{\to} A$$

$$\Rightarrow A_n x_n \to Ax, \text{ i. e. } x \in D(A) \text{ and } Ax = y.$$

Let us now consider two sequences of operators, namely $\{A_n\}$ and $\{C_n\}$, where A_n and C_n have the common domain $D_n \subset \mathring{X}$ $(n \in N)$, let also A and C be two operators with common domain $D(A) \subset \mathring{X}$, and assume that all these operators map their domains into $\mathring{Y}$ where Y is a linear space.

Let

$$B = A + C \ , \ B_n = A_n + C_n \quad (n = 1,2,\ldots).$$

The discrete closure of $[\{A_n\}, A]$ and of $[\{C_n\}, C]$ does not necessarily imply the discrete closure of $[\{B_n\},B]$ but the following theorem holds:

Theorem 2.4: Concerning the operators A_n, B_n, C_n, A, B, C $(n = 1,2,\ldots)$ of Remark 2.2, assume $[\{A_n\}, A]$ to be discretely closed with respect to $(X, D(A), \{D_n\})$, let $\{D_n\}^* \subset D(C) = D(A)$ [1] and assume

$$x_n \to x(x_n \in D_n, x \in X) \Rightarrow x \in D(C) \text{ and } C_n x_n \to Cx \text{ [2]}.$$

Then $[\{B_n\}, B]$ is discretely closed with respect to $(X, D(A), \{D_n\})$.

---

1) $\{x_n\}^*$ is the set of all the limits of convergent subsequences of $\{x_n\}$, and $\{D_n\}^*$ then is the union of all sets $\{x_n\}^*$ with $x_n \in D_n$ $(n = 1,2,\ldots)$.

2) e. g.: $D_1 = D_2 = \ldots \subset D(C)$ and $C_n \overset{c}{\to} C$ on D_1.

Proof:

Consider a sequence $\{x_n \mid x_n \in D_n;\ n = 1,2,\ldots\}$ with $x_n \to x$ and with $B_n x_n \to y$. Then $C_n x_n \to Cx$ and therefore $(B_n - C_n)x_n \to y - Cx$, i.e. $A_n x_n \to y - Cx$. Hence, $x \in D(A)$ and $Ax = y - Cx$, i.e. $(A + C)x = y$, thus $Bx = y$.

Remark 2.5: Theorem 2.4 is particularly of interest if one tries to show the discrete closure of a nonlinearily perturbed pair $[\{B_n\},B]$ of linear operators A_n, A $(n \in \mathbb{N})$ (cf. Example 3.4).

Remark 2.6: In the particular case $A_n = \Theta$ (zero-operator) $(n = 1,2,\ldots)$, $A = \Theta$, Theorem 2.4 is nothing other than Remark 2.4 applied to the case where Y is a linear space.

3. Examples

Example 3.1: The first example is very simple, but informative:

$$\mathring{X} = C^1[0,1] \subset \mathring{Y} = C^0[0,1] = \mathring{W},$$

$$d_X(u,v) := \|u-v\|_\infty + \|u'-v'\|_\infty, \quad d_Y(u,v) := \|u-v\|_\infty =: d_W(u,v)$$

$$A = A_n = \frac{d}{dt}, \quad \text{hence } D(A) = D_n = \mathring{X} \quad (n = 1,2,\ldots) \quad \text{and } Y = W.$$

Assumption (i) of Theorem 2.2 is fullfilled, and this also holds for assumption (ii) because of the Arzelà-Ascoli Theorem. If $x_n \to x$ in W $(x_n \in D_n)$ and $A_n x_n \to y$ in Y, i. e. $\|x_n' - y\|_\infty \to 0$, then from

$$x_n(t) - x_n(0) = \int_0^t x_n'(\tau)d\tau$$

for $n \to \infty$:

$$x(t) - x(0) = \int_0^t y(\tau)d\tau.$$

Hence, $x'(t) = y(t)$, i. e. $Ax = y$.
Thus, also the assumption (iii) of Theorem 2.2 is fullfilled such that $\frac{d}{dt}$ is (asymptotically) quasi-regular.

Example 3.2: Define $\mathring{X}$, $\mathring{W}$, $\mathring{Y}$, d_X, d_Y, A as in Example 3.1. Let A_n be defined by

$$[A_n w](t) = \begin{cases} \dfrac{w(t+\frac{1}{n}) - w(t)}{\frac{1}{n}} & \text{for} \quad 0 \le t \le 1 - \frac{1}{n} \\[2ex] \dfrac{w(1) - w(1-\frac{1}{n})}{\frac{1}{n}} & \text{for} \quad 1 - \frac{1}{n} \le t \le 1\,. \end{cases} \tag{11}$$

Herewith, let $w \in C^1[0,1]$, i. e. we consider the operators A_n only on $D_n := \mathring{X}$[1] $(n = 1,2,\ldots)$.
Again, the assumptions (i) and (ii) of Theorem 2.2 are fullfilled (by the same arguments as in Example 3.1).
Now, let $x_n \to x$ in W and $A_n x_n \to y$ $(n \in \mathbb{N}' \subset \mathbb{N})$, i. e.

$$||x_n - x||_\infty \to 0, \quad ||A_n x_n - y||_\infty \to 0.$$

Then, for $0 \le t < 1$ and for sufficiently large n (namely $n > \frac{1}{1-t}$):

$$\int_0^t [A_n x_n](\tau)d\tau = n\left\{\int_t^{t+\frac{1}{n}} x_n(\tau)d\tau - \int_0^{\frac{1}{n}} x_n(\tau)d\tau\right\}.$$

For $n\to\infty$, this leads to

$$\int_0^t y(\tau)d\tau = x(t) - x(0),$$

i. e.

$$x'(t) = y(t). \tag{12}$$

(12) also holds for $t = 1$ as can be shown easily.
Thus, from $x_n \to x$ in W and $A_n x_n \to y$ in Y we really get $x \in D(A)$ and $Ax = y$ such that also assumption (iii) of Theorem 2.2 is fullfilled.

Remark 3.1: The considerations in Example 3.2 concerning the forward difference quotients $A_n = \partial_n$ hold analogously for the backward difference quotients $\bar{\partial}_n$, even if $\bar{\partial}_n$ is at the ends of the interval not defined analogously to (11) but by other suitable rules.
So - e. g. - the results of Example 3.2 also hold if the operators ∂_n used there will be replaced by

$$[\bar{\partial}_n w](t) := \begin{cases} \dfrac{w(\frac{1}{n}) - w(0)}{\frac{1}{n}} & \text{for } 0 \le t \le \frac{1}{n} \\[2ex] \dfrac{w(t) - w(t-\frac{1}{n})}{\frac{1}{n}} & \text{for } \frac{1}{n} \le t \le 1-\frac{1}{n} \\[2ex] \dfrac{w(1-\frac{1}{n}) - w(1-\frac{2}{n})}{\frac{1}{n}} & \text{for } 1-\frac{1}{n} \le t \le 1. \end{cases} \tag{13}$$

These particular operators $\bar{\partial}_n$ $(n = 1,2,\ldots)$ will be helpful when treating the next example.

---

1) though (11) would make sense also for (e. g.) $w \in \mathring{Y}$.

Example 3.3: In view of boundary value problems, we ask for the validity of the assumptions of Theorem 2.2 in the case of $\mathring{X}=C^2[0,1] \cap \mathring{W}$ with $\mathring{W} = \{x \in C^1[0,1] \mid x(0) = x(1) = 0\} \subset \mathring{Y}$ where $\mathring{Y} = C^0[0,1]$,

$d_X(u,v) = ||u-v||_\infty + ||u'-v'||_\infty + ||u''-v''||_\infty$,

$d_W(u,v) = ||u-v||_\infty + ||u'-v'||_\infty$,

$d_Y(u,v) = ||u-v||_\infty$,

$A = \frac{d^2}{dt^2}$, $D(A) = \mathring{X}$,

$$[A_n\, w](t) := \frac{w(t+\frac{1}{n}) - 2w(t) + w(t-\frac{1}{n})}{\frac{1}{n^2}} \quad \text{for} \quad \frac{1}{n} \le t \le 1-\frac{1}{n}$$

and then

$$[A_n\, w](t) = [A_n\, w](\tfrac{1}{n}) \qquad \text{for} \quad 0 \le t \le \frac{1}{n}, \tag{14}$$

$$[A_n\, w](t) = [A_n\, w](1-\tfrac{1}{n}) \qquad \text{for} \quad 1-\frac{1}{n} \le t \le 1,$$

where only functions $w \in \mathring{X}$ are admitted [1)], i. e. $D_n = \mathring{X}$ $(n=1,2,\ldots)$. Again, the assumptions (i) and (ii) of Theorem 2.2 are fullfilled because the boundedness of a sequence $\{x_n\}$ in X implies the boundedness of $\{x'_n\}$ in the maximum-norm, hence the equicontinuity of the elements of $\{x_n\}$, as well as the boundedness of $\{x''_n\}$, hence the equicontinuity of the elements of $\{x'_n\}$. From the Arzelà-Ascoli Theorem,

$$\exists\, x \in \mathring{Y}, \quad \exists\, z \in \mathring{W}, \quad \exists\, \mathbb{N}' \subset \mathbb{N} \quad \text{such that}$$

$$||x_n - x||_\infty \to 0, \quad ||x'_n - z||_\infty \to 0 \qquad (n \in \mathbb{N}').$$

From Example 3.1, we get $z = x'$ such that $x \in \mathring{W}$ and $x_n \to x$ in W $(n \in \mathbb{N}')$.[2)]

In order to prove the validity of assumption (iii), consider a sequence $x_n \to x$ in W $(x_n \in D_n = \mathring{X})$, i. e.

$||x_n - x||_\infty + ||x'_n - x'||_\infty \to 0$ for a certain $x \in \mathring{W}$, with $||A_n\, x_n - y||_\infty \to 0$ for a certain $y \in \mathring{Y}$.

Obviously,

$$|[\partial_n\, x_n](t) - x'_n(t)| = n\, \left| \int_t^{t+\frac{1}{n}} \{x'_n(\tau) - x'_n(t)\}d\tau \right| \tag{15}$$

with the operators ∂_n from Example 3.2. The functions x'_n are equicontinuous because they are continuous and because they converge

---

1) though (14) would make sense also for (e. g.) $w \in \mathring{Y}$.

2) If the approximation of weak solutions of boundary value problems is under consideration, it should be realized that also in this case theorems concerning the compact embedding between Sobolev spaces are available (cf. [8] or [13]).

uniformly to the continuous function x'. Thus, (15) leads to

$$||\partial_n x_n - x'_n||_\infty \to 0.$$

This, together with $||x'_n - x'||_\infty \to 0$ yields

$$||\partial_n x_n - x'||_\infty \to 0, \tag{16}$$

i. e.

$$\partial_n x_n \to x' \quad \text{in} \quad C^0[0,1].$$

The particular choice of $\bar{\partial}_n$ at the ends of the interval [0,1] (cf. (13)) allows to write the operators A_n defined by (14) as

$$A_n = \bar{\partial}_n \partial_n.$$

If we replace in Example 3.2 the operators ∂_n by $\bar{\partial}_n$ (see Remark 3.1), the sequence $\{x_n\}$ by $\{\partial_n x_n\}$, and x by x', (16) leads to the existence of (x')' and to

$$y = x'' = Ax \text{ }^{1)}. \tag{17}$$

Thus, $x \in \overset{\circ}{X} = D(A)$, and (17) then stands for discrete closure of $[\{A_n\}, A]$ with respect to $(W, D(A), \{D_n\})$. Therefore, also assumption (iii) of Theorem 2.2 is fullfilled, indeed.

<u>Example 3.4</u>: X, W, Y, A_n, A, D(A), $\{D_n\}$ as in Example 3.3. Hence, the pair $[\{A_n\}, A]$ is discretely closed with respect to $(W, D(A), \{D_n\})$. Let

$$B = A + C, \quad B_n = A_n + C_n \quad (n = 1,2,...)$$

with

$$[Cx](t) := f(t, x, x'), \quad x \in D(C) := \overset{\circ}{X},$$

and with

$$\left.\begin{aligned} &[C_n x_n](t) = f\left(t, x_n(t), \frac{x_n(t+\frac{1}{n}) - x_n(t-\frac{1}{n})}{\frac{2}{n}}\right) \\ &\qquad\qquad \text{for } \frac{1}{n} \le t \le 1-\frac{1}{n} \\ &\text{and then} \\ &[C_n x_n](t) = [C_n x_n](\tfrac{1}{n}) \quad \text{for } 0 \le t \le \tfrac{1}{n} \\ &[C_n x_n](t) = [C_n x_n](1-\tfrac{1}{n}) \text{ for } 1-\tfrac{1}{n} \le 1 \le 1 \end{aligned}\right\} \tag{18}$$

where f(t, v, w) maps $[0,1] \times \mathbb{R} \times \mathbb{R}$ into $\mathbb{R}$ and is a uniformly lipschitz-continuous function (uniformly in t) with respect to v and w:

---

1) The linear homogeneous boundary conditions included in the definition of $\overset{\circ}{W}$ do not effect the arguments.

$$\exists\, L,K \in \mathbb{R}: |f(t,\tilde{v},\tilde{w}) - f(t,v,w)| \le L|\tilde{v}-v| + K|\tilde{w}-w|. \tag{19}$$

We consider the operators C_n only on the domains $D_n = \overset{\circ}{X}$ of the operators A_n $(n = 1,2,\ldots)$. Thus, also the operators B_n map D_n into $\overset{\circ}{Y}$ $(n = 1,2,\ldots)$, and $D(B) = D(C) = D(A) = \overset{\circ}{X} = C^2[0,1]$. Our aim is to show the discrete closure of $[\{B_n\}, B]$ by application of Theorem 2.4. In order to do this, it is sufficient to show $C_n \overset{c}{\to} C$ on W:

Indeed, if $x_n \to x$ (in W) $(x_n \in D_n)$ and for $\frac{1}{n} \le t \le 1-\frac{1}{n}$, we get from (19)

$$|[C_n x_n](t)-[Cx](t)| = |f(t,x_n(t), \frac{x_n(t+\frac{1}{n})-x_n(t-\frac{1}{n})}{\frac{2}{n}}) - f(t,x(t),x'(t))|$$

$$\le L|x_n(t)-x(t)| + K|\frac{x_n(t+\frac{1}{n})-x_n(t-\frac{1}{n})}{\frac{2}{n}} - x'(t)|$$

$$\le L||x_n-x||_\infty + K|x_n'(t-\frac{\vartheta}{n}) - x'(t)|$$

$$\le L||x_n-x||_\infty + K|x_n'(t-\frac{\vartheta}{n}) - x'(t-\frac{\vartheta}{n})|$$

$$+ K|x'(t-\frac{\vartheta}{n}) - x'(t)|$$

$$\le L||x_n-x||_\infty + K||x'_n-x'||_\infty + K|x'(t-\frac{\vartheta}{n}) - x'(t)|$$

with a certain $\vartheta = \vartheta(n) \in (-1, 1)$; analogous estimates hold on $0 \le t \le \frac{1}{n}$ and on $1-\frac{1}{n} \le t \le 1$, i.e. $C_n x_n \to Cx$ (in Y). Hence $C_n \overset{c}{\to} C$ on W.

Now - from Theorem 2.2 -: $\{B_n\}$ asymptotically quasi-regular.

Therefore - for $y_n \to y$ in the C^0-norm topology - the solutions x_n of $B_n x_n = y_n$ $(n \in \mathbb{N}''$ for a certain subset $\mathbb{N}'' \subset \mathbb{N})$ converge to a solution of $Bx = y$ in the C^1-norm topology.

<u>Remark 3.2</u>: A particular case of Example 3.4, namely the linear case $f(t,v,w) = a(t)v + b(t)w$ with continuous functions a and b, was treated by Vainikko in [12], but in a setting that describes the numerical procedure on a set of finite dimensional spaces. The example was again treated - but in the setting used here - in [2] where even asymptotic regularity of $\{B_n\}$ was shown - hence convergence in the C^2-norm topology -, but the set $\overset{\circ}{X}$ was restricted to a set of C^2-functions the elements of which have equitone second derivatives[1].

---

1) all x" are monotone increasing or they are all monotone decreasing

Example 3.4 shows that this monotonicity assumption can be omitted provided that we restrict ourselves to quasi-regularity which means in this case convergence of the descretization method in the C^1-norm only, but not necessarily in the C^2-norm though the limit is a solution of the problem, i. e. a C^2-function.

References

[1] Anselone, P. M. : Collectively compact operator approximation theory. Prentice Hall Publ., Englewood Cliffs 1971

[2] Anselone, P. M. and R. Ansorge : Compactness principles in nonlinear operator approximation theory. Numer. Funct. Anal. and Optimiz. 1 (1979), 589 - 618

[3] Ansorge, R. : Zur Frage der Verallgemeinerung des Äquivalenzsatzes von Lax. Numer. Math. 8 (1966), 178 - 185

[4] du Bois-Reymond, P. : Sitzungsber. Preußische Akad. der Wiss., Berlin (1886), 359 - 360

[5] Courant, R. : Über eine Eigenschaft der Abbildungsfunktionen bei konformer Abbildung. Gött. Nachr. (1914), 101 - 109

[6] Grigorieff, R. D. : Verallgemeinerte approximativ kompakte Operatoren. Frankfurt 1969

[7] Grigorieff, R. D. : Über diskrete Approximation nichtlinearer Gleichungen 1. Art. Math. Nachr. 69 (1975), 253 - 272

[8] Mazja, W. : Zur Theorie Sobolewscher Räume. Teubner, Leipzig 1981

[9] Reinhardt, H.-J. : Analysis of approximation methods for differential and integral equations. Springer, New York Berlin Heidelberg Tokyo 1985

[10] Rinow, W. : Die innere Geometrie der metrischen Räume. Springer, Berlin Göttingen Heidelberg 1961

[11] Stummel, F. : Discrete convergence of mappings. In: Proc. Conf. Numer. Anal. Dublin 1972 (ed.: J. Miller) Academic Press, New York London 1973

[12] Vainikko, G. : Funktionalanalysis der Diskretisierungsmethoden. Teubner, Leipzig 1976

[13] Wloka, J. : Partielle Differentialgleichungen. Teubner, Stuttgart 1982

ON THE INCLUSION OF SOLUTIONS IN INITIAL VALUE PROBLEMS FOR ORDINARY DIFFERENTIAL EQUATIONS BY MEANS OF INTERVALANALYTICAL METHODS

Hartmut Bauch[1], Volkmar Wiebigke[2]

1. Introduction

Given the initial value problem in a system of m ordinary differential equations of first order

$$Ty(t) := y'(t) - f(t,y(t)) = 0, \quad t \in [0,b] = B \subset R, \tag{1.1}$$

with either a fixed input vector $y(0) = c \in R^m$ or an arbitrary vector c within a given interval $C \in I(R^m)$. Moreover, there exists a continuous, interval-vector-valued function $Y \in \tilde{C}_m$ with the boundary functions $\underline{y},\bar{y}$: $B \rightarrow R^m$, which are continuously differentiable on B, such that all components f_d, d=1(1)m, of f possess sufficiently smooth partial derivatives w.r.t. y_j, j=1(1)m, and that f and the derivatives of f are continuous in

$$D_0(Y) = \{(t,y) : t \in B \text{ and } y \in [\underline{y}(t),\bar{y}(t)]\}.$$

Monotonicities of any kind are not required for f.

Tz denotes the defect of equation (1.1) w.r.t. z, and $\tilde{C}_m(B)$ the set of all continuous interval-vector-valued functions with m components, defined on the interval B. Under the given preconditions or similar ones the initial value problem (1.1) possesses (exactly) one continuously differentiable solution y_c^* in B for each $c \in C$.

For the set of all solutions $\{y_c^*\}_{c \in C}$ safe inclusions in $\tilde{C}_m(B)$ are to be found which can be determinated numerically with reasonable computing expenditure and high precision. Here special attention has to be paid to

- computer-implemented execution of construction such that all kinds of errors (input errors, procedural errors, truncation errors, round-off errors) are safely taken into account. (For the employment of a rounded (machine) interval arithmetic see MOORE (1966) or the paper by ADAMS, CORDES, LOHNER in this volume.)
- the blowing-up of intervals (the so-called "wrapping" effect), i.e. the presentation errors and
- the applicability of an algorithm also in the case of selected (difficult) mathematical models.

----

1) Paedagogische Hochschule "K.F.W. Wander" Dresden, Sekt. Mathematik
DDR - 8060 Dresden Wigardstr. 17

2) Technische Hochschule "Carl Schorlemmer" Leuna-Merseburg, Sekt. Mathematik
DDR - 4200 Merseburg Otto-Nuschke-Strasse

In this paper the following notations are used:
$A = [\underline{a}, \bar{a}]$ represents an interval of elements, where $\underline{a}, \bar{a}$ are real numbers, vectors and matrices, resp. The set of interval numbers, interval vectors with m components or interval (m,n)-matrices is denoted by $I(R)$, $I(R^m)$, $I(R^{m,n})$, resp. The ordering used is always the natural (componentwise) ordering. Arithmetic operations among these quantities are defined as usual (a working knowledge of interval arithmetics is presupposed) as well as interval evaluations of real functions $f(x)$ which are denoted by $f(X)$ and which are defined componentwise for vector functions. The width of an interval quantity is denoted by

$$w(A) = w([\underline{a}, \bar{a}]) := \bar{a} - \underline{a}.$$

Remarks:

1) The more general problem, e.g. investigated by WIEBIGKE (1983), LOHNER (1985) and ADAMS, CORDES, LOHNER (in this volume)

$$\{T_\lambda x(t) = x'(t) - f(t,x(t),\lambda) = 0, \quad t \in B \subset R, \qquad x(0) = c \in C \in I(R^n), \quad \lambda \in \Lambda \in I(R^k)\} \tag{1.2}$$

which also contains a pertubation parameter $\lambda \in \Lambda \in I(R^k)$ in the differential equations can be reduced to problem (1.1) by transferring Λ into an initial condition. However then the dimension of the problem increases by k, since by means of

$$y^T := (x,\lambda)^T \in R^M, \; M := m+k$$

$$g(t,y(t)) = g(t,x(t),\lambda) = (f(t,x(t),\lambda),0)^T \tag{1.3}$$

$$g: B \times R^m \rightarrow R^M$$

the equivalence of (1.2) with the original problem (1.1) extended to the dimension m+k

$$y'(t) = g(t,y(t)), \; t \in B \subset R, \; y(0) \in (C,\Lambda)^T \in I(R^M) \tag{1.4}$$

can be proved.

2) An autonomous form of Problem (1.1) is obtained by the extension

$$y_{m+1} := t\,, \quad y'_{m+1} = 1, \quad y_{m+1}(0) = 0\,. \tag{1.5}$$

3) In the present paper a review is given of some of the authors' results on constructing iterative methods for solving the problem (1.1). The relationship with enclosure methods for problem (1.1) given in related papers by other authors is not discussed here. The same applies to the numerical treatment of the characteristic initial-value problem with hyperbolic differential equation $z_{st} = f(s,t,z)$ and boundary value problems.

For solving problem (1.1) and (1.2), resp., several methods using interval techniques have been developed up to now, which may not be classified uniquely into a few groups. However, there are three classes of methods for proving the existence of a solution to the initial-value problem (1.1). With any of these a class of corresponding numerical approximation methods can be associated. This correspondence is shown in the following table (see NICKEL (1983)):

Existence Theorem	Numerical Method
Peano	Finite Differences
Picard-Lindeloef	Numerical Quadrature and Iteration
Power Series Method	Taylor Series

Due to lack of space no further statements are made on the following numerical principles (methods):

a) Interval difference methods resp. discretization methods (bounds are computed by inclusion of the local discretization error of the approximation method, see e.g. MOORE (1966,1979), KRUECKEBERG (1969), SHOKIN (1981), BACHMANN (1983), WIEBIGKE (1983), ADAMS (1982), LOHNER (1985) ADAMS, CORDES, LOHNER (1986).

b) Power series method (e.g. REICHMANN (1980), NICKEL (1983)).

c) The defect method (a posteriori bounds by using e.g. inclusion theorems by SCHROEDER (1961), WALTER (1964), SCHOLZ (1968), BAUCH (1973), MARCOWITZ (1975), CONRADT (1980), BOEHMER, HEMKER, STETTER (1984) and others).

d) Combined forms of a) to c) or special techniques, e.g. for fighting the wrapping effect (see NICKEL (1985), ADAMS et al. (in this volume)).

Since the set of all solutions of the initial value problem (1.1) is in general of such a complex nature that its numerical computation causes difficulties, solution bounds in form of so-called interval (outer) hulls (or parallelepipeds and simplices, resp.) are determined.

The following explanations are confined to fixed point theorems and iterative methods for determining bounds. In general, the iterative improvement of bounds is carried out in an integral equation of Volterra-type equivalent to the original problem (1.1). One gets bounds for a solution y_c, if the integrand is estimated. The determination of extreme values necessary for this (but generally not feasible) may be avoided by using either

(a) additional demands on f, e.g.

- monotonicity (see e.g. WALTER (1964), ADAMS, SCHEU (1975), ADAMS (1982) and others),
- monotonicity and a special form of f which leads to alternating upper and lower bounds (see e.g. AMES, GINSBERG (1974)),
- use of monotone decomposible operators (see e.g. COLLATZ and

others),

or

(b) strict employment of weighting functions resp. a Lipschitz condition on f

- TSCHAPLYGIN (1950) defined a tangent/secant-method for determining lower and upper bounds,
- RAITH (1981) made strict use of the Lipschitz constant in order to guarantee simple computable bounds,
- WIEBIGKE (1983) extends RAITH's method for problems of the type (1.2),

or

(c) interval extension of the integrand

- BAUCH (1973), (1980), (1982) made use of a Taylor expansion of f w.r.t. y, which leads to a contractive interval-integral-operator.
- ADAMS (1982), LOHNER (1985), ADAMS et al. (in this volume) made use of the mean value theorem, which provides a representation of a solution y_o^* in terms of parameters standing for (local) input or procedural errors.

For all iterative methods (and also for some other procedures) a priori bounds are required. In order to obtain such (rough) bounds fixed-point theorems can be applied.
For the iterative improvement of bounds using the fixed-point theorem of BANACH and an interval-Picard iteration see e.g. BAUCH (1973). (The numerical problems of this method are considered to be solved. Even the simpliest version of this method guarantees safe bounds, if the existence of an inclusion isotonic interval extension is guaranteed for the right-hand side of (1.1). "One big advantage in using Picard-iteration is that no stability problems occur. One of the disadvantages, however, is the slow (linear) convergence. This may be the reason why, up to now, this class of metods has been used only occasionally" (NICKEL (1983)).)
Because of the slow convergence of PICARD's method it is an obvious step to develop an implementable interval extension of the (real) Newton method for solving problem (1.1). But "sharp" bounds for a solution y_c^*, $c \in C$ fixed, of problem (1.1) can only be guaranteed, if in the iteration steps no estimations are performed for the first derivatives of f. For nonlinear equations ALEFELD (1974) constructed an interval-Newton method, which does not require an interval extension of the first derivatives. It is possible to generalize this idea and so one gets the following interval-Newton method for solving problem(1.1).

2. An interval-Newton-algorithm

Under the usual conditions of differentiability on f in (1.1), we obtain according to the Taylor expansion of f w.r.t. y_c^*, $c\in C$ is fixed, the theoretical foundation for the following interval-Newton-method.

Theorem 1:

For the initial-value problem (1.1) with a fixed initial value $c \in C$, the following conditions are to hold:

- There exists a continuous, interval-valued function $Y \in \tilde{C}_m(B)$, with the boundary functions $\underline{y}$, $\bar{y}$ which are continuously differentiable on B, such that all components f_d, $d = 1(1)m$, of f possess sufficiently smooth partial derivatives w.r.t. y_j, $j = 1(1)m$, and that f and the derivatives of f are continuous in $D_0(Y)$. Let $D_m(B)$ denote the set of all continuously differentiable vector-valued functions on B contained in $D_0(Y)$.
- There exist continuous interval-valued functions F_1 and F_2, which enclose the partial derivatives Df and D^2f of f for all arguments considered, i.e.

$$Df(t,y) := \left(\frac{\partial f_d(t,y)}{\partial y_k}\right)_{d,k=1(1)m} \in F_1(t), \quad F_1 \in \tilde{C}_{m,m}(B), \quad \text{and}$$

$$D^2f(t,y) := (D^2 f_2(t,y))_{d=1(1)m} \in F_2(t), \; F_2 \in \tilde{C}_{m\ m,m}(B), \; \forall t \in B, \; y \in Y.$$

The iteration rule (formulated implicitly for reasons of stability; see BAUCH (1980)) is defined as follows:

$$\text{(I)} \quad Q_n(t) := \{c + \int_0^t [f(s,y_n)+F_1(s)\ (Y_n(s)-y_n(s))]ds\} \cap Y_n(t),$$

$$y_n \in Y_n \text{ arbitrary, e.g. } y_n = m(Y_n)$$

$$\text{(II)} \quad Z_{n+1}(t) := c + \int_0^t [p_n(s)\ Z_{n+1}(s) + w_n(s)]ds,$$

$$q_n \in Q_n \text{ arbitrary, e.g. } q_n = m(Q_n),$$

$$p_n(t) := \begin{cases} f_y(t,q_n), & \text{for } m = 1 \\ Df(t,q_n), & \text{for } m > 1 \end{cases}$$

$$w_n(t) := f(t,q_n) - p_n(t)\ q_n(t)$$

$$W_n(t) := w_n(t) + \begin{cases} \tfrac{1}{2}F_2(t)\ (Q_n(t) - q_n(t))^2, & \text{for } m = 1 \\ \bar{F}(t, Q_n - q_n\ F_2),^{1)} & \text{for } m > 1. \end{cases}$$

$$\text{(III)} \quad Y_{n+1}(t) = Z_{n+1}(t) \cap Q_n(t), \; n \in N, \; t \in B,$$

---

1) $\bar{F}(t,Z,F_2) := (Z(t)^T F_{2,d}(t)\ Z(t))_{d=1(1)m}, \; Z \in \tilde{C}_m(B)$

where $Y_0 \subset Y$ is a first candidate for an enclosure of y_c^*, $c \in C$, with $w(Y_0)$ sufficiently small for all $t \in B$, and $y_0 = m(Y_0)$.

The rule (2.1) provides a sequence of inclusion functions $(Y_n)_{n \in N}$ for $y_c^* \in D_m(B)$, $c \in C$ is fixed, of (1.1) with the following properties for all $t \in B$ and $n \in N$ (for the proof see below):

(1) $y_c^* \in Y_n$ (2.2)

(2a) $Y_{n+1} \subset Y_n$ (2.3)

(2b) $\lim_{n \to \infty} w_m(Y_n) = 0$ [1)] (2.4)

(3) $w_m(Y_{n+1}) \leqq K\, w_m^2(Y_n)$. (2.5)

Thus the Method (2.1) represents a globally convergent interval-iteration procedure of convergence of second order according to ALEFELD (1974). As in practical computation a rounded machine interval arithmetic is used, the property 2b) as given in (2.4) is, above all, of theoretical interest.

Under appropriate conditions of differentiability on f, interval iteration methods of convergence order $p \geqq 3$, $p \in N$, may bedirectly derived from (2.1). However, the practical application of such methods is not very efficient because of the high computing expenditure.

For $w(C) > 0$, i.e. interval initial conditions are given, the initial value $c \in R^m$ in (2.1) has to be replaced (enclosed) by $C \in I(R^m)$. Thus, for the iteratives we obtain the following properties of convergence for all $t \in B$ and $n \in N$ (those properties naturally differ from the ones in (2.4) and (2.5)):

(1) $y_c^* \in Y_n$ (2.6)

(2a) $Y_{n+1} \subset Y_n$ (2.7)

(3) $w_m(Q_n) \leqq \max \{w(C), \frac{K_1}{\alpha} w_m(Y_n)\}$, $0 \leqq K_1 < \alpha$ (see proof) (2.8)

---

1) By means of

$$w_m(X) := \sup_{t \in B} (e^{-\alpha t} w(X(t)))$$

$$:= \sup_{t \in B} (e^{-\alpha t} \max_{1 \leqq k \leqq m} w(X_k(t)))$$

$$:= \sup_{t \in B} (e^{-\alpha t} \max_{1 \leqq k \leqq m} (\bar{x}_k(t) - \underline{x}_k(t)))$$

In BAUCH (1980) an adapted definition for the width of interval functions of the complete space $(\tilde{C}_m(B), r)$ is introduced, where

$$r : \tilde{C}_m(B) \times \tilde{C}_m(B) \to R^+,$$

$$r(X,Y) := \sup_{t \in B} (e^{-\alpha t} \max_{1 \leqq k \leqq m} \{ |\underline{x}_k(t) - \underline{y}_k(t)|, |\bar{x}_k(t) - \bar{y}_k(t)| \} \quad (2.9)$$

denotes a generalized MORGENSTERN-metric with the parameter $\alpha > 0$, arbitrary but not variable. To the choice of α see proof.

(4) $$w_m(Y_{n+1}) \leq \begin{cases} w_m(Q_n), & \text{if } w_m(Q_n) \leq \varepsilon \\ K\, w_m^2(Q_n), & \text{otherwise} \end{cases} \qquad (2.10)$$

with

$$\varepsilon := (\alpha\, w(C))^{1/2} \cdot (K_{10}\, K)^{-1/2} \quad \text{and} \qquad (2.11)$$

$$K := \frac{m^2\, K_2}{(\alpha - K_{10})} \sinh(b)\ . \qquad (2.12)$$

In the special case $w(C) = 0$, the relations (2.4) and (2.5) do hold instead of (2.8) and (2.10). The proof for this special case is contained in BAUCH (1980).

<u>On the proof of the properties (2.6) - (2.8), (2.10):</u>

Given interval initial conditions, i.e. $w(C) > 0$ and a first enclosure $Y_0 \subset Y$ for y_c^*:

(1) Using The Tailor expansion of f w.r.t. y_c^* we get

$$y_c^* = c + \int_0^t (f(s,y_0)+Df(s,\eta)(y_c^*(s)-y_0(s)))ds \in Y_0(t),$$

with the unknown intermediate vektor η and $c \in C$ fixed. According to the subset property, follows $y_c^* \in Q_0$, for all $t \in B$ and $c \in C$ fixed. Using Lemma H1 (see BAUCH (1980), p. 141) with the values $Q := Q_0$, $Z := Z_1$, we get for $c \in C$, arbitrary but fixed the inclusion $y_c^* \in Z_1$, for all $t \in B$. Thus $y_c^* \in Q_0 \cap Z_1 = Y_1$, for all $t \in B$ and $c \in C$. (Property (2.6)).

(2a) Because of $Y_1 \subset Q_0 \subset Y_0$, there follows the property (2.7).

(3) For Q_0 according to (2.1,I) we get:

$$w_m(Q_0) = \sup_{t \in B} (e^{-\alpha t} (w(C) + \int_0^t w_m(F_1(s)(Y_0(s)-y_0(s)))ds))$$

<u>Additional notes :</u>

$$w_m(\underbrace{F_1(s)}_{(m,m)} \cdot \underbrace{(Y_0(s)-y_0(s))}_{(m,1)}) = \max_{1 \leq d \leq m} w(\sum_{k=1}^{m} F_{1,dk}(s)\ (Y_0(s)-y_0(s))_k)$$

$$= \max_{1 \leq d \leq m} \sum_{k=1}^{m} (w(F_{1,dk}(s))\ \underbrace{|Y_0(s)-y_0(s)|_k}_{=w(Y_0(s)-y_0(s))_k \,\leq\, w_m(Y_0(s)-y_0(s))} + |F_{1,dk}(s)|\, w(Y_0(s)-y_0(s))_k)$$

$$\leq \max_{1 \leq d \leq m} \sum_{k=1}^{m} (w(F_{1,dk}(s)) + |F_{1,dk}(s)|) \cdot \overbrace{w_m(Y_0(s)-y_0(s))}^{= w_m(Y_0(s))}$$

$= K_1 \cdot w_m(Y_0(s))$, with

$$K_1 := m \cdot \max_{t \in B} \max_{1 \leq d,k \leq m} (w(F_{1,dk}(t)) + |F_{1,dk}(t)|)\ . \qquad (2.13)$$

Hence, it follows for Q_0 in (3)

$$w_m(Q_0) \leqq \sup_{t\in B} (e^{-\alpha t} (w(C)+K_1 \int_0^t e^{\alpha s} \sup_{s\in B} (e^{-\alpha s} w_m(Y_0(s)))ds))$$

$$\leqq \sup_{t\in B} (e^{-\alpha t} (w(C)+K_1 \cdot w_m(Y_0) \frac{1}{\alpha} (e^{\alpha t}-1)))$$

$$\leqq \sup_{t\in B} (e^{-\alpha t} (w(C) - \frac{K_1}{\alpha} w_m(Y_0))) + \frac{K_1}{\alpha} w_m(Y_0)$$

$$\leqq \max \{ w(C), \frac{K_1}{\alpha} w_m(Y_0) \} \quad \text{(property (2.8))}.$$

(4) Estimation of Z_1 in (2.1,II):

$$w_m(Z_1) = \sup_{t\in B} (e^{-\alpha t} w_m(C+ \int_0^t (p_0(s)\cdot Z_1(s)+W_0(s))ds))$$

Additional notes:

$$\int_0^t w_m(p_0(s)\cdot Z_1(s) + W_0(s))ds$$

$$\leqq \int_0^t \underbrace{w_m(Df(s,q_0)\ (Z_1(s)-q_0(s)))}_{:= I_1(s)}ds + \int_0^t \underbrace{w_m(\bar{P}(s,Q_0-q_0,F_2))}_{:= I_2(s)}ds$$

(a1) $I_1(s) = K_{10}\, w_m(Z_1(s))$, with

$$K_{10} := m \cdot \sup_{(t,q_0)\in D_0(Y)} \max_{1\leqq d,k\leqq m} |Df(t,q_0)|_{dk}) \leqq K_1 \tag{2.14}$$

(a2) $$A_1(t) := \int_0^t I_1(s)ds = \frac{K_{10}}{\alpha} w_m(Z_1)\ (e^{\alpha t}-1). \tag{2.15}$$

(b1) $I_2(s) = w_m(\bar{P}(s,Q_0-q_0,F_2))$

$$= \max_{1\leqq d\leqq m} w((Q_0(s)-q_0(s))^T \cdot F_{2d}(s) \cdot (Q_0(s)-q_0(s)))$$

type: (1,m) (m,m) (m,1)

$$\leqq \max_{1\leqq d\leqq m} \sum_{k,l=1}^{m} (2|F_{2d,kl}(s)|+w(F_{2d,kl}(s)))w(Q_0(s))_k w(Q_0(s))_l$$

$\leqq K_2\, m^2 \cdot w_m^2(Q_0(s))$, with

$$K_2 := \max_{t\in B} \max_{1\leqq d,k,l\leqq m} (2|F_{2d,kl}(t)|+w(F_{2d,kl}(t))). \tag{2.16}$$

(b2) $A_2(t) := \int_0^t I_2(s)ds \leqq K_2 \cdot m^2 \int_0^t e^{2\alpha s} \sup_{s \in B} (e^{-\alpha s} w_m^2(Q_0(s)))ds$

$$\leqq K_2 m^2 \frac{1}{2\alpha} w_m^2(Q_0)(e^{2\alpha t}-1).$$

Finally we get:

$$w_m(Z_1) = \sup_{t \in B} (e^{-\alpha t} (w(C) + \int_0^t w_m(p_0(s) \cdot Z_1(s) + W_0(s))ds))$$

$$\leqq \sup_t (e^{-\alpha t} (w(C) + A_1(t) + A_2(t)))$$

$$\leqq \sup_t (e^{-\alpha t}(w(C) + \frac{K_{10}}{\alpha} w_m(Z_1)(e^{\alpha t}-1)) + \frac{K_2 m^2}{2\alpha}(e^{\alpha t}-e^{-\alpha t}) w_m^2(Q_0))$$

$$\leqq \sup_t (e^{-\alpha t}(w(C) - \frac{K_{10}}{\alpha} w_m(Z_1))) + \underbrace{\frac{K_2 m^2}{\alpha} \sinh(b)}_{:= K_3} \cdot w_m^2(Q_0) + \frac{K_{10}}{\alpha} w_m(Z_1)$$

and with $\alpha > K_1 \geqq K_{10}$ there follows

$$w_m(Z_1) \leqq \frac{\alpha}{\alpha - K_{10}} (\sup_t (e^{-\alpha t}(w(C) - \frac{K_{10}}{\alpha} w_m(Z_1))) + K_3 w_m^2(Q_0))$$

$$= \frac{\alpha}{\alpha - K_{10}} \sup_t (e^{-\alpha t}(w(C) - \frac{K_{10}}{\alpha} w_m(Z_1))) + K \cdot w_m^2(Q_0). \tag{2.17}$$

<u>Case distinction:</u>

Case 1: $w(C) \leqq \frac{K_{10}}{\alpha} w_m(Z_1)$.

The second statement in (2.10) is due to

$w_m(Y_1) \leqq w_m(Z_1) \leqq K \cdot w_m^2(Q_0)$.

Case 2: $w(C) > \frac{K_{10}}{\alpha} w_m(Z_1)$. Thus we get

$w_m(Z_1) \leqq \frac{\alpha}{\alpha - K_{10}} (w(C) - \frac{K_{10}}{\alpha} w_m(Z_1)) + K \cdot w_m^2 (Q_0)$, i.e.

$w_m(Z_1) \leqq w(C) + \frac{K}{\alpha} (\alpha - K_{10}) w_m^2(Q_0)$.

Thus we get:

$$w_m(Z_1) \leqq \min \{ w(C) \frac{\alpha}{K_{10}} , w(C) + \frac{K}{\alpha} (\alpha - K_{10}) w_m^2(Q_0) \} .$$

A further case distinction yields for $K_2 \neq 0$:

(a) If

$$w(C) \cdot \frac{\alpha}{K_{10}} = w(C) + \frac{K}{\alpha}(\alpha - K_{10}) w_m^2(Q_0), \quad \text{i.e.} \quad w_m(Q_0) \geqq \varepsilon ,$$

then because of $w_m(Z_1) < w(C) \cdot \frac{\alpha}{K_{10}}$ and (2.1,III)

$$w_m(Y_1) \leqq \min \{ w_m(Q_0), \frac{\alpha}{K_{10}} w(C) \} = K \cdot w_m^2(Q_0).$$ (Second statment in (2.10))

(b) If $w_m(Q_0) \leqq \varepsilon$, then

$$w_m(Y_1) = \min \{ w_m(Q_0), w(C) + \frac{K}{\alpha}(\alpha - K_{10}) w_m^2(Q_0) \} \leqq w_m(Q_0).$$

First statment in (2.10))

By means of a complete induction on n, the validity of the statements (2.6) to (2.10) follows; q.e.d.

<u>Remarks</u>

1. An extension of method (2.1) to characteristic initial-value problems with hyperbolic differential equation $z_{st} = f(s,t,z)$ is possible (see the latest edition of [28] in English).
2. In each iteration step of (2.1,II), a fixed-point equation of the form Z = SZ, with

$$SZ(t) = c + \int_0^t (p(\hat{t})\, Z(\hat{t}) + W(\hat{t}))d\hat{t}$$

has to be solved for the determination of the enclosures Z_{n+1}. The operator S is contracting, because we get, for arbitrary inclusion functions X and Z:

$$r(SX,SZ) = \sup_{t \in B}(e^{-\alpha t} \max_{1 \leqq k \leqq m} \{ |\underline{(SX(t))}_k - \underline{(SZ(t))}_k|, |\overline{(SX(t))}_k - \overline{(SZ(t))}_k| \})$$

$$\leqq \sup_{t \in B}(e^{-\alpha t} \mid \int_0^t \underbrace{\sup_{\substack{s \in B \\ q \in Y}} |f_y(s,q(s))|}_{\leqq :K} e^{\alpha s} \cdot \underbrace{\sup_{s \in B}(e^{-\alpha s} \max_{1 \leqq k \leqq m} \{ |\underline{x}_k(s) - \underline{z}_k(s)|, |\overline{x}_k(s) - \overline{z}_k(s)| \} ds)}_{\leqq r(X,Z)}$$

$$= \frac{K}{\alpha} r(X,Z).$$

With α sufficiently large, e.g. $\alpha = 2K$, we obtain the assertation. (Note that for the practical computation no concrete value for α is required.) Since S is contracting, (2.1,II) may also be solved iteratively (PICARD-LINDELOEF-iteration). So we obtain a linear convergent sequence of enclosure functions $Z_{n+1}^{(\nu)}$, $\nu = 0,1,2,...$ for the fixed-point Z_{n+1}^{*} of of (2.1,II). Because of the inclusion monotony of the iterates Z_{n+1}, the iteration can be stopped at any ν ; the corresponding $Z_{n+1}^{(\nu)}$ is always an enclosure for Z_{n+1}^{*}.

On the other hand, the iterates Z_{n+1} may also be given explicitly by an interval-analytical evaluation of the formula for the solution of linear differential equations; to this see e.g. BAUCH (1982) and NICKEL (1985), resp. This leads to better numerical results than the integral estimations proposed by AVENHAUS (1970). (Rule (2.1,II) has to be appropriately modified.)

If we drop the interval concept for the determination of the iterates Z_{n+1} and in place of (2.1,II) consider the corresponding real differential problems, then traditional methods can be used for its solution. But in each new step of method (2.1) a continuous enclosure has to be constructed from the data obtained by the solution of the linear auxiliary problems. This causes extraordinary difficulties since, with the progress of the iteration, not only enclosures are required but also sharper and sharper bounds. Finally we mention that at present a combination of known (real) approximation methods with interval-analytical construction of the bounds is designed in order to reduce the computing expenditure of method (2.1) and to fight the wrapping effects.

3. If a partition of B into subintervals is nessesary, then there occurs a corresponding number of initial-value problems with "perturbed" (interval) initial conditions. Because of $C_i := Y_{n,i-1}(t_i)$, the initial values $C_i \in I(R^m)$, i=1,2,... are directly influenced by the width of the available enclosures $Y_{n,i-1}$, defined on $[t_{i-1},t_i]$, i=1,2,... . Therefore NICKEL (1985) investigated under which conditions optimal enclosures Y_{i-1} for $t \in [t_{i-1},t_i]$, i=1,2,... can be constructed. He devoted special attention to the wrapping effect of interval methods for the numerical solution of (linearized) systems of differential equations. If a linear system is given in (1.1) and the initial set C is a simplex, parallelepiped, ellipsoid and convex set, resp., then the set of solutions $\{y_c^*\}_{c \in C}$ belongs to the same corresponding class (see NICKEL (1985)). Thus, in principle, it is possible according to the special situation to avoid wrapping effects. But note that, in general,

wrapping effects are unavoidable, if the boundary of the set of solutions is not of a very special type. (In general, in practical computation only a reduction of the influence of the wrapping effects on the quality of the bounds is possible.)

4. The implementation of the interval-iteration methods mentioned above requires

- an efficient machine-interval arithmetic (rounded interval arithmetic) and
- an appropriate approximation concept in order to guarantee e.g. the computer presentation and processing of (interval) bounding functions.

(Differential or integral equations are posed in function spaces; however, the practical computation is confined to finite dimensional approximations. This is the principal difficulty in the construction of methods yielding safe (guaranteed) bounds.)

3. Some remarks on RAITH's method

Generalizing the secant-tangent method used by CHAPLYGIN (1950), RAITH (1981) developed an explicit single-step method. Proceeding from bounds at an initial point this method simultaneously generates lower and upper bounding functions, which for $h \to 0$ in B converge uniformly towards y^* with an arbitrarily given order. It is assumed that f in (1.1) is a continuous function that satisfies a Lipschitz condition

$$|f_d(t,y) - f_d(t,z)| = \sum_{j=1}^{m} l_{dj}\,|y_j - z_j|, \quad \text{for } t \in B;\; y,z \in R^m; \quad 0 \leqq l_{dj} \in R,\; d=1(1)m. \tag{3.1}$$

if the l_{dj} are collected in a matrix L and the step size h is fixed such that

$$0 \leqq h \leqq b \text{ and } h \cdot \| L+2\,(L-d(L)) \| < 1, \tag{3.2}$$

then bounding functions for the set of all solutions of (1.1) in $B' = [0,h] \subset B$ may be determined as follows:

I. a priori-bounds

Let $\underline{\alpha} := f(0,\underline{c})$; $\bar{\alpha} := f(0.c)$

$$\underline{z}(t) := \underline{c} + \underline{\alpha}\, t; \quad \bar{z}(t) := \bar{c} + \bar{\alpha}\, t, \quad t \in B' \tag{3.3}$$

$$\beta := (\bar{\beta}, \underline{\beta}), \text{ with } \bar{\beta} := (\beta_1, \beta_2, \ldots, \beta_m);\; \underline{\beta} := (\beta_{m+1}, \ldots, \beta_{2m})$$

be the solution of the linear system $A\,\beta = b$, which

$$A := \begin{pmatrix} A_1 & A_2 \\ A_2 & A_1 \end{pmatrix}, \quad \begin{matrix} A_1 := E - h(2L - d(L)) \\ A_2 := - h(L - d(L)) \end{matrix} \quad \text{of the type } (m,m)\ ,$$

$$b := \begin{pmatrix} \left(-\bar{\alpha} + \max\limits_{t\in B'} f(t,\bar{z}) + (L-d(L)) \max\limits_{t\in B'} (\bar{z}(t)-\underline{z}(t))\right) \\ \left(\underline{\alpha} - \min\limits_{t\in B'} (f(t,\underline{z}) + (L-d(L)) \max\limits_{t\in B'} (\bar{z}(t)-\underline{z}(t))\right) \end{pmatrix} \qquad (3.4)$$

$u_0(t) := \underline{c} + (\underline{\alpha} - \beta)\, t; \quad v_0(t) := \bar{c} + (\bar{\alpha}+\bar{\beta})\, t$.

Therefore (see RAITH (1981), Theorem 4.1, p.30)

$u_0(t) \leqq y^*(t) \leqq v_0$, for all $t \in B'$ and all y^* with $y^*(0) \in C$.

II. Iterative Improvement of the bounding functions

Iteration rule (RAITH (1981)):

$u_{n+1}(t) := u_n(t) + r_{n+1}(t)$

$v_{n+1}(t) := v_n(t) - s_{n+1}(t)$, with

$$r'_{n+1}(t) + d(L)\cdot r_{n+1}(t) = -Tu_n(t) - (L-d(L))(v_n(t) - u_n(t))$$
$$r_{n+1}(0) = 0$$
$$s'_{n+1}(t) + d(L)\cdot s_{n+1}(t) = Tv_n(t) - (L-d(L))(v_n(t) - u_n(t))$$
$$s_{n+1}(0) = 0.$$

The solutions may be given explicitly (see for RAITH (1981), p.37).

For all $t \in B' \subset B$ and all $y(0) \in C$:

$r_{n+1}, s_{n+1} \geqq 0$, i.e. $u_n \leqq u_{n+1} \leqq \ldots \leqq y^* \leqq \ldots \leqq v_{n+1} \leqq v_n$, for all $n \in N$.

The continuation of the method in B is possible, if (3.2) is satisfied. The properties of the bounds can be guaranteed, even if "small perturations" (rounding errors, computation of extreme values and definite integrals) occur, for this see RAITH (1981, p.40). Especially the bounding functions remain globally bounded for $h \to 0$. This property is of significance for the convergence of the sequences (u_n), (v_n), see RAITH (1981, Theorem 5.4, p.53). Because of expenditure considerations it seems to be advisable to generalize RAITH's method for problems of the form (1.2). Thus the transformation (1.3) can be avoided, and, consequently, there is no increase of dimension m of the problem; see for this WIEBIGKE (1983). Here, let the function f be Lipschitz bounded and possess for all a continuous intervall extension F on $B \times I(R^m) \times \Lambda$. Thus in (3.3) and (3.4)) the parameter λ is to be added to the arguments of f and is to be estimated in the same sense to get the extremes.

---

1) This statement remains valid, if the vector b has to be replaced by $\tilde{b} \geqq b$ (computation of the extremes!); see RAITH (1981, remark p.34).

Due to the generalized problem (1.2) the correction funktions r_{n+1} and s_{n+1} are constructed by means of two auxiliary interval polynomials

$$\varrho_{n+1}(t) := -T_\Lambda u_n(t) - (L-d(L))\,(v_n(t)) \tag{3.5}$$

$$\sigma_{n+1}(t) := T_\Lambda v_n(t) - (L-d(L))\,(v_n(t)) \tag{3.6}$$

So we have

$$r_{n+1,d}(t) := \begin{cases} \int_0^t (1+l_{dd}(t-s))\,\varrho_{n+1,d}(s)ds, & \text{if } \min_{t\in B''} \varrho_{n+1,d}(t)=0 \\ 0, \text{ otherwise} & \end{cases} \tag{3.7}$$

$$s_{n+1,d}(t) := \begin{cases} \int_0^t (1+l_{dd}(t-s))\,\underline{\sigma}_{n+1,d}(s)ds, & \text{if } \min_{t\in B''} \underline{\sigma}_{n+1,d}(t)=0 \\ 0, \text{ otherwise} & \end{cases} \tag{3.8}$$

Under the condition $(L-d(L))r_{n+1}>0$, $(L-d(L))s_{n+1}>0$ no deterioration of the bounds occurs (for growing n), i.e. we have

$$u_n \leqq u_{n+1} \leqq \ldots \leqq y^* \leqq \ldots \leqq v_{n+1} \leqq v_n, \quad \text{for all } t\in B''\subseteq B'\subset B \text{ and } y^* \text{ with } y^*(0)\in C.$$

Example (WIEBIGKE (1983, p.80)):

The following interval-initial-value problem is given:

$$\begin{aligned} y_1'(t) &= \lambda_1\, y_2(t) + \lambda_2\, y_3(t) + t^2, & c_1=y_1(0) &\in [1;\ 1.01] \\ y_2'(t) &= -0.2\, y_2(t) - 0.3\, y_3(t) - t, & c_2=y_2(0) &\in [2;\ 2.01] \\ y_3'(t) &= \lambda_3\, y_1(t) - 0.5\, y_2(t) + 1, & c_3=y_3(0) &\in [0.99;\ 1] \end{aligned} \tag{3.9}$$

for all $t\in[0;\ 1]$, $\lambda_1\in[0.3;\ 0.31]$, $\lambda_2\in[0.2;\ 0.21]$, $\lambda_3\in[0.1;\ 0.0.11]$.

Using the method by WIEBIGKE (step size h=0.5) we get after three iteration steps solution bounds Y with rel $w(Y_1(1.0))$ = 0.073 86 rel $w(Y_2(1.0))$ = 0.090 01, rel $w(Y_3(1.0))$ = 0.090 47, where rel $w(Y(t)) := w(Y(t))/m(Y(t))$, $m(Y(t)) \neq 0$. As to practical computation see WIEBIGKE (1983).

If we apply the transformation (1.3) to problem (3.9), we obtain analogous results by using RAITH's method and the Newton method (2.1), resp. (Deviations occur only in the last two digits.)

Without doubt, for avoiding the "wrapping effect" the parallelepiped enclosures used by the authors mentioned above have in general a positive effect on the width of the computed bounds.

4. References

[1] ADAMS, E.: Invers-Monotonie, direkte und indirekte Intervalmethoden. Ber. Math.-Statist. Sekt. Forsch. Graz 185 (1982).

[2] ADAMS, E., CORDES, D., LOHNER, R.: Enclosure of Solution of Ordinary Initial Value Problems and Applications. Lecture held at the International Seminar on Applied Mathematics ISAM-86, Weissig, GDR (1986) - in this volume.

[3] ADAMS, E. SCHEU, G.: Zur numerischen Konstruktion konvergenter Schrankenfolgen fuer Systeme nichtlinearer, gewoehnlicher Anfangswertaufgaben. Interval Mathematics, Ed. by K. Nickel, Lecture Notes in Computer Science 29, Springer-Verlag, S. 279 - 287 (1975).

[4] AMES, W.F., GINSBERG, M.: Bilateral Algoritms and their Applications. Computational Mechanics. Lecture Notes in Mathmatics 461, ed. by A. Dold, B. Eckmann. International Conference on Computational Methods in Nonlinear Mechanics, Austin, Texas. Springer-Verlag, S. 1 - 32 (1974).

[5] ALEFELD, G.: Stets konvergente Verfahren hoeherer Ordnung zur Berechnung von reellen Nullstellen. Computing 13, 55 - 65 (1974).

[6] AVENHAUS, J.: Zur numerischen Behandlung des Anfangswertproblems mit exakter Fehlererfassung. Dissertation, Universitaet Karlsruhe (1970).

[7] BACHMANN, K.-H.: Fehlereinschliessung fuer Naeherungsloesungen von Systemen gewoehnlicher Differentialgleichungen. ZAMM 63, 63 - 65 (1983).

[8] BAUCH, H.: Loesungseinschliessung bei Anfangswertaufgaben gewoehnlicher Differentialgleichungen mit Hilfe intervallanalytscher Methoden. Dissertation, Universitaet Dresden (1973).

[9] BAUCH, H.: Zur Loesungseinschliessung bei Anfangswertaufgaben gewoehnlicher Differentialgleichungen nach der Defektmethode. ZAMM 57, 387 - 396 (1977).

[10] BAUCH, H.: Zur iterativen Loesungseinschliessung bei Anfangswertproblemen mittels Intervallmethoden. ZAMM 60, 137 - 145 (1980).

[11] BAUCH, H.: Zur iterativen Loesung von Anfangswertaufgaben bei gewoehnlichen Differentialgleichungen. ZH. vychisl. mat. i mat. fiz. 22, 309 - 321 (1982). See also (engl. transl.): U.S.S.R. Comput. Maths. Math. Phys. 22, 58 - 72 (1982).

[12] BOEHMER, K.; HEMKER, P.; STETTER, H.J.: The Defect Correction Approach. Computing, Suppl. 5, 1 - 32 (1984).

[13] CHAPLYGIN, S.A. see TSCHAPLYGIN, S.A.

[14] CONRADT, J.: Ein Intervallverfahren zur Einschliessung des Fehlers einer Naeherungsloesung bei Anfangswertaufgaben fuer Systeme von gewoehnlichen Differentialgleichungen. Freiburger Intervall-Berichte 80/1, Universitaet Freiburg i. Br. (1980).

[15] KRUECKEBERG, F.: Ordinary Differential Equations. Topics in Interval Analysis. Ed. by E. Hansen, Oxford University Press, 91 - 97 (1969).

[16] LOHNER, R.: Enclosing the Solutions of Ordinary Initial- and Boundary-value Problems. In: Proceedings of the 11th IMACS World Congress, Oslo, 1985, Vol. 1, 99 - 102 (1985).

[17] MARCOWITZ, U.: Fehlerabschaetzung bei Anfangswertaufgaben fuer Systeme von gewoehnlichen Differentialgleichungen mit Anwendung auf das Reentry-Problem. Numer. Math. 24, 249 - 275 (1975).

[18] MOORE, R.E.: Interval Analysis. Prentice-Hall Inc., Englewood Cliffs, N. J. (1966).

[19] MOORE, R.E.: Methods and Applications of Interval Analysis. SIAM, Philadelphia (1979).

[20] NICKEL K.: Using Interval Methods for the Numerical Solution of ODEs. Freiburger Intervall-Berichte 83/10, Universitaet Freiburg i. Br. (1983).

[21] NICKEL, K.: How to Fight the Wrapping Effect. In: Lecture Notes in Computer Science 212, Interval Mathematics 1985, Proc. of the Intern. Symp., Freiburg i. Br., Sept. 1985, Springer-Verlag, Berlin, Heidelberg, New York, Tokyo, S. 121 -132 (1986).

[22] OELSCHLAEGEL, D.; WIEBIGKE, V.: Untersuchungen zu numerischen Verfahren fuer Intervallanfangswertaufgaben. Freiburger Intervall-Berichte 85/2, Universitaet Freiburg i. Br. (1985).

[23] RAITH, M.: Ein Vervahren zur Berechnung von Schranken beliebiger Konvergenzordnung bei Anfangswertproblemen mit Systemen gewoehnlicher Differentialgleichungen. Freiburger Intervall-Berichte 81/1. Dissertation, Universitaet Freiburg (1981).

[24] REICHMANN, K.: Die Konvergenz von Intervall-(Potenz-)Reihen mit Anwendung auf Intervall-Anfangswertprobleme. Freiburger Intervall-Berichte 80/4, Universitaet Freiburg i. Br. (1980).

[25] SCHOLZ, S.: Fehlerabschaetzung und Genauigkeitssteigerung bei Runge-Kutta-Verfahren. Dissertation, TU Dresden (1970).

[26] SHOKIN, YU.I.: Interval Analysis (in Russian). Nauka, Novosibirsk (1981).

[27] SCHROEDER, J.: Fehlerabschaetzung mit Rechenanlagen bei gewoehnlichen Differentialgleichungen 1. Ordnung. Numer. Math. 3, 39 - 61 (1961)

[13] TSCHAPLYGIN, S.A.:Eine neue Methode zur naeherungsweisen Integration von Differentialgleichungen (russ.). Gostechisdat, Moskau (1950).

[28] WALTER, W.: Differential- und Integral-Ungleichungen und ihre Anwendung bei Abschaetzungs- und Eindeutigkeitsproblemen. Springer-Verlag, Berlin, Heidelberg, New York (1964).

[29] WIEBIGKE, V.: Intervallmathematische Behandlung von Anfangswertproblemen bei gewoehnlichen Differentialgleichungssystemen unter Beruecksichtigung von Datenfehlern. Dissertation, TH Leuna-Merseburg (1983).

SOLUTION OF A CLASS OF WEAKLY COUPLED PROGRAMS VIA DUALIZATION, AND APPLICATIONS

W.Burmeister, S.Dietze, W.Heß, J.W.Schmidt[1)]

1. Introduction

In this paper programming problems of the following special structure will be studied:

$$\text{Program PA: Minimize } F(m) = \sum_{i=1}^{n} F_i(m_{i-1},m_i) \qquad (1.1)$$
$$\text{s.t. } (m_{i-1},m_i) \in W_i \ (i=1,\dots,n).$$

Here $F_i : R^\ell \times R^\ell \to R$ denotes a strictly convex function and $W_i \subset R^\ell \times R^\ell$ a closed convex set. Programs of type (1.1) arise in several applications, e.g. in the shape preserving spline interpolation, in certain discrete variational problems with side conditions and in tridiagonal linear complementarity problems.

For solving (1.1), the special structure should be utilized by practical algorithms in order to be efficient. Here it is proposed to consider the dual program:

$$\text{Program DA: Maximize } H(p) = -\sum_{i=1}^{n} H_i^*(p_{i-1},-p_i) \qquad (1.2)$$
$$\text{with } p_0 = p_n = 0 .$$

The so-called conjugate function $H_i^* : R^\ell \times R^\ell \to R \cup \{+\infty,-\infty\}$ is therein to be defined by

$$H_i^*(\xi,\eta) = \sup\{\xi^T x + \eta^T y - F_i(x,y) : (x,y) \in W_i\}. \qquad (1.3)$$

If always $H_i^*(\xi,\eta) < +\infty$ then program DA is unconstrained and, moreover, the Hessian of the objective function is tridiagonal. These properties are indeed very convenient for numerical purposes. Further, if a solution p of DA is known then, under assumptions, a solution m of PA is explicitly given by means of the partial derivatives of H_i^*,

$$m_{i-1} = \partial_1 H_i^*(p_{i-1},-p_i)^T$$
$$m_i = \partial_2 H_i^*(p_{i-1},-p_i)^T \quad (i=1,\dots,n). \qquad (1.4)$$

---

1) Technische Universität Dresden, Sektion Mathematik, Wissenschaftsbereich Numerische Mathematik, DDR-8027 Dresden, Mommsenstr. 13

This concept of solving PA via dualization is developed by the authors in recent years, see Burmeister/Heß/Schmidt [1] and Dietze/Schmidt [2], and several applications are given. The main purpose of the present paper is to give a unified representation of these results.

The plan of the paper is as follows. Section 2 contains concrete examples of problems PA. Especially they are coming from the shape preserving spline interpolation. By these the present investigations have been originally stimulated, and the results have been obtained step by step, see [1],[2],[3],[4] and [7]. Now, they are derived more deductively. In section 3 an abstract algorithm due to Schmidt/Heß [6] is described by means of which it can be decided whether the specially structured feasible domain of PA is empty or not. In section 4 the pair PA,DA of programming problems is shown to be dual, and strong duality statements, including the important formula (1.4), are verified. This is now done by means of the well-known Lagrange theory while in [2] Fenchel's theory is used. Section 5 is concerned with the application of the general results of section 4 to concrete problems. Especially several programs DA are stated explicitly by computing the necessary conjugate functions directly. In the last section several numerical aspects are discussed.

2. Problems of Structure PA

In this section concrete problems are named which have structure PA.

2.1 Convex Spline Interpolation

A set of points

$$(x_0,y_0),(x_1,y_1),\dots,(x_n,y_n) \text{ with } x_0 = 0 < x_1 < \dots < x_n = 1 \qquad (2.1)$$

is said to be in convex position if

$$\tau_1 \le \tau_2 \le \dots \le \tau_n \qquad (2.2)$$

where $\tau_i = (y_i-y_{i-1})/h_i$, $h_i = x_i-x_{i-1}$ denotes the slope of the secant passing through the points (x_{i-1},y_{i-1}) and (x_i,y_i). The aim is to construct interpolants which are also convex. This shall be done by means of spline functions.
Let $Sp_m(k)$ be the set of $C^k[0,1]$-polynomial splines s of degree $\le m$, that is

$$s \in C^k[0,1],\ s \in \operatorname{span}\{1,x,\dots,x^m\} \text{ for } x_{i-1} \le x \le x_i,$$

and let $ESp(k;\lambda)$ respectively $CSp(k;\lambda)$ be the sets of $C^k[0,1]$-exponential splines s defined by

$$s \in C^k[0,1],\ s \in \operatorname{span}\{1,x,\exp(\lambda_i x),\exp(-\lambda_i x)\} \text{ for } x_{i-1} \le x \le x_i$$

respectively by

$$s \in C^k[0,1], s \in \operatorname{span}\{1,x,\cosh(\lambda_i x)\} \text{ for } x_{i-1} \le x \le x_i,$$

where $\lambda = (\lambda_1,\dots,\lambda_n)$ is a given vector with positive components. It is natural to set $ESp(k;0) = Sp_3(k)$ and $CSp(k;0) = Sp_2(k)$.

Denote by $m_i = s'(x_i)$ the first derivative of a spline s in the node x_i. Then the convexity of interpolants s, $s(x_i) = y_i$ $(i=1,\dots,n)$, is characterized as follows.

Proposition 1[19]: $s \in Sp_3(1)$ is convex on $[0,1]$ if and only if

$$(m_{i-1},m_i) \in K_i \quad (i=1,\dots,n), \qquad K_i = \{(x,y) \in R^2:\ 2x+y \le 3\tau_i,\ x+2y \ge 3\tau_i\}. \tag{2.3}$$

Proposition 2: $s \in Sp_2(1)$ is convex on $[0,1]$ if and only if

$$(m_{i-1},m_i) \in \tilde{K}_i \quad (i=1,\dots,n), \qquad \tilde{K}_i = \{(x,y) \in R^2:\ x+y = 2\tau_i,\ x \le \tau_i\}. \tag{2.4}$$

Proposition 3[3]: $s \in ESp(1;\lambda)$ is convex on $[0,1]$ if and only if

$$(m_{i-1},m_i) \in K_i^{\lambda_i} \quad (i=1,\dots,n), \qquad K_i^{\lambda_i} = \{(x,y) \in R^2:\ \alpha_i x+\beta_i y \le \tau_i,\ \beta_i x+\alpha_i y \ge \tau_i\}, \tag{2.5}$$

where

$$\alpha_i = 1-\beta_i,\ \beta_i = (\sinh\mu_i-\mu_i)/(\mu_i(\cosh\mu_i-1)),\ \mu_i = h_i\lambda_i. \tag{2.6}$$

A necessary and sufficient condition for the convexity of a spline interpolant $s \in ESp(2;\lambda)$ is given in [22].

For the mentioned cases convex spline interpolants do not necessarily exist. An effective algorithm by which the existence can be decided is described in section 3. If at all, there is either one or an infinite number of convex spline interpolants. Therefore, the question arises which of these splines shall be chosen. According to Holladay's theorem, a strategy in $Sp_3(1)$ is to select that spline which minimizes

$$\int_0^1 s''(x)^2dx = \sum_{i=1}^n \frac{4}{h_i}\{m_{i-1}^2+m_{i-1}m_i+m_i^2-3\tau_i(m_{i-1}+m_i)+3\tau_i^2\}, \tag{2.7}$$

see [1]. But there are also other acceptable objective functions, e.g.

$$\sum_{i=1}^n w_i \int_{x_{i-1}}^{x_i} s''(x)^2dx \text{ with } w_i = 1/(1+\tau_i^2)^3 \tag{2.8}$$

or

$$\max_{i=1,..,n} w_i \int_{x_{i-1}}^{x_i} s''(x)^2dx, \quad \max_{i=1,..,n} \sqrt{w_i} \max_{x_{i-1}\le x\le x_i} |s''(x)|.$$

A widely used objective function in $ESp(1;\lambda)$ reads

$$\int_0^1 \{s''(x)^2 + \Lambda(x)^2 s'(x)^2\}dx = \tag{2.9}$$

$$= \sum_{i=1}^{n} \frac{\gamma_i}{h_i} \{m_{i-1}^2 + \frac{2\beta_i}{\alpha_i} m_{i-1}m_i + m_i^2 - \frac{2\tau_i}{\alpha_i}(m_{i-1}+m_i) + \delta_i \tau_i^2\},$$

where $\Lambda(x) = \lambda_i$ for $x_{i-1} \leq x \leq x_i$ and α_i, β_i according to (2.6),

$$\gamma_i = \frac{\mu_i^2 \cosh\mu_i - \mu_i \sinh\mu_i}{2+\mu_i \sinh\mu_i - 2\cosh\mu_i}, \quad \delta_i = \frac{\mu_i^2 \sinh\mu_i}{\mu_i \cosh\mu_i - \sinh\mu_i}, \tag{2.10}$$

see [3]. In all cases that belong to an objective function (2.7),(2.8) or (2.9), a programming problem PA arises.

2.2 Monotone Spline Interpolation

A set of points (2.1) is said to be in isotone position if

$$\tau_1 \geq 0, \tau_2 \geq 0, \ldots, \tau_n \geq 0 . \tag{2.11}$$

Conditions under which this property is preserved by interpolants s are the following.

Proposition 4[13]: $s \in Sp_3(1)$ is isotone on $[0,1]$ if and only if

$$\begin{aligned} &(m_{i-1}, m_i) \in D_i \cup E_i \quad (i=1,\ldots,n), \\ &D_i = \{(x,y) \in R^2 : 0 \leq x \leq 3\tau_i, 0 \leq y \leq 3\tau_i\}, \\ &E_i = \{(x,y) \in R^2 : x^2+xy+y^2-6\tau_i(x+y)+9\tau_i^2 \leq 0\}. \end{aligned} \tag{2.12}$$

Proposition 5: $s \in Sp_2(1)$ is isotone on $[0,1]$ if and only if

$$\begin{aligned} &(m_{i-1}, m_i) \in \tilde{E}_i \quad (i=1,\ldots,n), \\ &\tilde{E}_i = \{(x,y) \in R^2 : x+y = 2\tau_i, x \geq 0, y \geq 0\}. \end{aligned} \tag{2.13}$$

It is easily shown that there are always isotone spline interpolants $s \in Sp_3(1)$ but, in general, they are not uniquely determined. When selecting one of them by means of a choice function (2.7) or (2.8) one is led to a program PA.

2.3 Discrete Variational Problems with Side Conditions

Let the following variational problem be given:

$$\begin{aligned} &\int_0^1 \{a(t)x'(t)^2+b(t)x(t)^2-2r(t)x(t)\}dt \to \min! \\ &\text{s.t. } x(0) = \alpha, \ x(1) = \beta, \ d(t) \leq x(t) \leq c(t) \ (0<t<1), \end{aligned} \tag{2.14}$$

where $a(t) \geq \bar{a} > 0$, $b(t) \geq \bar{b} > 0$. The usual discretization yields

$$\sum_{i=1}^{n} \{\frac{a_{i-1/2}}{h}(m_{i-1}-m_i)^2 + \frac{h}{2}(b_{i-1}m_{i-1}^2 + b_i m_i^2) - h(r_{i-1}m_{i-1} + r_i m_i)\} \to \min!$$
$$\text{s.t. } m_0 = \alpha,\ m_n = \beta,\ d_i \leq m_i \leq c_i \ (i=1,\dots,n-1). \tag{2.15}$$

Here $h > 0$ denotes the step-size, $t_i = i\,h$ $(i=0,\dots,n)$ the nodes, m_i the approximation to $x(t_i)$, $a_{i-1/2} = a(t_i - h/2)$, $b_i = b(t_i)$ and so on. Of course, (2.15) is a program PA, but in addition boundary conditions occur. The sets $W_1,\dots,W_n$ may now be e.g. equal to

$$W_i = \{(x,y) \in R^2;\ d_{i-1} \leq x \leq c_{i-1},\ d_i \leq y \leq c_i\}. \tag{2.16}$$

2.4 Tridiagonal Linear Complementarity Problems

Here the problem is that of finding a vector m such that

$$Am+b \geq 0,\ m \geq c,\ (m-c)^T(Am+b) = 0 \tag{2.17}$$

holds, where the $(n+1,n+1)$-matrix A and the vectors $b,c \in R^{n+1}$ are given. The matrix A is assumed to be symmetric positive definite and tridiagonal. Then, (2.17) is equivalent to

$$F(m) = m^T A m + 2b^T m \to \min!$$
$$\text{s.t. } m \geq c. \tag{2.18}$$

Let $\varepsilon > 0$ be such that the matrix $A-\varepsilon I$ is positive definite, and L be the Cholesky factor to $A-\varepsilon I$,

$$L = \begin{pmatrix} u_0 & & & \\ v_0 & u_1 & & \\ & \ddots & \ddots & \\ & & v_{n-1} & u_n \end{pmatrix},\ u_0 > 0,\dots,u_n > 0.$$

Then it follows that

$$F(m) = m^T L\, L^T m + \varepsilon\, m^T m + 2b^T m$$
$$= \sum_{i=1}^{n} \{(u_{i-1}m_{i-1} + v_{i-1}m_i)^2 + \varepsilon\, m_{i-1}^2 + 2b_{i-1}m_{i-1}\} \tag{2.19}$$
$$+ u_n^2 m_n^2 + \varepsilon m_n^2 + 2b_n m_n .$$

Thus, the tridiagonal complementarity problem (2.17) is equivalent to a program PA, and the sets $W_1,\dots,W_n$ may now be

$$W_i = \{(x,y) \in R^2;\ x \geq c_{i-1}, y \geq c_i\}, \tag{2.20}$$

see [5]. For example, discretizations of free boundary value problems occuring in journal bearings can be transformed into a complementarity problem (2.17), see [29],[30].

3. Existence and Determination of Feasible Solutions

3.1 The Basic Algorithm

In order to test the feasibility of the preceding problems always the following question arises: For given nonempty subsets $W_1,\dots,W_n \subset R^\ell \times R^\ell$ are there vectors $m_0,m_1,\dots,m_n \in R^\ell$ such that

$$(m_{i-1},m_i) \in W_i \ (i=1,\dots,n) \tag{3.1}$$

is valid?

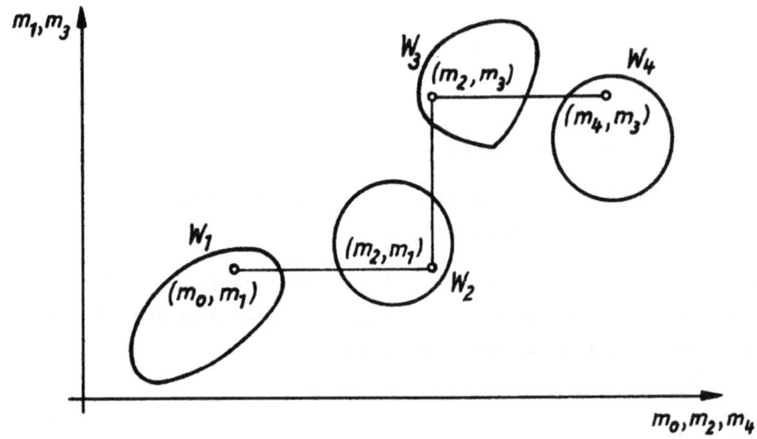

Fig.1: Does there exists a "staircase" on $W_1,\dots,W_n$?

For solving this weakly coupled problem the following general algorithm and the corresponding criteria are developed by Schmidt/Heß [6].

Algorithm 6: Set $P_0 = R^\ell$ and then successively for i=1,2,..,n:

$$P_i = \{y \in R^\ell: \text{there exists a } x \in P_{i-1} \text{ with } (x,y) \in W_i\}. \tag{3.2}$$

Theorem 7: Problem (3.1) is solvable if and only if

$$P_i \neq \emptyset \ (i=1,\dots,n) \tag{3.3}$$

is valid, and then each solution of (3.1) can be determined as follows: choose $m_n \in Q_n = P_n$ and then successively for $i=n,n-1,\dots,1$:

$$m_{i-1} \in Q_{i-1} = Q_{i-1}(m_i) \quad = P_{i-1} \cap \{x \in R^\ell: (x,m_i) \in W_i\}. \tag{3.4}$$

Proof: Suppose that a solution $m_0,m_1,\dots,m_n$ of (3.1) exists. Then one gets $m_i \in P_i$ (i=0,..,n); this implies (3.3). Indeed, $m_0 \in P_0$ trivially holds. And if $m_{i-1} \in P_{i-1}$ is valid, $(m_{i-1},m_i) \in W_i$ already assures $m_i \in P_i$. Further, every solution satisfies (3.4). Conversely, let now (3.3) be true. Whenever $m_j \in Q_j$ has been selected, $m_j \in P_j$ holds by (3.4).

Because of (3.2), an $x \in P_{j-1}$ exists such that $(x, m_j) \in W_j$. The choice $m_{j-1} = x$ assures that $m_{j-1} \in Q_{j-1}$. Using this the existence of a solution of (3.1) follows by induction. Thus, theorem 7 is proven.

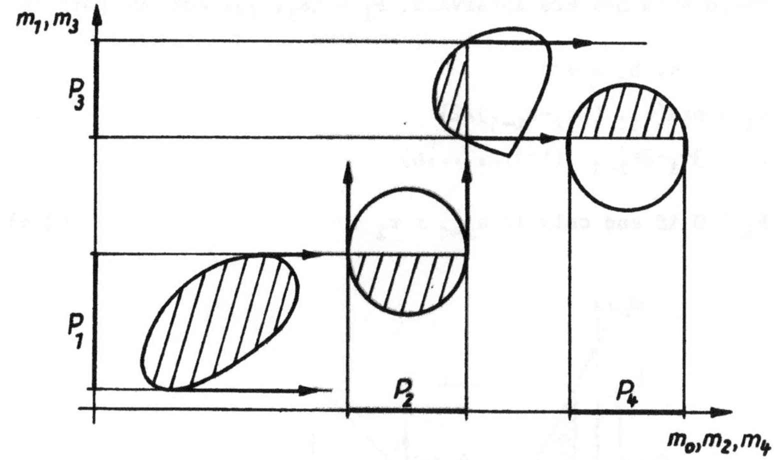

Fig.2: Existence of "staircases"; test (3.3)

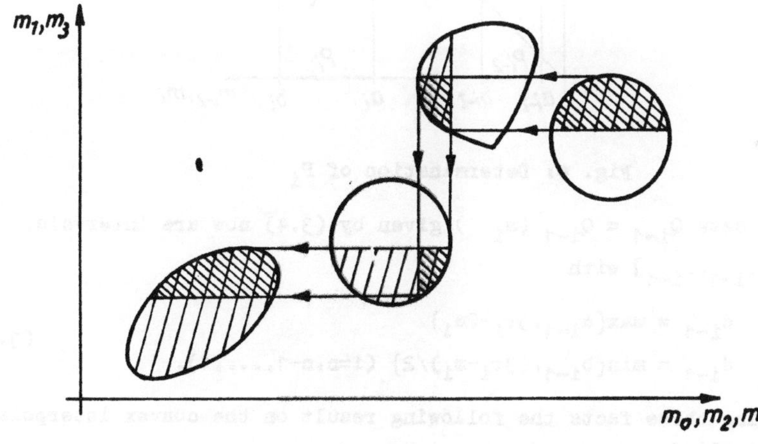

Fig.3: Construction of all "staircases"; algorithm (3.4)

3.2 Applications of the Basic Algorithm

3.2.1 In view of proposition 1, for the convex interpolation in $Sp_3(1)$ the sets P_i defined by (3.2), which correspond to $W_i = K_i$, are of interest. These sets now are intervals, $P_i = [a_i, b_i]$, and one gets inductively

$$\begin{aligned} a_0 &= -\infty,\ b_0 = +\infty, \\ a_i &= \max\{\tau_i,\ (3\tau_i - b_{i-1})/2\} \qquad (3.5) \\ b_i &= 3\tau_i - 2a_{i-1} \quad (i=1,2,\dots,n) \end{aligned}$$

and

$$P_i \neq \emptyset \text{ if and only if } a_{i-1} \leq \tau_i \,. \qquad (3.6)$$

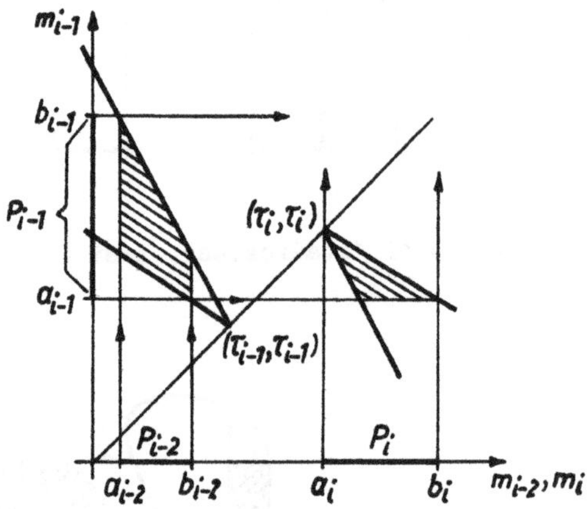

Fig. 4: Determination of P_i

Also the sets $Q_{i-1} = Q_{i-1}(m_i \quad)$ given by (3.4) now are intervals, $Q_{i-1} = [c_{i-1}, d_{i-1}]$ with

$$\begin{aligned} c_{i-1} &= \max\{a_{i-1}, 3\tau_i - 2m_i\} \\ d_{i-1} &= \min\{b_{i-1}, (3\tau_i - m_i)/2\} \quad (i=n, n-1, \dots, 1). \end{aligned} \qquad (3.7)$$

Summarizing these facts the following result on the convex interpolation is implied.

Theorem 8[6]: To a data set (2.1) being in convex position there exist convex spline interpolants $s \in Sp_3(1)$ if and only if

$$a_{i-1} \leq \tau_i \quad (i=1,\dots,n) \qquad (3.8)$$

where the numbers a_i are recursively defined by (3.5). If (3.8) is met all these splines can be determined as follows:

$$s(x) = y_{i-1} + m_{i-1}(x - x_{i-1}) + r_i(x - x_{i-1})^2 + s_i(x - x_{i-1})^3 \text{ for } x_{i-1} \leq x \leq x_i \quad (3.9)$$

with

$$r_i = (3\tau_i - 2m_{i-1} - m_i)/h_i, \quad s_i = (m_{i-1} + m_i - 2\tau_i)/h_i^2 \tag{3.10}$$

and

$$m_n \in [a_n, b_n], \quad m_{i-1} \in [c_{i-1}, d_{i-1}] \quad (i=n,n-1,..,1). \tag{3.11}$$

For $n \leq 3$ the test (3.8) is, for all data sets (2.1) satisfied, but for $n > 3$ this test may fail. Therefore, the problem of convex interpolation is not always solvable in $Sp_3(1)$.

Note that the total amount in arithmetic operations and in comparisons for performing (3.8)-(3.11) is only $O(n)$. Another algorithm for this problem which, however, requires $O(n^2)$ operations is proposed in [17].

3.2.2 The situation for the convex interpolation in $Sp_2(1)$ is analoguous. With regard to proposition 2 the sets P_i and Q_{i-1} belonging to $W_i = \tilde{K}_i$ are to be determined. One gets $P_i = [a_i, b_i]$ and $Q_{i-1} = [c_{i-1}, d_{i-1}]$ with

$$\begin{aligned} a_i &= \max\{\tau_i, 2\tau_i - b_{i-1}\} \\ b_i &= 2\tau_i - a_{i-1} \quad (i=1,2,\ldots,n), \end{aligned} \tag{3.12}$$

$$c_{i-1} = d_{i-1} = 2\tau_i - m_i \quad (i=n,n-1,..,1) \tag{3.13}$$

and

$$P_i \neq \emptyset \text{ if and only if } a_{i-1} \leq \tau_i. \tag{3.14}$$

Theorem 8 holds verbatim in $Sp_2(1)$ if one takes into account that $s \in Sp_2(1)$ means

$$s(x) = y_{i-1} + m_{i-1}(x - x_{i-1}) + r_i(x - x_{i-1})^2 \text{ for } x_{i-1} \leq x \leq x_i \tag{3.15}$$

with

$$r_i = (\tau_i - m_{i-1})/h_i. \tag{3.16}$$

The equality occuring in criteria (2.4) allows also another existence test for convex $Sp_2(1)$-splines. The resulting question whether there are numbers $m_o, m_1, .., m_n$ such that

$$m_{i-1} + m_i = 2\tau_i, \quad m_{i-1} \leq \tau_i \quad (i=1,\ldots,n) \tag{3.17}$$

can be, as easily seen, equivalently transformed into

$$\delta_* \leq \delta^* \tag{3.18}$$

with

$$\delta_* = \max\{\delta_i : i \text{ even}\}, \quad \delta^* = \min\{\delta_i : i \text{ odd}\}, \tag{3.19}$$

$$\delta_1 = \tau_1, \quad \delta_{i+1} = \delta_i + (-1)^i(\tau_{i+1} - \tau_i) \quad (i \geq 1). \tag{3.20}$$

Every $m_o \in [\delta_*, \delta^*]$ leads to a convex spline interpolant $s \in Sp_2(1)$. Moreover, the problem of selecting one of these splines by means of the objective function (2.8) turns out to be a simple one-dimensional optimization problem,

$$4 \sum_{i=1}^{n} w_i(m_0-\delta_i)^2/h_i \to \min! \text{ s.t. } \delta_* \le m_0 \le \delta^* . \tag{3.21}$$

For extensions to splines $s \in CSp(1;\lambda)$ see [7].

3.2.3 Exponential splines $s \in ESp(1;\lambda)$ are introduced in order to avoid the mentioned negative results in convex interpolation. For sufficient large parameter vectors λ there are always interpolating exponential splines which are convex if the given data set is in strict convex position, see [3]. For verifying this the sets P_i and Q_{i-1} from (3.2) and (3.4) are computed for $W_i = K_i^{\lambda_i}$ where $K_i^{\lambda_i}$ is defined by (2.5). Also in this case one gets $P_i = [a_i,b_i]$ and $Q_{i-1} = [c_{i-1},d_{i-1}]$, now with

$$a_i = \max\{\tau_i,(\tau_i-\beta_i b_{i-1})/\alpha_i\}$$

$$b_i = (\tau_i-\alpha_i a_{i-1})/\beta_i \quad (i=1,2,\ldots,n), \tag{3.22}$$

$$c_{i-1} = \max\{a_{i-1},(\tau_i-\alpha_i m_i)/\beta_i\}$$

$$d_{i-1} = \min\{b_{i-1},(\tau_i-\beta_i m_i)/\alpha_i\} \quad (i=n,n-1,\ldots,1), \tag{3.23}$$

and

$$P_i \neq \emptyset \text{ if and only if } a_{i-1} \le \tau_i . \tag{3.24}$$

Because of

$$(\tau_i-\beta_i b_{i-1})/\alpha_i \to \tau_i \text{ for } \mu_i = h_i\lambda_i \to +\infty \tag{3.25}$$

it follows the essential

Proposition 9: Assume that $\tau_{i-1} < \tau_i$ holds. Then

$$a_{i-1} \le \tau_i \quad (P_i \neq \emptyset)$$

is valid for sufficiently large λ_i.

Therefore, for data sets (2.1) in strictly convex position, the problem of convex interpolation is always solvable in $ESp(1;\lambda)$ if λ is sufficiently large, and then convex spline interpolants $s \in ESp(1;\lambda)$ can be computed as follows:

For $x_{i-1} \le x \le x_i$, with $t = (x-x_{i-1})/h_i$ and $\mu_i = h_i\lambda_i$, set

$$s(x) = y_{i-1}+h_i m_{i-1}t+r_i(\cosh\mu_i t-1)/\lambda_i+s_i(\sinh\mu_i t-\mu_i t)/\lambda_i \tag{3.26}$$

where

$$r_i = \frac{\mu_i(\cosh\mu_i-1)\tau_i-(\mu_i\cosh\mu_i-\sinh\mu_i)m_{i-1}-(\sinh\mu_i-\mu_i)m_i}{2+\mu_i\sinh\mu_i-2\cosh\mu_i}$$

$$s_i = \frac{-\mu_i(\sinh\mu_i)\tau_i+(1+\mu_i\sinh\mu_i-\cosh\mu_i)m_{i-1}-(1-\cosh\mu_i)m_i}{2+\mu_i\sinh\mu_i-2\cosh\mu_i} \tag{3.27}$$

and, with (3.22) and (3.23),

$$m_n \in [a_n,b_n],\ m_{i-1} \in [c_{i-1},d_{i-1}]\ (i=n,n-1,\dots,1). \tag{3.28}$$

3.2.4 It is more promising to get a convex spline s if the usual interpolation condition

$$s(x_i) = y_i\ (i=0,\dots,n) \tag{3.29}$$

is weakened to

$$s(x_i) \in [\underline{y}_i,\bar{y}_i]\ (i=0,\dots,n) \tag{3.30}$$

with given numbers $\underline{y}_i$ and $\bar{y}_i$. For the interval interpolation introduced in this way convexity preserving conditions for cubic splines are derived in [4].

According to proposition 1 the sets P_i and Q_{i-1} defined by (3.2) and (3.4), respectively, are to be determined for

$$W_i = \{(f,x,g,y) \in R^4\colon 2x+y \le 3(g-f)/h_i \le x+2y,\ \underline{y}_{i-1} \le f \le \bar{y}_{i-1},\ \underline{y}_i \le g \le \bar{y}_i\}. \tag{3.31}$$

It turns out that

$$P_i = \{(g,y) \in R^2\colon a_i(g) \le y \le b_i(g),\ A_i \le g \le B_i\} \tag{3.32}$$

with

$$b_i(g) = 3(g-A_{i-1})/h_i-2a_{i-1}(A_{i-1}),\qquad a_i(g) = \max_{j\in I_i}\{p_j^i g + q_j^i\} \tag{3.33}$$

where $I_i = I_{i-1} \cup \{i_1,i_2\}$, and for $j \in I_{i-1}$:

$$p_j^i = p_j^{i-1}/(1+h_ip_j^{i-1}),\quad q_j^i = q_j^{i-1}/(1+h_ip_j^{i-1}),$$

for $j = i_1$:

$$p_j^i = 1/h_i,\quad q_j^i = -B_{i-1}/h_i,$$

and for $j = i_2$:

$$p_j^i = 3/(2h_i),\quad q_j^i = -3B_{i-1}/(2h_i)-b_{i-1}(B_{i-1})/2.$$

For a somewhat sharper choice of I_i see [4]. The numbers A_i and B_i are to be declared by

$$A_i = \max\{\underline{y}_i,A_{i-1}+h_ia_{i-1}(A_{i-1})\},\quad B_i = \bar{y}_i. \tag{3.34}$$

At the beginning one has to set

$$a_0(g) = -\infty,\ b_0(g) = +\infty,\ A_0 = \underline{y}_0,\ B_0 = \bar{y}_0,\ I_0 = \emptyset\ . \tag{3.35}$$

The important test (3.3), now relevant to the existence of convex spline interpolants $s \in Sp_3(1)$, reads by using the introduced quantities

$$a_{i-1}(A_{i-1}) \le (\tilde{y}_i - A_{i-1})/h_i \quad (i=1,\dots,n). \tag{3.36}$$

The construction concrete convex spline interpolants of the form (3.9), (3.10) can be done by choosing

$$(y_{i-1}, m_{i-1}) \in Q_{i-1} \quad (i=n+1,n,\dots,1) \tag{3.37}$$

where

$$\begin{aligned} Q_{i-1} &= \{(f,x) \in R^2 \colon c_{i-1}(f) \le x \le d_{i-1}(f), \\ &\max\{A_{i-1}, y_i - h_i m_i, \gamma_{i-1}\} \le f \le \min\{B_{i-1}, \delta_{i-1}\}\} \end{aligned} \tag{3.38}$$

with

$$\begin{aligned} c_{i-1}(f) &= \max\{a_{i-1}(f),\ 3(y_i-f)/h_i - 2m_i\} \\ d_{i-1}(f) &= \min\{b_{i-1}(f),\ (3(y_i-f)/h_i - m_i)/2\} \end{aligned} \tag{3.39}$$

while the numbers γ_{i-1} and δ_{i-1} are the solutions f of the equations

$$b_{i-1}(f) = 3(y_i-f)/h_i - 2m_i, \quad a_{i-1}(f) = (3(y_i-f)/h_i - m_i)/2, \tag{3.40}$$

respectively.

3.2.5 The numbers $m_o = m_1 = \dots = m_n = 0$ obviously satisfy condition (2.12). Thus, in view of proposition 4, the problem of monotone spline interpolation is always solvable in $Sp_3(1)$. In $Sp_2(1)$ there may exist no solution. Indeed, the sets $W_i = \tilde{E}_i$ from (2.13) define by (3.2) the sets $P_i = [a_i, b_i]$ as follows:

$$\begin{aligned} a_o &= 0,\ b_o = +\infty, \\ a_i &= \max\{0, 2\tau_i - b_{i-1}\} \\ b_i &= 2\tau_i - a_{i-1} \quad (i=1,\dots,n), \end{aligned} \tag{3.41}$$

and the condition

$$a_{i-1} \le 2\tau_i \quad (i=1,\dots,n)$$

which is equivalent to (3.3) may fail.

3.2.6 In the problems described in sections 2.3 and 2.4 the feasible domains are always nonempty if $d_i \le c_i$ $(i=1,\dots,n-1)$ or $c_i \ge 0$ $(i=0,\dots,n)$, respectively.

4. Duality Statements via Lagrange's Theory

In order to construct the problem DA corresponding to PA the latter is restated as

PA1: $$f(u) = \sum_{i=1}^{n} F_i\,(m_{i-1},M_i) \to \inf!$$

where

$$u = (m_0,M_1,\ldots,m_{n-1},M_n) \in W_1 \times \ldots \times W_n$$

and

$$h_i(u) = M_i - m_i = 0 \quad (i=1,\ldots,n-1).$$

Here a change of notation $M_n := m_n$ is applied, so that m_n no longer appears in PA1.

The problem PA1, due to the introduction of auxiliary variables $M_1,\ldots,M_{n-1}$, is separable, which enables the decomposition principle to be used (see e.g.[28], § 28,[27]). For simplicity of notation it will be assumed that $m_i, M_i \in R$ and $W_i \subset R^2$, the extension to the vector case is immediate.

Now, the required facts from Lagrange's theory will be stated. Let $U \subset R^n$ be convex, $U \neq \emptyset$, $f: R^n \to R$ a convex and $g_i: R^n \to R^{n_i}$ $(i=1,\ldots,r)$, $h_j: R^n \to R$ $(j=1,\ldots,s)$ affine functions. Further, let

$$G = \{u \in R^n:\ g_i(u) \le 0\ (i=1,\ldots,r),\ h_j(u) = 0\ (j=1,\ldots,s)\}.$$

Consider

P : $\inf P := \inf\{f(u) : u \in U \cap G\}$.

With the Lagrangian

$$L(u,v) = f(u) + \sum_{i=1}^{r} q_i g_i(u) + \sum_{j=1}^{s} p_j h_j(u),$$

$$u \in R^n,\ v = (q_1,\ldots,q_r,p_1,\ldots,p_s) \in R^m,$$

$$m = n_1 + \ldots + n_r + s$$

one gets the dual objective function

$$\varphi : R^m \to R \cup \{-\infty\},\ \varphi(v) = \inf\{L(u,v) : u \in U\}.$$

The domain of the dual problem is

$$V = \{v \in R^m:\ q_1 \ge 0,\ldots,q_r \ge 0\}.$$

The dual of P is then

D : $\sup D := \sup\{\varphi(v) : v \in V\}$.

The following duality theorem of Lagrange type for the pair P,D results e.g. from [26], theorems 5.7 and 5.12, or [28], § 28. The symbol ri denotes the relative interior of a convex set.

Theorem 10: Let U, f, g_i, h_j satisfy the above assumptions. Then

a) if $G \cap \text{ri}\, U \neq \emptyset$ and $\inf P > -\infty$, problem D is solvable and

$$\inf P = \max D.$$

b) if D is solvable and $\inf P = \max D$ and v is a solution of D, then u solves P if and only if

$$u \in U \cap G \tag{4.1}$$

$$L(u,v) = \varphi(v) \tag{4.2}$$

$$q_i g_i(u) = 0 \quad (i=1,\dots,r). \tag{4.3}$$

Let us return to problem PA1 which is of type P. Denote by DA1 the corresponding Lagrange dual problem. Its objective function φ takes the form, with regard to (1.3),

$$\begin{aligned}\varphi(v) &= \inf\{f(u) + \sum_{i=1}^{n-1} p_i(M_i - m_i):\ u \in W_1 \times \dots \times W_n\} = \\ &= \sum_{i=1}^{n} \inf\{F_i(m_{i-1}, M_i) - p_{i-1} m_{i-1} + p_i M_i:\ (m_{i-1}, M_i) \in W_i\} = \\ &= -\sum_{i=1}^{n} H_i^*(p_{i-1}, -p_i)\end{aligned} \tag{4.4}$$

with

$$v = (p_1, \dots, p_{n-1}) \in R^{n-1} \text{ and } p_0 = p_n = 0 .$$

Thus, DA1 is identical with DA.

The assumptions about W_i and F_i stated in the introduction as well as further required assumptions are collected as

Assumption (V): $F_i : R^2 \to R$ are strictly convex, $W_i \neq \emptyset$ convex, the auxiliary problems (1.3) are solvable for all $(\xi, \eta) \in R^2$ with optimal value in R, problem PA is feasible and $\inf PA > -\infty$. The unique solution of (1.3) for $(\xi,\eta) \in R^2$ will be denoted by

$$(\bar{x}_i(\xi,\eta),\ \bar{y}_i(\xi,\eta)). \tag{4.5}$$

Proposition 11: Suppose that (V) holds. Then

a) if
$$\{(m_0, \dots, m_n) \in R^{n+1}:\ (m_{i-1}, m_i) \in \text{ri}\, W_i \text{ for } i=1,\dots,n\} \neq \emptyset \tag{4.6}$$
then DA is solvable and $\inf PA = \max DA$.

b) if the W_i are given by a finite set of affine inequalities

$$W_i = \{(x,y) \in R^2:\ a_i x + b_i y + c_i \le 0\}$$

with fixed $a_i, b_i, c_i \in R^{n_i}$ then DA is solvable and $\inf PA = \max DA$.

c) the functions H_i^* and the objective function $-\varphi$ of DA are convex and continuously differentiable. The relations

$$\partial_1 H_i^*(\xi,\eta) = \bar{x}_i(\xi,\eta) \qquad (4.7)$$
$$\partial_2 H_i^*(\xi,\eta) = \bar{y}_i(\xi,\eta)$$

hold.

d) if DA is solvable and inf PA = max DA and $(p_1,\dots,p_{n-1})$ is a solution of DA then PA has a unique solution given by

$$m_o = \partial_1 H_1^*(0,-p_1)$$
$$m_i = \partial_1 H_{i+1}^*(p_i,-p_{i+1}) = \partial_2 H_i^*(p_{i-1},-p_i) \quad (i=1,\dots,n-1) \qquad (4.8)$$
$$m_n = \partial_2 H_n^*(p_{n-1},0).$$

Proof:

a) Condition (4.6) is obviously true if and only if

$$\{u = (m_o,M_1,\dots,m_{n-1},M_n) \in R^{2n}:$$
$$M_i-m_i = 0 \ (i=1,\dots,n-1) \text{ and } u \in ri(W_1 \times \dots \times W_n)\} \neq \emptyset .$$

Now, part a) of theorem 10 applied to PA1, DA together with the equivalence of PA and PA1 imply the result.

b) PA is rewritten as

PA2: $f(u) \to \inf$ with $u = (m_o,M_1,\dots,m_{n-1},M_n) \in R^{2n}$,
$g_i(u) = a_i m_{i-1} + b_i M_i + c_i \le 0 \quad (i=1,\dots,n)$
$h_i(u) = M_i - m_i = 0 \qquad (i=1,\dots,n-1).$

Since this is a problem of type P, theorem 10, part a) yields

$$\inf PA = \inf PA2 = \max DA2 := \max\{\phi(v): q_1 \ge 0,\dots,q_n \ge 0\} \qquad (4.9)$$

where

$$v = (q_1,\dots,q_n,p_1,\dots,p_{n-1}) \in R^m, \quad m = n_1 + \dots + n_n + n-1,$$

and the dual objective function reads

$$\phi(v) = \inf\{f(u) + \sum_{i=1}^{n} q_i g_i(u) + \sum_{i=1}^{n-1} p_i h_i(u): u \in R^{2n}\}.$$

It will be shown that if v is a solution of DA2 then $w = (p_1,\dots,p_{n-1})$ is a solution of DA with the same optimal value, equal to inf PA. In fact,

$$\phi(v) \le \inf\{f(u) + \sum_{i=1}^{n} q_i g_i(u) + \sum_{i=1}^{n-1} p_i h_i(u): u \in W_1 \times \dots \times W_n\}$$
$$\le \inf\{f(u) + \sum_{i=1}^{n-1} p_i h_i(u): u \in W_1 \times \dots \times W_n\} = \varphi(w)$$
$$\le \inf PA$$

whenever $q_1,\dots,q_n \ge 0$. With regard to (4.4) and (4.9)

$$\max\{\phi(v): q_1 \ge 0,\dots,q_n \ge 0\} = \max DA = \inf PA \qquad \text{follows.}$$

c) The convexity is a property of conjugate functions. The pair (x_i,y_i) is in $\partial H_i^*(\xi,\eta)$, the subdifferential of H_i^*, if and only if (x_i,y_i) is a maximizer in (1.3). Since this maximizer is unique, H_i^* is differentiable, and (4.7) follows. Moreover, $\partial_1 H_i^*$ and $\partial_2 H_i^*$ are continuous (see e.g. [28], § 23 and § 25).

d) Let $w = (p_1,\ldots,p_{n-1})$ be a solution of DA = DA1 and define $u = (m_0,M_1,\ldots,m_{n-1},M_n)$ by

$$m_{i-1} = \partial_1 H_i^* \, (p_{i-1},-p_i)$$

$$M_i = \partial_2 H_i^* \, (p_{i-1},-p_i)$$

with $p_0 = p_n = 0$. It remains to prove that u is the unique solution of (4.1),(4.2) ((4.3) is vacuous). By this, because of the equivalence of PA and PA1, the assertion of d) would be established.

By means of (4.7) and the definition of $(\bar{x}_i,\bar{y}_i)$ as a solution of (1.3) it follows that $(m_{i-1},M_i) \in W_i$, thus $u \in U = W_1 \times \ldots \times W_n$. Since w is a solution of DA,

$$0 = \partial_i \varphi(w) = \partial_2 H_i^*(p_{i-1},-p_i) - \partial_1 H_{i+1}^*(p_i,-p_{i+1})$$

$$= M_i - m_i,$$

so that (4.1) and (4.8) are true. Moreover, it results from (4.7) again that the infimum in (4.4) is attained at u, this implies (4.2). The uniqueness of u is a consequence of the strict convexity of F_i.

5. Applications of the General Duality Theory

Here the general duality statements derived in section 4 shall be used in order to treat the concrete problems of structure PA given in section 2.

5.1 Convex Spline Interpolation

For computing convex cubic spline interpolants which minimize the functional (2.8) one has to solve the

Program PC: $$4\sum_{i=1}^{n} \frac{w_i}{h_i}\{(m_{i-1}-\tau_i)^2+(m_{i-1}-\tau_i)(m_i-\tau_i)+(m_i-\tau_i)^2\} \to \min!$$
$$\text{s.t.}\quad 2m_{i-1}+m_i \le 3\tau_i,\ m_{i-1}+2m_i \ge 3\tau_i \quad (i=1,\ldots,n). \tag{5.1}$$

Because the Hessian of the objective function is positive definite program PC has a unique solution if the feasible domain is nonempty. Now, a corresponding dual program is stated. To this end, the subprogram (1.3),

$$d_i(x,y) = \xi x+\eta y-4\frac{w_i}{h_i}\{(x-\tau_i)^2+(x-\tau_i)(y-\tau_i)+(y-\tau_i)^2\} \to \max!$$
$$\text{s.t. } 2x+y \le 3\tau_i,\; x+2y \ge 3\tau_i \tag{5.2}$$

is to be solved for all $(\xi,\eta) \in R^2$. Especially the optimal value $d_{i\,max} = H_i^*(\xi,\eta)$ is of interest.

Proposition 12: The optimal value of program (5.2) is equal to

$$\begin{aligned} H_i^*(\xi,\eta) &= \tau_i(\xi+\eta) + \frac{h_i}{12w_i}(\xi^2-\xi\eta+\eta^2) \text{ for } \xi \le 0,\ \eta \ge 0 \\ &= \tau_i(\xi+\eta) + \frac{h_i}{12w_i}\left(\frac{\xi}{2}-\eta\right)^2 \quad \text{for } 0 \le \xi \le 2\eta \\ &= \tau_i(\xi+\eta) + \frac{h_i}{12w_i}\left(\xi-\frac{\eta}{2}\right)^2 \quad \text{for } 2\xi \le \eta \le 0 \\ &= \tau_i(\xi+\eta) \quad \text{for } \xi \ge 2\eta,\ 2\xi \ge \eta. \end{aligned} \tag{5.3}$$

Proof: By means of the Lagrangian to (5.2),

$$\Phi(x,y,\lambda,\mu) = -d_i(x,y) + \lambda(2x+y-3\tau_i) + \mu(3\tau_i-x-2y), \tag{5.4}$$

the Kuhn-Tucker conditions to (5.2), now necessary and sufficient, read as follows:

$$\begin{aligned} &\Phi_x = 0,\ \Phi_y = 0 \\ &\Phi_\lambda \le 0,\ \lambda \ge 0,\ \lambda\Phi_\lambda = 0 \\ &\Phi_\mu \le 0,\ \mu \ge 0,\ \mu\Phi_\mu = 0\,. \end{aligned} \tag{5.5}$$

System $\Phi_x = \Phi_y = 0$ has the solution

$$x = \tau_i + \frac{h_i}{12w_i}(2\xi-\eta-3\lambda),\quad y = \tau_i + \frac{h_i}{12w_i}(-\xi+2\eta+3\mu). \tag{5.6}$$

Case 1: Let $\lambda = 0$, $\mu = 0$. Then (x,y) according to (5.6) is the maximizer of (5.2) if $\Phi_\lambda \le 0$, $\Phi_\mu \le 0$; this occurs exactly for $\xi \le 0$, $\eta \ge 0$. The optimal value $H_i^*(\xi,\eta) = d_i(x,y)$ is as given in (5.3) at the first place.

Case 2: Let $\Phi_\lambda = 0$, $\mu = 0$. Then $\lambda = \xi/2$ follows in view of $\Phi_\lambda = 0$, and using this expression for λ the maximizer of (5.2) is given once more by (5.6) if $\lambda \ge 0$, $\Phi_\mu \le 0$; this happens if and only if $\xi \ge 0$, $\xi \le 2\eta$. And $H_i^*(\xi,\eta) = d_i(x,y)$ with $H_i^*(\xi,\eta)$ equal to the second part in (5.3) is easily verified.

Case 3: Let $\Phi_\mu = 0$, $\lambda = 0$. This case is symmetric to case 2.

Case 4: Let $\Phi_\lambda = 0$, $\Phi_\mu = 0$. Then (x,y) with $x = y = \tau_i$ is the maximizer of (5.2) if $3\lambda = 2\xi - \eta \geq 0$, $3\mu = \xi - 2\eta \leq 0$, and $H_i^*(\xi,\eta) = d(\tau_i,\tau_i) = \tau_i(\xi+\eta)$ follows.

Thus the proof of proposition 12 is complete.

Using the functions H_i^* defined by (5.3) the dual program (1.2) reads now as follows:

$$\text{Program DC:} \qquad -\sum_{i=1}^{n} H_i^*(p_{i-1},-p_i) \to \max! \text{ with } p_o = p_n = 0 \,. \tag{5.7}$$

Proposition 11 yield immediately the following strong duality statements.

<u>Theorem 13</u>: Both programs PC and DC are solvable if and only if the feasible domain of PC is nonempty; this can be checked by theorem 8. Then the optimal values are equal. Furthermore, if a solution $(p_o,p_1,\ldots,p_n)$ of DC is known the solution $(m_o,m_1,\ldots,m_n)$ of PC can be computed by formula (1.4), where now H_i^* is to be chosen according to (5.3).

Theorem 13 is due to Burmeister/Heß/Schmidt [1]. There this result is derived in another way by starting from the Kuhn-Tucker conditions corresponding to PC.

Analogous results are valid for the convex interpolation in $ESp(1;\lambda)$, see [3]. Now, functional (2.9) is minimized, leading to

Program PCE:

$$\sum_{i=1}^{n} \frac{\gamma_i}{h_i}\{m_{i-1}^2 + \frac{2\beta_i}{\alpha_i} m_{i-1}m_i + m_i^2 - \frac{2\tau_i}{\alpha_i}(m_{i-1}+m_i) + \delta_i\tau_i^2\} \to \min! \tag{5.8}$$

$$\text{s.t. } \alpha_i m_{i-1} + \beta_i m_i \leq \tau_i, \beta_i m_{i-1} + \alpha_i m_i \geq \tau_i \quad (i=1,\ldots,n).$$

Here the constants $\alpha_i,\beta_i,\gamma_i,\delta_i$ are given by (2.6) and (2.10).

As before, the conjugate function H_i^* is determined to be

$$\begin{aligned} H_i^*(\xi,\eta) &= \tau_i(\xi+\eta) + \frac{h_i}{4\delta_i}(\xi^2 - \frac{2\beta_i}{\alpha_i}\xi\eta + \eta^2) + h_i\lambda_i^2\tau_i^2 && \text{for } \xi \leq 0,\ \eta \geq 0 \\ &= \tau_i(\xi+\eta) + \frac{h_i}{4\delta_i}(\frac{\beta_i}{\alpha_i}\xi - \eta)^2 + h_i\lambda_i^2\tau_i^2 && \text{for } 0 \leq \xi \leq \frac{\alpha_i}{\beta_i}\eta \\ &= \tau_i(\xi+\eta) + \frac{h_i}{4\delta_i}(\xi - \frac{\beta_i}{\alpha_i}\eta)^2 + h_i\lambda_i^2\tau_i^2 && \text{for } \frac{\alpha_i}{\beta_i}\xi \leq \eta \leq 0 \\ &= \tau_i(\xi+\eta) + h_i\lambda_i^2\tau_i^2 && \text{for } \xi \geq \frac{\alpha_i}{\beta_i}\eta,\ \frac{\alpha_i}{\beta_i}\xi \geq \eta. \end{aligned} \tag{5.9}$$

With these functions the dual program (1.2) to PCE reads

Program DCE: $-\sum_{i=1}^{n} H_i^*(p_{i-1},-p_i) \to \max!$ with $p_0 = p_n = 0$, (5.10)

and for the pair PCE, DCE theorem 13 holds verbatim.

5.2 Monotone Spline Interpolation

The problem of determining isotone cubic spline interpolants which minimize functional (2.8) leads to

Program PM:
$$4\sum_{i=1}^{n} \frac{w_i}{h_i} \{(m_{i-1}-\tau_i)^2+(m_{i-1}-\tau_i)(m_i-\tau_i)+(m_i-\tau_i)^2\} \to \min!$$
$$\text{s.t.} \quad (m_{i-1},m_i) \in D_i \cup E_i \ (i=1,\dots,n) \tag{5.11}$$

where the sets D_i and E_i are given by (2.12). Here the feasible domain is always nonempty. Therefore, in view of the positive definiteness of the Hessian of the objective function, program PM possesses always a unique solution. For formulating the corresponding dual program (1.2) the subprogram

$$d_i(x,y) = \xi x+\eta y - \frac{4w_i}{h_i} \{(x-\tau_i)^2+(x-\tau_i)(y-\tau_i)+(y-\tau_i)^2\} \to \max!$$
$$\text{s.t.} \quad (x,y) \in D_i \cup E_i \tag{5.12}$$

is to be considered for all $(\xi,\eta) \in R^2$.

Proposition 14: The optimal value $d_{i\,\max} = H_i^*(\xi,\eta)$ of program (5.12) is, with $\sigma_i = 12\tau_i w_i/h_i$, given by

$$\begin{aligned}
\frac{12w_i}{h_i} H_i^*(\xi,\eta) &= \sigma_i(\xi+\eta)+\xi^2-\xi\eta+\eta^2 && \text{for } (\xi,\eta) \in A_i \cup B_i \\
&= -\tfrac{1}{4}\sigma_i^2 + \tfrac{3}{2}\sigma_i\xi + \tfrac{3}{4}\xi^2 && \text{for } (\xi,\eta) \in S_{1i} \\
&= -\tfrac{1}{4}\sigma_i^2 + \tfrac{3}{2}\sigma_i\eta + \tfrac{3}{4}\eta^2 && \text{for } (\xi,\eta) \in S_{2i} \\
&= -\sigma_i^2 && \text{for } (\xi,\eta) \in T_i \\
&= -2\sigma_i^2 + 2\sigma_i(\xi+\eta) + \\
&\quad + 2\sigma_i\sqrt{\xi^2-\xi\eta+\eta^2-\sigma_i(\xi+\eta)+\sigma_i^2} && \text{for } (\xi,\eta) \in C_i
\end{aligned} \tag{5.13}$$

where

$$A_i = \{(\xi,\eta) \in R^2: -\sigma_i \le 2\xi-\eta \le 2\sigma_i, -2\sigma_i \le \xi-2\eta \le \sigma_i\},$$

$$B_i = \{(\xi,\eta) \in R^2: \xi^2-\xi\eta+\eta^2-\sigma_i(\xi+\eta) \le 0\},$$

$$S_{1i} = \{(\xi,\eta) \in R^2: \xi-2\eta \ge \sigma_i, -\sigma_i \le \xi \le \sigma_i\},$$

$$S_{2i} = \{(\xi,\eta) \in R^2: 2\xi-\eta \le -\sigma_i, -\sigma_i \le \eta \le \sigma_i\},$$

$$T_i = \{(\xi,\eta) \in R^2: \xi \le -\sigma_i, \eta \le -\sigma_i\},$$

$$C_i = R^2 \setminus \{A_i \cup B_i \cup S_{1i} \cup S_{2i} \cup T_i\}.$$

Using the fact that

$$D_i \cup E_i = \{(x,y) \in R^2: x-\sqrt{xy} + y \le 3\tau_i, x \ge 0, y \ge 0\} \qquad (5.14)$$

(5.13) can, in principle, be proven analogously to (5.3). For another proof it is referred to [2].

With H_i^* according to (5.13) the following program DM is obtained which is dual to PM:

$$\text{Program DM:} \quad -\sum_{i=1}^{n} H_i^*(p_{i-1},-p_i) \to \max! \text{ with } p_o = p_n = 0. \qquad (5.15)$$

If $\tau_j = 0$ for some $j \in \{1,\dots,n\}$, then $m_{j-1} = m_j = 0$ follows and, therefore, PM can be decomposed into subprograms of lower dimensions. Thus, without loss of generality, $\tau_1 > 0,\dots,\tau_n > 0$ can be assumed, and by means of proposition 11 one gets immediately

Theorem 14 [2]: Both programs PM and DM possess optimal solutions and the optimal values are equal. Moreover, if a solution $(p_o,p_1,\dots,p_n)$ of DM is known the solution $(m_o,m_1,\dots,m_n)$ of PM can be computed by formula (1.4), where now H_i^* is to be defined by (5.13).

5.3 Discrete Variational and Complementarity Problems

Under assumptions described in sections 2.3 and 2.4 now one is led to problems of the type

$$\text{Program PV:} \quad \sum_{i=1}^{n}\{(u_i m_{i-1}+v_i m_i)^2+\varepsilon_i m_{i-1}^2+\varkappa_i m_i^2+r_i m_{i-1}+s_i m_i\} \to \min! \qquad (5.16)$$
$$\text{s.t. } m_i \ge c_i \ (i=0,\dots,n).$$

The constraints may also be $c_i \le m_i \le d_i$ $(i=0,\dots,n)$.

Suppose that the inequalities

$$u_i^2+\varepsilon_i > 0, \ v_i^2+\varkappa_i > 0, \ D_i = \varkappa_i u_i^2+\varepsilon_i v_i^2+\varepsilon_i\varkappa_i > 0 \qquad (5.17)$$

are valid. Then the subprogram (1.3),

$$d_i(x,y) = \xi x+\eta y-(u_ix+v_iy)^2-\varepsilon_ix^2-\varkappa_iy^2-r_ix-s_iy \to \max!$$

$$\text{s.t. } x \ge c_{i-1},\ y \ge c_i, \tag{5.18}$$

is uniquely solvable, and the optimal value $d_{i\,max} = H_i^*(\xi,\eta)$ is equal to

$$H_i^*(\xi,\eta) = \frac{1}{4D_i}\{(v_i^2+\varkappa_i)(\xi-r_i)^2-2u_iv_i(\xi-r_i)(\eta-s_i)+(u_i^2+\varepsilon_i)(\eta-s_i)^2\}$$
$$\text{for } q_1(\xi,\eta) \ge 2c_{i-1}D_i,\ q_2(\xi,\eta) \ge 2c_iD_i$$

$$= \frac{1}{4(v_i^2+\varkappa_i)}\{(\eta-s_i)^2+4(v_i^2+\varkappa_i)c_{i-1}(\xi-r_i)-4u_iv_ic_{i-1}(\eta-s_i)-4D_ic_{i-1}^2\}$$
$$\text{for } q_1(\xi,\eta) \le 2c_{i-1}D_i,\ \eta-s_i \ge q_4 \tag{5.19}$$

$$= \frac{1}{4(u_i^2+\varepsilon_i)}\{(\xi-r_i)^2-4u_iv_ic_i(\xi-r_i)+4(u_i^2+\varepsilon)c_i(\eta-s_i)-4D_ic_i^2\}$$
$$\text{for } q_2(\xi,\eta) \le 2c_iD_i,\ \xi-r_i \ge q_3$$

$$= c_{i-1}(\xi-r_i)+c_i(\eta-s_i)-(u_i^2+\varepsilon_i)c_{i-1}^2-2u_iv_ic_{i-1}c_i-(v_i^2+\varkappa_i)c_i^2$$
$$\text{for } \xi-r_i \le q_3,\ \eta-s_i \le q_4,$$

where

$$q_1(\xi,\eta) = (v_i^2+\varkappa_i)(\xi-r_i)-u_iv_i(\eta-s_i),$$

$$q_2(\xi,\eta) = -u_iv_i(\xi-r_i)+(u_i^2+\varepsilon_i)(\eta-s_i),$$

$$q_3 = 2(u_i^2+\varepsilon_i)c_{i-1}+2u_iv_ic_i,\quad q_4 = 2(v_i^2+\varkappa_i)c_i+2u_iv_ic_{i-1}.$$

Proposition (5.19) is proven analogously to (5.3).

To PV belongs the dual

$$\text{Program DV: } -\sum_{i=1}^{n} H_i^*(p_{i-1},-p_i) \to \max! \text{ with } p_0 = p_n = 0, \tag{5.20}$$

where H_i^* is now defined by (5.19), and for the pair PV,DV theorem 14 holds verbatim, see [5].

6. Numerical Aspects

Several numerical tests are performed on the problems described above, see [1],[3],[5] and [7]. For solving a program PA in these papers always the following general procedure is used:

Step 1: By means of theorem 7 is tested whether the feasible domain W of PA is nonempty or not.

Step 2: If $W \neq \emptyset$ holds, the unconstrained dual program DA is numerically treated by the usual Newton's method

$$H'(p^{(\nu)}) + H''(p^{(\nu)})(p^{(\nu+1)}-p^{(\nu)}) = 0 \quad (\nu = 0,1,\ldots). \tag{6.1}$$

Step 3: The solution m of PA is computed via the return-formula (1.4) with $p = \lim p^{(\nu)}$.

Step 4: In the spline problems the optimal interpolant s is determined by means of corresponding formulas, e.g. by (3.9),(3.10) if $s \in Sp_3(1)$. Also s may be plotted.

For the complexity of this procedure step 2 is essential. It turns out that a good choice for the starting vector $p^{(o)}$ is

$$p^{(o)} = (0,0,\dots,0,0) \tag{6.2}$$

for the monotone $Sp_3(1)$-interpolation and

$$p^{(o)} = (0,\tau_1-\tau_2,\dots,\tau_{n-1}-\tau_n,0) \tag{6.3}$$

for the convex $Sp_3(1)$-interpolation. Then it is observed that Newton's method terminates after a small number of iteration steps, say after two, three or four steps for $n \leq 100$. Thus, since the Hessian H" in the linear system (6.1) is tridiagonal, for the spline problems the total amount of arithmetical operations for computing the optimal splines by performing the steps 1-4 is only equal to

$$O(n) \text{ additions, multiplications, comparisons,} \tag{6.4}$$

if the mentioned finiteness of Newton's method generally holds. Of course, using bad starting vectors Newton's method may even be divergent. Further, application of the gradient method in step 2 cannot be recommended since then the number of iteration steps is in general very large.

It may be pointed out that the advantage of the proposed procedure is mainly based on the following properties:

(i) Program DA is unconstrained
(ii) The dimension of DA is not greater than the dimension of PA
(iii) The Hessian of the objective function in DA is tridiagonal
(iv) A return-formula (1.4) is valid.

Finally, it shall be noticed that in the spline problems the choice of the objective function must be undertaken with some care. Tests are performed with the objective function (2.8) where the weights are set

$$w_i = 1 \ (i=1,\dots,n) \tag{6.5}$$

respectively

$$w_i = 1/(1+\tau_i^2)^3 \ (i=1,\dots,n). \tag{6.6}$$

It turns out that in the test examples the weights (6.6) lead to the better results, compare fig.5.

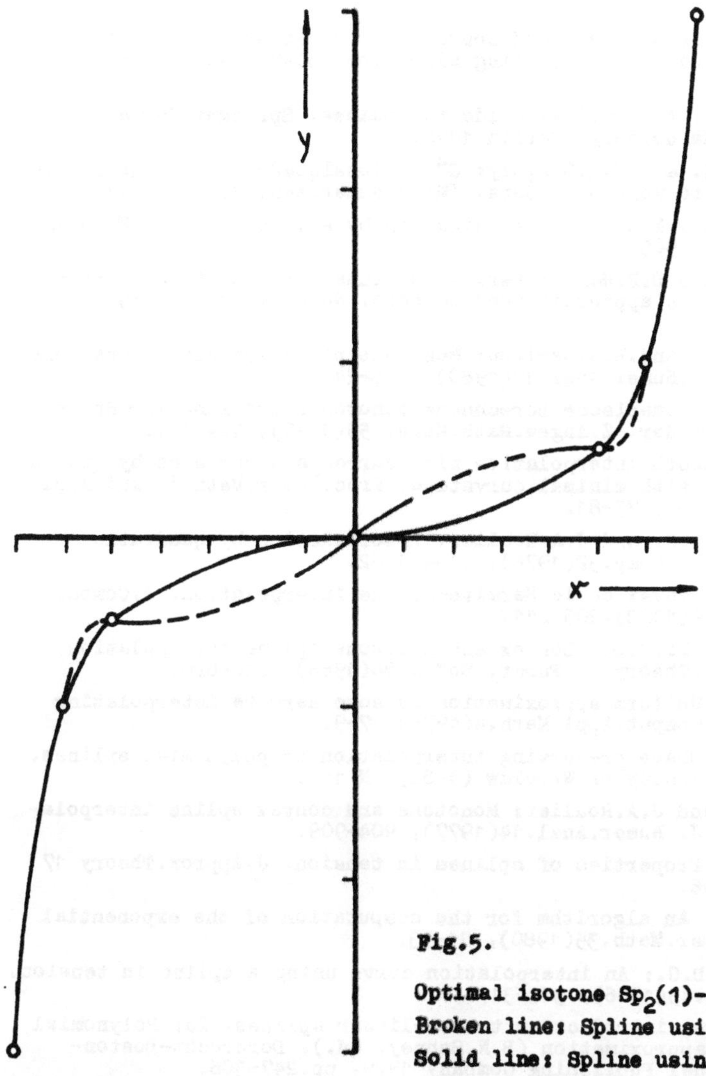

Fig.5.

Optimal isotone $Sp_2(1)$-splines.
Broken line: Spline using weights (6.5)
Solid line: Spline using weights (6.6)

References

[1] Burmeister,W., W.Heß and J.W.Schmidt: Convex spline interpolants with minimal curvature. Computing 35(1985), 219-229.

[2] Dietze,S. and J.W.Schmidt: Determination of shape preserving spline interpolants with minimal curvature via dual programs. J.Approx.Theory (to appear).

[3] Heß,W. and J.W.Schmidt: Convexity preserving interpolation with exponential splines. Computing 36(1986), 335-342.

[4] Schmidt,J.W.: Convex interval interpolation with cubic splines. BIT (to appear).

[5] Schmidt,J.W.: On tridiagonal complementarity problems. Numer.Math. (submitted).

[6] Schmidt,J.W. and W.Heß: Schwach verkoppelte Ungleichungssysteme und konvexe Spline-Interpolation. Elem.Math. 39(1984), 85-95.

[7] Schmidt,J.W. and W.Heß: Quadratic and related exponential splines in shape preserving interpolation. J.Comp.Appl.Math. (submitted).

[8] Beatson,R.K.: Monotone and convex approximation by splines: error estimates and a curve fitting algorithm. SIAM J.Numer.Anal.19(1982), 1278-1285.

[9] de Boor, C.: A practical guide to splines. Springer-Verlag, New York, Heidelberg, Berlin 1978.

[10] Delbourgo,R. and J.A.Gregory: C^2 rationalquadratic spline interpolations to monotonic data. IMA J.Numer.Anal.3(1983), 141-152.

[11] De Vore,R.A.: Monotone approximation by splines. SIAM J.Math.Anal. 8(1977), 891-905.

[12] Dott,S.L. and D.F.McAllister: Algorithms for computing shape preserving spline approximations to data. Numer.Math.46(1985), 159-174.

[13] Fritsch,F.N. and.R.E.Carlson: Monotone piecewise cubic interpolation. SIAM J.Numer.Anal.17(1980), 238-246.

[14] Hornung,U.: Numerische Berechnung monotoner und konvexer Spline-Interpolierender. Z.Angew.Math.Mech. 59(1979), T64-T65.

[15] Maeβ,G.: Smooth interpolation with curves and surfaces by quadratic splines with minimal curvature. Proc.Numer.Methods and Appl. Sofia 84(1985), 75-81.

[16] McAllister,D.F. and J.A.Roulier: Interpolation by quadratic splines. Math.Comp.32(1978), 1154-1162.

[17] Mettke,H.: Convex cubic Hermite-spline interpolation. J.Comput. Appl.Math. 9(1983),205-211.

[18] Miroshnichenko,V.L.: Convex and monotone spline interpolation. Proc.Constr.Theory of Funct. Sofia 84(1985), 610-620.

[19] Neuman,E.: Uniform approximation by some Hermite interpolating splines. J.Comput.Appl.Math.4(1978), 7-9.

[20] Neuman,E.: Shape preserving interpolation by polynomial splines. Report University of Wroclaw (1982), N 112.

[21] Passow,E. and J.A.Roulier: Monotone and convex spline interpolation. SIAM J. Numer.Anal.14(1977), 904-909.

[22] Pruess,S.: Properties of splines in tension. J.Approx.Theory 17 (1976),86-96.

[23] Rentrop,P.: An algorithm for the computation of the exponential spline. Numer.Math.35(1980), 81-93.

[24] Schweikert,D.G.: An interpolation curve using a spline in tension. J.Math.Phys.45(1966), 312-317.

[25] Werner,H.: An introduction to non-linear splines. In: Polynomial and spline approximation (B.N.Sahney, ed.), Dordrecht-Boston-London: Reidel Publishing Company 1979, pp.247-306.

[26] Elster,K.-H., R.Reinhardt, M.Schäuble and G.Donath: Einführung in die nichtlineare Optimierung. Math.-naturw.Bibliothek. Bd.63, Teubner Verlagsgesellsch., Leipzig 1977.

[27] Lasdon,L.S.: Optimization theory for large systems. Macmillan, London 1971.

[28] Rockafellar,R.T.: Convex Analysis. Princeton University Press, Princeton, NJ. 1970.

[29] Cottle,R.W. and R.S.Sacher: On the solution of large, structured linear complementarity problems: The tridiagonal case. Appl.Math. Optim.3(1977), 321-340.

[30] Cryer,C.W.: The solution of a quadratic programming problem using systematic overrelaxation. SIAM J.Control 9(1971), 385-392.

[31] Lüthi,H.-J.: Komplementaritäts- und Fixpunktalgorithmen in der mathematischen Programmierung, Spieltheorie und Ökonomie. Springer-Verlag, Berlin-Heidelberg-New York 1976.

[32] Mandel,J.: A multilevel iterative method for symmetric positive definite linear complementarity problems. Appl.Math.Optim.11(1984), 77-95.

FINITE ELEMENT SOLUTION OF A NONLINEAR DIFFUSION PROBLEM WITH A MOVING BOUNDARY

Libor ČERMÁK [1)]

Abstract. The above problem is a boundary value problem simulating the redistribution of impurities in semiconductor device structures in the oxidation - diffusion technological process, see [5]. In [1] the problem was formulated in a variational form and a fully discrete finite element solution was constructed. If the diffusion is comparatively very small with respect to the velocity of the moving boundary the developed scheme does not work well. Therefore we propose here its modification based on the streamline artificial diffusion technique, see [3]. This new scheme is applicable independently on the magnitude of the diffusion coefficient. Stability of the scheme is proved and an error estimate derived.

1. Introduction

We consider the physical model of the oxidation - diffusion process introduced, for example, in Maldonado [5]. It consists in solving the following boundary value problem:

$$\frac{\partial u}{\partial t} = \nabla \cdot [\, D(u) \nabla u \,] \qquad \text{in } \Omega(t),\ 0<t<T, \tag{1}$$

$$\Omega(t) = \{(x,y) \mid \varphi(y,t) < x < L_0,\ 0<y<B\},$$

$$D(u)\frac{\partial u}{\partial n} = \gamma \dot{\varphi}_n u \qquad \text{on } \Gamma(t),\ 0<t<T, \tag{2}$$

$$\Gamma(t) = \{(x,y) \mid x = \varphi(y,t),\ 0<y<B\},$$

$$\frac{\partial u}{\partial n} = 0 \qquad \text{on } \partial\Omega(t) - \Gamma(t), \tag{3}$$

$$u(x,y,0) = u^*(x,y) \qquad \text{in } \Omega(0). \tag{4}$$

Here, u is the unknown concentration of an impurity, D(u) is the diffusion coefficient and we assume that $D(w) \in C^0((-\infty,\infty))$ and

$$0 < d_L \leqq D(w) \leqq d_U . \tag{5}$$

Further, φ is a given function of y and t belonging to $C^1(\langle 0,B\rangle \times \langle 0,T\rangle)$, $\frac{\partial}{\partial n}$ is the derivative in the direction of the outward normal, γ is a

1) VUT - OVC ČSSR - 602 00 Brno, třída Obránců míru 21

constant, $\dot{\varphi}_n$ is the rate of the motion of $\Gamma(t)$ in the direction of the outward normal ($\dot{\varphi}_n = -\frac{\partial\varphi}{\partial t}\left[1+\left(\frac{\partial\varphi}{\partial y}\right)^2\right]^{-\frac{1}{2}}$) and in the problem considered

$$\frac{\partial\varphi}{\partial t} \geqq 0 \, , \quad 0 \leqq \varphi \leqq \frac{1}{2} L_0 \, , \tag{6}$$

$u^*(x,y)$ is the given initial concentration. For more details we refer the reader to [5].

The motion of the boundary $\Gamma(t)$ introduces a convection into our diffusion problem. The mutual relation between convection and diffusion is usually expressed by means of the Peclet number which, in our case, can be represented by the ratio

$$Pe = \frac{\partial\varphi}{\partial t} L_0 / (2D) \, . \tag{7}$$

If the Peclet number is large then due to the boundary condition (2) there exists a boundary layer along $\Gamma(t)$, in which the concentration u changes significantly. In the problem (1)-(4) the Peclet number belongs typically to the interval (10,1000), but it may be even larger. For large Peclet numbers the scheme introduced in [1] produces unsatisfactory solutions. Therefore, in the present paper we construct a fully discrete finite element solution which behaves well independently on the magnitude of the diffusion coefficient D. We prove stability and derive an error estimate.

2. A finite element solution

Let Ω be a bounded open set in R^2. We denote by $H^{m,p}(\Omega)$, $m = 0, 1, \ldots$, $1 \leqq p \leqq \infty$, the Sobolev space $H^{m,p}(\Omega) = \{v \mid D^\alpha v \in L^p(\Omega) \ \forall \alpha \leqq m\}$ normed by $\|v\|_{m,p,\Omega} = \sum_{|\alpha| \leqq m} \|D^\alpha v\|_{L^p(\Omega)}$. For p=2 the index p will be omitted. The scalar product in $H^m(\Omega)$ is denoted by $(\cdot,\cdot)_{m,\Omega}$. For m=0 we have $H^0(\Omega) = L^2(\Omega)$ and instead of $\|\cdot\|_{0,\Omega}$ and $(\cdot,\cdot)_{0,\Omega}$ we use the notation $\|\cdot\|_\Omega$ and $(\cdot,\cdot)_\Omega$, respectively.

If u is a sufficiently smooth solution of (1)-(4) then by multiplying (1) by $v \in H^1(\Omega(t))$ and integrating over $\Omega(t)$ we get

$$\forall t \in (0,T) \quad \left(\frac{\partial u}{\partial t}, v\right)_{\Omega(t)} + a(u,t,f;u,v) = 0 \qquad \forall v \in V(t) \equiv H^1(\Omega(t)). \tag{8}$$

Here

$$a(w,t,s;u,v) = \int_{\Omega(t)} D(w)\nabla u \cdot \nabla v \, dx\,dy - s \int_{\Gamma(t)} \dot{\varphi}_n u v \, d\Gamma . \tag{9}$$

The equation (8) will be used for defining a semidiscrete finite element solution. To this end, we need the one-to-one mapping of the rectangle $\bar{Q} = \langle 0,L_o\rangle \times \langle 0,B\rangle$ on $\bar{\Omega}(t)$:

$$x = F(\alpha,\beta,t) = \varphi(\beta,t) + \alpha\left[1 - L_o^{-1}\varphi(\beta,t)\right], \quad y = \beta. \tag{10}$$

We cover $\bar{\Omega}(0)$ by triangles completed along $\Gamma(0)$ by curved elements in a manner described in Zlámal [7]. Let us denote this triangulation by $\mathcal{T}_h$ where h is the discretisation parameter denoting the greatest side of all elements. Let $P_k = (x_k,y_k)$, $k = 1,\dots,d$, be the nodes of $\mathcal{T}_h$ and let $Q_k = (\alpha_k,\beta_k)$ be their inverse immages in the mapping (10), i.e.

$$\alpha_k = \left[x_k - \varphi(\beta_k,0)\right]\cdot\left[1 - L_o^{-1}\varphi(\beta_k,0)\right]^{-1}, \quad \beta_k = y_k. \tag{11}$$

The triangulation $\mathcal{T}_h(t)$ of $\bar{\Omega}(t)$ is determined for $t > 0$ by the nodes

$$P_k(t) = (x_k(t), y_k), \quad x_k(t) = F(\alpha_k,\beta_k,t), \quad 0 < t \leq T. \tag{12}$$

The elements of $\mathcal{T}_h(t)$ are again triangles or curved elements. Let K(t) be an arbitrary element of $\mathcal{T}_h(t)$. At this moment, we use a local notation $P_1(t)$, $P_2(t)$, $P_3(t)$ of the vertices of K(t). We map K(t) on the (time independent) reference element $\hat{K}$ with vertices $R_1 = (0,0)$, $R_2 = (1,0)$, $R_3 = (0,1)$ in the ξ,η plane. We have (see [7])

$$\left.\begin{aligned} x &= x(\xi,\eta,t) = \sum_{j=1}^{3} x_j(t)N_j(\xi,\eta) + (1-\xi-\eta)\Phi(\eta,t) \\ y &= y(\xi,\eta) = \sum_{j=1}^{3} y_j N_j(\xi,\eta) \end{aligned}\right\} \tag{13}$$

where

$$N_1 = 1-\xi-\eta, \quad N_2 = \xi, \quad N_3 = \eta$$

and

$$\Phi(\eta,t) = 0 \qquad \text{for triangles,}$$

$$\Phi(\eta,t) = (1-\eta)^{-1}\left\{\varphi(y_1 + (y_3-y_1)\eta, t) - x_1(t) - \left[x_3(t) - x_1(t)\right]\eta\right\}$$

for elements with a curved side $\widehat{P_1P_3}$ ($x_1(t) = \varphi(y_1,t)$, $x_3(t) = \varphi(y_3,t)$). The trial functions are on each element K(t) of the form

$$v(x,y,t) = \hat{v}(\xi,\eta) \equiv \sum_{j=1}^{3} v_j N_j(\xi,\eta), \quad \xi = \xi(x,y,t), \quad \eta = \eta(x,y,t) \tag{14}$$

where $\xi = \xi(x,y,t)$, $\eta = \eta(x,y,t)$ is the inverse mapping to the mapping (13). We denote by $V_h(t)$ the set of all trial functions. We have $V_h(t) \subset V(t)$, $0 \leq t \leq T$. We denote by $w_k(x,y,t)$, $k = 1,\dots,d$, the basis

functions of $V_h(t)$. w_k is uniquely determined by $w_k(x,y,t) \in V_h(t)$, $w_j(x_j(t),y_j,t) = \delta_k^j$, $0 \leq t \leq T$.
The notation $\hat{v}$ will be used also for functions which are not trial functions. If v is defined on an element $K(t)$, then $\hat{v}(\xi,\eta,t)$ is defined on $\hat{K}$ by $\hat{v}(\xi,\eta,t) = v(x(\xi,\eta,t), y(\xi,\eta),t)$.

The semidiscrete solution of the problem (1)-(4) is assumed in the form

$$U(x,y,t) = \sum_{j=1}^{d} U_j(t)\, w_j(x,y,t) . \tag{15}$$

In [1] we have supposed that U satisfies

$$\left.\begin{aligned} &\forall t \in (0,T) \quad \left(\frac{\partial U}{\partial t}, v\right)_{\Omega(t)} + a(U,t,\gamma; U, v) = 0 \quad \forall v \in V_h(t) \\ &\qquad\qquad\qquad\qquad\qquad U(x,y,0) = U^*(x,y) \end{aligned}\right\} \tag{16}$$

where $U^* \in V_h(0)$ is a suitable approximation of u^*. Let us express the term $\frac{\partial U}{\partial t}$. We have

$$\frac{\partial U}{\partial t} = \sum_{j=1}^{d} \left[\dot{U}_j(t)\, w_j(x,y,t) + U_j(t) \frac{\partial w_j(x,y,t)}{\partial t} \right] \quad \text{where} \quad \dot{U}_j(t) = \frac{dU_j(t)}{dt} . \tag{17}$$

By differentiating on the reference element we obtain

$$0 = \frac{d\hat{w}_j(\xi,\eta)}{dt} = \frac{\partial w_j(x(\xi,\eta,t),y(\xi,\eta),t)}{\partial x} \frac{\partial x(\xi,\eta,t)}{\partial t} + \frac{\partial w_j(x(\xi,\eta,t),y(\xi,\eta),t)}{\partial t}$$

and, consequently,

$$\frac{\partial w_j}{\partial t} = -G \frac{\partial w_j}{\partial x} \quad \text{where} \quad G(x,y,t)\Big|_{K(t)} = \frac{\partial x(\xi,\eta,t)}{\partial t}\Big|_{\substack{\xi=\xi(x,y,t)\\ \eta=\eta(x,y,t)}} . \tag{18}$$

From the last two relations we get

$$\frac{\partial U}{\partial t} = \dot{U} - G \frac{\partial U}{\partial x} \quad \text{where} \quad \dot{U} = \sum_{j=1}^{d} \dot{U}_j w_j . \tag{19}$$

If we set (19) into (16), we see that U satisfies for each $t \in (0,T)$ the convection - diffusion equation

$$(\dot{U},v)_{\Omega(t)} - (G\frac{\partial U}{\partial x},v)_{\Omega(t)} + a(U,t,\gamma;U,v) = 0 \qquad \forall v \in V_h(t). \tag{20}$$

From (13) and (10) we easily derive

$$\left.\begin{aligned} &G \geq 0, \\ &G(x_j(t),y_j,t) = \frac{\partial\varphi(y_j,t)}{\partial t}\,\frac{L_0 - x_j}{L_0 - \varphi(y_j,0)}, \\ &G = \frac{\partial\varphi}{\partial t} \text{ on } \Gamma(t), \quad G = 0 \text{ for } x = L_0. \end{aligned}\right\} \tag{21}$$

Hence, we deduce that G is proportional to $\frac{\partial\varphi}{\partial t}$. Thus the grid Peclet number of the semidiscrete problem (16) can be characterized by

$$Pe_g = \frac{\partial\varphi}{\partial t}\,h/(2D) = Pe \cdot h/L_0 . \tag{22}$$

We observe that Pe_g remains large if Pe is large and h is not unrealisticaly small. Therefore for large Peclet numbers the scheme (16) will, in general, produce severely oscillating discrete solutions U, which are not close to the exact solution u in, e.g., the L^2-norm, cf.e.f. [6], [2]. This assertion was also confirmed by the numerical experiment. If we increase the diffusion coefficient by adding an artificial diffusion, which acts in the direction of the x-axis only and is of approximate magnitude $h\frac{\partial\varphi}{\partial t}$, we obtain a modification of the scheme (16) with the modified grid Peclet number less then one. This idea, which was proposed by Hughes and Brooks in [3], enables us to introduce a modified semidiscrete formulation

$$\left.\begin{aligned} &\forall t \in (0,T) \quad (\dot{U},v)_{\Omega(t)} - (G\frac{\partial U}{\partial x},v)_{\Omega(t)} + A(U,t,\gamma;U,v) = 0 \\ &\qquad\qquad\qquad\qquad\qquad \forall v \in V_h(t) \\ &\qquad\qquad U(x,y,0) = U^*(x,y). \end{aligned}\right\} \tag{23}$$

Here

$$A(w,t,s;u,v) = a(w,t,s;u,v) + \int_{\Omega(t)} GH \frac{\partial u}{\partial x}\frac{\partial v}{\partial x}\,dx\,dy \tag{24}$$

and H is an arbitrary function satisfying

$$0 < \varkappa_L h \leqq H(x,y,t) \leqq \varkappa_U h \tag{25}$$

where $\varkappa_L, \varkappa_U$ are some constants independent of h.

We discretize (23) in time. For simplicity, we use a uniform partition of the interval $\langle 0,T\rangle$: $t_i = i\cdot\Delta t$, $i = 1,\ldots,q$ (hence $q\cdot\Delta t = T$). In the sequel $U^i, \Omega^i, \ldots$ means $U(t_i), \Omega(t_i), \ldots$. Now we set $t = t_{i+1}$ in (23), replace $\dot{U}_j^{i+1}$ by $\Delta t^{-1}(U_j^{i+1} - U_j^i)$, $j=1,\ldots,d$, and linearize the nonlinear term in (23). We get

$$\left.\begin{aligned} &(U^{i+1}-\tilde{U}^i, v)_{\Omega^{i+1}} - \Delta t\left(G^{i+1}\frac{\partial U^{i+1}}{\partial x}, v\right)_{\Omega^{i+1}} + \Delta t\, A(\tilde{U}^i, t_{i+1}, \mathcal{F}; U^{i+1}, v) = 0 \\ &\qquad \forall v \in V_h^{i+1},\ i = 0,\ldots,q-1, \\ &U^0 = U^* \end{aligned}\right\} \tag{26}$$

where

$$\tilde{U}^i = \sum_{j=1}^{d} U_j^i w_j^{i+1}. \tag{27}$$

For practical computations it is necessary to do one more step: to replace curved elements by triangles and to compute all integrals numerically.

3. Properties of the triangulation

We will consider a family $\{\mathcal{T}_h^0\}$ of the triangulations of $\bar{\Omega}^0$ such that h approaches zero and that the minimum angle condition is satisfied (see Zlámal [7]), i.e. $\vartheta_h \geqq \vartheta_0$, $\vartheta_0 = \text{const.} > 0$ where ϑ_h is the smallest angle of all elements of $\mathcal{T}_h^0$. We remind that

$$h = \max_{K^0 \in \mathcal{T}_h^0} h_{K^0} \tag{28}$$

where h_{K^0} is the greatest side of an element K^0. From $\{\mathcal{T}_h^0\}$ we construct $\{\mathcal{T}_h(t)\}$ for all t as described in the preceeding section. In [1] we have proved the following assertions.

<u>Lemma 1.</u> Let the family $\{\mathcal{T}_h^0\}$ of triangulations satisfy the minimum angle condition and let $\frac{\partial^{k+1}\varphi}{\partial y^{k+1}}$, $k \geqq 1$, and $\frac{\partial^3\varphi}{\partial t\,\partial y^2} \in C^0(\langle 0,B\rangle \times \langle 0,T\rangle)$.

If h is sufficiently small, $h \leq h_0$ where h_0 does not depend on K(t), then (13) maps $\hat{K}$ one-to-one on K(t) for any $t \in \langle 0,T \rangle$. Further, the mapping (13) and its Jacobian determinant $J(\xi,\eta,t)$ are bounded on $\langle 0,T \rangle$ in this way:

$$C_0 h_{K^0}^2 \leq |J(\xi,\eta,t)| \leq C_0^{-1} h_{K^0}^2, \quad \left|\frac{\partial J(\xi,\eta,t)}{\partial t}\right| \leq C_0 h_{K^0}^2, \tag{29}$$

$$\left.\begin{aligned} &|D^\alpha x(\xi,\eta,t)| \leq C_0 h_{K^0}^{|\alpha|}, \quad |D^\alpha y(\xi,\eta)| \leq C_0 h_{K^0}^{|\alpha|}, \quad |\alpha| = 1,\dots,k, \\ &|D^\beta \xi(x,y,t)| \leq C_0 h_{K^0}^{-1}, \quad |D^\beta \eta(x,y,t)| \leq C_0 h_{K^0}^{-1}, \quad |\beta| = 1. \end{aligned}\right\} \tag{30}$$

In addition, the family $\{\mathcal{T}_h(t)\}$ satisfies the minimum angle condition uniformly on $\langle 0,T \rangle$, i.e. the minimum angle $\vartheta_h(t)$ of $\mathcal{T}_h(t)$ satisfies $\vartheta_h(t) \geq \vartheta_1$, $t \in \langle 0,T \rangle$. Here, C_0 and ϑ_1 are positive constants.

Lemma 2. Let the assumptions of lemma 1 with k = 2 be satisfied and let u(x,y,t) belongs for each $t \in \langle 0,T \rangle$ to $H^{2,p}(\Omega(t))$, $p \geq 1$. Then

$$\left.\begin{aligned} &\|u-u_I\|_{L^p(\Omega(t))} + h\|u-u_I\|_{H^{1,p}(\Omega(t))} \leq C_0 h^2 \|u\|_{H^{2,p}(\Omega(t))} \\ &\qquad 0 \leq t \leq T \end{aligned}\right\} \tag{31}$$

where u_I is the interpolate of u.

Lemma 3. Let the assumptions of lemma 1 with k = 2 be satisfied. Then for each $t \in \langle 0,T \rangle$ the function G belongs to $H^{1,\infty}(\Omega(t))$ and

$$\|G\|_{H^{1,\infty}(\Omega(t))} \leq C \tag{32}$$

where C does not depend on h, t.

4. Stability and error estimates

In the sequel C denotes a positive constant independent on h, Δt, D and not necessarily the same in any two places.

Theorem 1. Let the assumptions of lemma 1 with k = 2 be satisfied and let $\gamma \geq -\frac{1}{2}$. Then for Δt sufficiently small, $\Delta t \leq \Delta t_0$ where Δt_0 does not depend on i,h,D, the fully discrete solutions U^i, i=1,...,q, of the systems (26) are uniquely determined. Furthermore, the scheme (26) is unconditionally stable in the following sense: there exists a constant C_S independend of h, Δt, D such that it holds

$$\max_{1 \leq i \leq q} \|U^i\|_{L^2(\Omega^i)} + \left[\Delta t \sum_{i=1}^{d} A(\tilde{U}^{i-1}, t_i, \gamma + \tfrac{1}{2}; U^i, U^i)\right]^{\frac{1}{2}} \leq C_S \|U^0\|_{L^2(\Omega^0)} . \tag{33}$$

The proof of this theorem is similar to the proof of theorem 1 in [1]. Before introducing the error estimate we formulate the assumptions on the data φ and D, on the function H and on the exact solution u:

A: (i) $\frac{\partial^2 \varphi}{\partial t^2}, \frac{\partial^3 \varphi}{\partial t \partial y^2}, \frac{\partial^3 \varphi}{\partial y^3} \in C^0(\langle 0,B\rangle \times \langle 0,T\rangle)$;

(ii) D(w) and D'(w) are bounded for $w \in (-\infty, \infty)$,

$$D'(w) \leq d'_U \quad w \in (-\infty, \infty);$$

(iii) $H \in H^{1,\infty}(\Omega(t)), \ \|H\|_{1,\infty,\Omega(t)} \leq Ch \quad \forall t \in \langle 0,T\rangle$;

(IV) $\frac{\partial u}{\partial x}, \frac{\partial u}{\partial y} \in L^\infty(\{(x,y,t) \mid t \in (0,T), (x,y) \in \Omega(t)\})$;

(V) $\|u\|_{2,\Omega(t)} + \|\frac{\partial u}{\partial x}\|_{2,\Omega(t)} + \|\frac{\partial u}{\partial t}\|_{2,\Omega(t)} + \|\frac{\partial^2 u}{\partial t^2}\|_{\Omega(t)} \leq C \quad \forall t \in (0,T)$.

Theorem 2. Let the family $\{\mathcal{T}^0_h\}$ of triangulations satisfy the minimum angle condition, let $\gamma > -\frac{1}{2}$ and the assumption A be fulfilled. Then for Δt sufficiently small, $\Delta t \leq \Delta t_0$ where Δt_0 does not depend on i,h,D, there holds

$$\max_{1 \leq i \leq q} \|u^i - U^i\|_{L^2(\Omega^i)} + \left[\Delta t \sum_{i=1}^{d} A(\tilde{U}^{i-1}, t_i, \gamma + \tfrac{1}{2}; u^i - U^i, u^i - U^i)\right]^{\frac{1}{2}} \leq$$

$$C_E \left[\|u^0 - U^0\|_{L^2(\Omega^0)} + h + \Delta t\right]. \tag{34}$$

Here C_E is a constant, $C_E = \exp\{C \cdot C_D^2\}$ where $C_D = \max(1, d_U) . (d_U + d'_U)/d_L$.
Remark. If $D(w) = \varepsilon < 1$ then $C_D = 1$ and C_E is a constant independent on h, Δt, D.
Proof. Applying the Green theorem on the term $(G^{i+1} \frac{\partial U^{i+1}}{\partial x}, v)_{\Omega^{i+1}}$ in (26) we obtain

$$(U^{i+1} - \tilde{U}^i, v)_{\Omega^{i+1}} + \Delta t (U^{i+1}, \frac{\partial}{\partial x}[G^{i+1} v])_{\Omega^{i+1}} + \Delta t\, A(\tilde{U}^i, t_{i+1}, \gamma + 1; U^{i+1}, v) = 0$$

$$\forall v \in V_h^{i+1}. \tag{35}$$

Introducing the operator $D_t = G \frac{\partial}{\partial x} + \frac{\partial}{\partial t}$ we obtain from (8)

$$(D_t u, v)_{\Omega(t)} - (G \frac{\partial u}{\partial x}, v)_{\Omega(t)} + a(u, t, \gamma; u, v) = 0 \quad \forall v \in V(t).$$

Further, using the Green theorem we get

$$(D_t u, v)_{\Omega(t)} + \left(u, \frac{\partial}{\partial x}[Gv]\right)_{\Omega(t)} + A(u, t, j+1; u, v) = \left(GH\frac{\partial u}{\partial x}, \frac{\partial v}{\partial x}\right)_{\Omega(t)} \quad \forall v \in V(t). \tag{36}$$

The equations (35) and (36) are the starting relations for deriving the error estimate. We introduce one more notation. If $z(x,y,t_i)$ is defined on Ω^i we denote by $\tilde{z}^i$ the function defined on Ω^{i+1} by

$$\tilde{z}^i(x,y)\Big|_{K^{i+1}} = \hat{z}(\xi(x,y,t_{i+1}), \eta(x,y,t_{i+1}), t_i).$$

This definition is in agreement with the notation $\tilde{U}^i$ introduced in (27). We decompose the error $u^i - U^i$ in the following way:

$$u^i - U^i = e^i + \varepsilon^i, \quad e^i = u^i - u_I^i, \quad \varepsilon^i = u_I^i - U^i \in V_h^i \tag{37}$$

(we remind that u_I denotes the interpolate of u). Multiplying (36) by Δt and choosing $t = t_{i+1}$ we obtain after some arrangements

$$(u_I^{i+1} - \tilde{u}_I^i, v)_{\Omega^{i+1}} + \Delta t\left(u_I^{i+1}, \frac{\partial}{\partial x}[G^{i+1}v]\right)_{\Omega^{i+1}} + \Delta t A(\tilde{U}^i, t_{i+1}, j+1; u_I^{i+1}, v) = F(u^{i+1}, \tilde{u}^i, \tilde{U}^i; v) \quad \forall v \in V_h^{i+1} \tag{38}$$

where

$$F(u^{i+1}, \tilde{u}^i, \tilde{U}^i; v) = \Delta t\left(G^{i+1}H^{i+1}\frac{\partial u^{i+1}}{\partial x}, \frac{\partial v}{\partial x}\right)_{\Omega^{i+1}} - \Delta t\left(e^{i+1}, \frac{\partial}{\partial x}[G^{i+1}v]\right)_{\Omega^{i+1}} +$$
$$(u^{i+1} - \tilde{u}^i - \Delta t D_t u^{i+1}, v)_{\Omega^{i+1}} - (e^{i+1} - \tilde{e}^i, v)_{\Omega^{i+1}} - \Delta t A(u^{i+1}, t_{i+1}, j+1; e^{i+1}, v) -$$
$$\Delta t \int_{\Omega^{i+1}} [D(u^{i+1}) - D(\tilde{U}^i)] \nabla u_I^{i+1} \cdot \nabla v \, dx\, dy. \tag{39}$$

Substracting (35) from (38) and choosing $v = \varepsilon^{i+1}$ we get

$$(\varepsilon^{i+1} - \tilde{\varepsilon}^i, \varepsilon^{i+1})_{\Omega^{i+1}} + \Delta t\left(\varepsilon^{i+1}, \frac{\partial}{\partial x}[G^{i+1}\varepsilon^{i+1}]\right)_{\Omega^{i+1}} + \Delta t A(\tilde{U}^i, t_{i+1}, j+1; \varepsilon^{i+1}, \varepsilon^{i+1}) = F(u^{i+1}, \tilde{u}^i, \tilde{U}^i; \varepsilon^{i+1}). \tag{40}$$

Using lemmas 1,2,3 and a technique similar to that used in the proof of Theorem 2 in [1] we find

$$|F(u^{i+1}, \tilde{u}^i, \tilde{U}^i; \varepsilon^{i+1})| \leq C \cdot C_D \cdot \Delta t \{ \lambda^{-1} [h^2 + \Delta t^2 + \| \tilde{\varepsilon}^i \|^2_{\Omega^{i+1}}] + \lambda [\| \varepsilon^{i+1} \|^2_{\Omega^{i+1}} + A(\tilde{U}^i, t_{i+1}, \gamma+1; \varepsilon^{i+1}, \varepsilon^{i+1})] \} \tag{41}$$

where λ is an arbitrary positive number. The proof of the assertion (41) is omitted here, however it will appear in the next paper. Arguing similarly as in the proof of Theorem 1 we get from (40) and from (41) with λ sufficiently small

$$\| \varepsilon^{i+1} \|^2_{\Omega^{i+1}} + \Delta t A(\tilde{U}^i, t_{i+1}, \gamma+\tfrac{1}{2}; \varepsilon^{i+1}, \varepsilon^{i+1}) \leq \tilde{C} \Delta t (h^2 + \Delta t^2) + (1 + \tilde{C} \Delta t) \| \varepsilon^i \|^2_{\Omega^i} ,$$

where $\tilde{C} = C.C_D^2$. Using the discrete Gronwall lemma we obtain

$$\max_{1 \leq i \leq q} \| \varepsilon^i \|^2_{\Omega^i} + \Delta t \sum_{i=1}^{q} A(\tilde{U}^i, t_{i+1}, \gamma+\tfrac{1}{2}; \varepsilon^{i+1}, \varepsilon^{i+1}) \leq C_E \cdot [h^2 + \Delta t^2 + \| \varepsilon^0 \|^2_{\Omega^0}].$$

From this inequality and (31) we get the assertion (34) of Theorem 2.

Remark. The constant C_E depends on the norm $\| \frac{\partial u}{\partial x} \|_{2, \Omega(t)}$ which, in general , is large if D is small. Therefore it would be worthwhile to derive a local error estimate on subdomains $\tilde{\Omega}(t) \subset \Omega(t)$, on which the above mentioned norm is no more large, see [6].

REFERENCES

[1] Čermák, L., Zlámal, M.: Finite Element Solution of a Nonlinear Diffusion Problem with a Moving Boundary. To appear in R.A.I.R.O. M²AN.

[2] Heinrich, J.C., Huyakorn, P.S., Zienkiewicz, O.C., Mitchell, A.R.: An "Upwind" Finite Element Scheme for Two-Dimensional Convective Transport Equation. Int.J.Numer.Meth.Engng.11 (1977), 131-143.

[3] Hughes, T.J.R., Brooks, A.N.: A Multidimensional Upwind Scheme with no Crosswind Diffusion. In AMD vol.34, Finite Element Methods for Convection Dominated Flows, T.J.R.Hughes (ed.), ASME, New York,1979.

[4] Lees, M.: A Priori Estimates for the Solutions of Difference Approximations to Parabolic Differential Equations. Duke Math.J. 27 (1960), 287-311.

[5] Maldonado, C.D.: ROMANS II, A Two-Dimensional Process Simulator for Modeling and Simulation in the Design of VLSI Devices. Applied Physics A31 (1983), 119-138.

[6] Nävert, U.: A Finite Element Method for Convection - Diffusion Problems. The Ph.D. dissertation thesis. Chalmers University of Tech - nology, Göteborg (1982).

[7] Zlámal, M.: Curved Elements in the Finite Element Method I. SIAM J. Numer.Anal 10 (1973), 229-240.

ERROR ESTIMATES FOR SOLUTION OF HYPERBOLIC DIFFERENTIAL EQUATIONS

Michael Fröhner [1)]

1. Introduction

In many time dependent simulations, the solution in most of the domain will be fairly smooth, with discontinuities or highly oscillatory phenomena occuring over a small fraction of the domain. With problems such as these, an error estimation and step size control can be an efficient, and often the only practical, solution method. Refined steps or grids with smaller and smaller mesh spacing are placed only where they are needed. Since we are solving a time dependent problem, the regions needing refinement will change, and therefore our grids must adapt to time as well.

In this paper we present an error estimation based on the Runge principle and on the predictor-corrector technique for the solution of hyperbolic partial differential equations using explicit finite difference techniques. We estimate the local truncation error. Other authors (see, for example, /1;2,3,6,12,14/) have considered enclosing methods and adaptive mesh refinements for solution of evolutionary problems in partial differential equations.

We consider the Euler equations for compressible inviscid flow in a two dimensional domain Ω

$$\frac{\partial U}{\partial t} + \sum_{i=1}^{2} \frac{\partial f_i(U)}{\partial x_i} = 0, \tag{1}$$

which forms a quasi-linear hyperbolic system of conservation laws. The state of the fluid at a point (x_1, x_2) is given by

$$U(x_1,x_2,t) = (\rho, \rho u, \rho v, E)^T .$$

The fluxes in the x_1- and x_2-direction are

$$f_1(U) = (\rho u, \rho u^2 + p, \rho uv, u(E+p))^T$$
$$f_2(U) = (\rho v, \rho uv, \rho v^2 + p, v(E+p))^T .$$

Additionally we have an equation of state

$$p = p(\rho, u, v, E)$$

and initial and boundary conditions.

For the general case, there exist only weak solutions, which are not necessarily unique. If we in addition assume the validity of entropy conditions, then the existence of an uniquely determined and physically

1) Ingenieurhochschule Cottbus, Karl-Marx-Straße 17, 7500 Cottbus, DDR

interpretable solution can be shown. Such entropy conditions are automatically fulfilled for many problems of gas dynamics. In the sequel let us suppose that all required conditions of regularity are satisfied. We shall give some theoretical results in the one-dimensional case and for one scalar equation. Then our model problem gets the form

$$\frac{\partial u}{\partial t} + \frac{\partial f(u)}{\partial x} = 0 \,. \qquad (2)$$

If we set $f = f(u) = au$, $a = \text{const}$, then we have the linear transport equation.

2. Discretization and local error estimation

2.1 Two-level difference schemes

We are going to the classical two-level schemes of Friedrichs and Lax-Wendroff. Our aim is to solve the equation (2) with stepsize τ and h in time and space, respectively, and to compare it with the results obtained with stepsize 2τ and $2h$. The Richardson extrapolation gives a solution of higher order accuracy and we can estimate the local truncation error. The procedure is similar to the classical Runge-Kutta method with control of the local error for ordinary differential equations.

Let $y_i^n = u(x_i, t_n)$ denote the approximate solution of $u(x,t)$ of equation (2). Furthermore, let T be the shift operator and $t_{n+1} - t_n = \tau$, $x_{i+1} - x_i = h$:

$$y_{i+1} = T_h \, y_i \,.$$

If function $u(x,t)$ is sufficiently smooth, we have for the semidiscrete function $\underline{u}(t)$

$$\|\underline{u}(t+\tau) - T_h \underline{u}(t)\| = \tau \left\{\tau^{q_1} a(t) + h^{q_2} b(t)\right\} + O(\tau^{q_1+2} + \tau h^{q_2+1})$$
$$\|\underline{u}(t+2\tau) - T_h^2 \underline{u}(t)\| = 2\tau \left\{(2\tau)^{q_1} \tilde{a}(t) + h^{q_2} \tilde{b}(t)\right\} + O(\tau^{q_1+2} + \tau h^{q_2+1})$$
$$\|\underline{u}(t+2\tau) - T_{2h} \underline{u}(t)\| = 2\tau \left\{(2\tau)^{q_1} \tilde{\tilde{a}}(t) + (2h)^{q_2} \tilde{\tilde{b}}(t)\right\} + O(\tau^{q_1+2} + \tau h^{q_2+1}).$$

Assume that for small τ and h the relations

$$a(t) \approx \tilde{a}(t) \approx \tilde{\tilde{a}}(t) \quad , \quad b(t) \approx \tilde{b}(t) \approx \tilde{\tilde{b}}(t)$$

hold. In the cases of the Friedrichs and the Lax scheme, we get $q_1 = q_2 = 1$, and for the 2-stage Lax-Wendroff scheme we have $q_1 = q_2 = 2$. If we compare the norm terms formulated above, a simple calculation leads to the local truncation error

$$l_e = \tau \left\{\tau^q a(t) + h^q b(t)\right\} = \left(T_{2h} \underline{u}(t) - T_h^2 \underline{u}(t) \right)/(2^{q+1} - 2) \,.$$

Note that four steps are needed for an error estimation at the point (x_i, t_n); see Fig. 1, too.

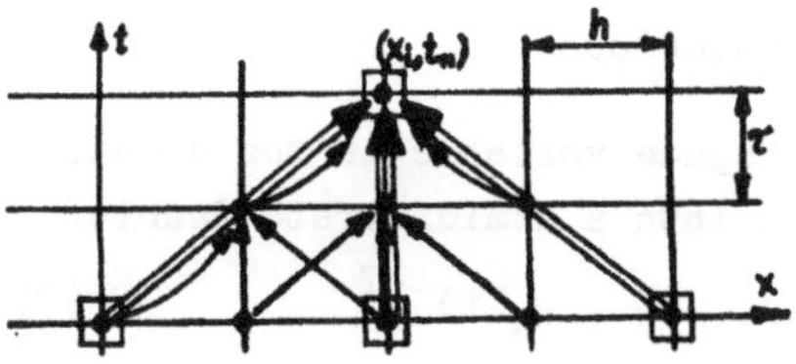

Fig. 1: The computational grid in the x,t - plane

Example: $\begin{cases} u_t - u_x = 0 \\ u(x,0) = \exp(i\xi x) \quad \text{(single wave number)}\ ,\ x \in R^1. \end{cases}$

The 2-stage Lax-Wendroff scheme has the following form

$$y_j^{n+1} = y_j^n + \frac{\lambda}{2}\left(y_{j+1}^n - y_{j-1}^n\right) + \frac{\lambda^2}{2}\left(y_{j+1}^n - 2\, y_j^n + y_{j-1}^n\right) ,$$

where $\lambda := \tau/h$, and the Courant condition of stability requires $\lambda \leq 1$. We can compute the amplification factor of the solution at

$$(h\ ,\tau)\ :\ r(\xi) = 1 + i\lambda \sin(\xi h) - 2\,\lambda^2 \sin^2(\xi h/2)$$

$$(2h,2\tau)\ :\ \tilde{r}(\xi) = 1 + i\lambda \sin(2\xi h) - 2\,\lambda^2 \sin^2(\xi h)\ .$$

Then the error function (q = 2) writes as

$$E\,(\xi) = \frac{1}{6}\left(r^2(\xi) - \tilde{r}(\xi)\right)\ .$$

On the other hand, the truncation error for the function u can be calculated as follows:

$$\underline{u}_j(t+\tau) - T_h\,\underline{u}_j(t) = \frac{\tau}{6}\left(\tau^2 - h^2\right)\frac{\partial^3}{\partial x^3}\,\underline{u}_j(t) + \tau \cdot O(\tau^3 + h^3).$$

A few numerical results for h = 0.05 (N = 20 in the interval [0, 1]), τ = 0.025, are given in Table 1. Since we have chosen λ = 0.5, the condition of stability is fulfilled. Table 2 gives results for the non-linear transport equation

$$u_t \; - \; u\, u_x = 0\ .$$

All the conditions stated here (stepsize in time and space, initial condition etc.) coincide with those of the linear equation.

Table 1: Results for the linear transport equation (see example)

No. of the time step	Method and its order of approx.	Estimation (Numerical res.)	Exact truncation error
10 (t = .25)	LAX, 1-stage (q=1)	1.6851 E-3	1.7272 E-3
	L-W, 2-stage (q=2)	4.2829 E-5	4.2705 E-5
100 (t = 2.5)	LAX, 1-stage (q=1)	1.6002 E-2	1.6387 E-2
	L-W, 2-stage (q=2)	3.8672 E-4	4.0517 E-4

Table 2: Results for the (non-linear) inviscid Burgers equation

100 (t = 2.5)	LAX, 1-stage (q=1)	4.9931 E-2	1.3181 E-2
	L-W, 2-stage (q=2)	5.1735 E-2	2.8624 E-2

2.2 A predictor - corrector - like technique [1)]

Now reconsider equation (2) for one space variable in the domain $G = [0, 1]\times[0, T]$. We shall transform it into a semidiscrete fashion

$$\frac{d\vec{u}}{dt} = \vec{F}(\vec{u}, t) , \quad \vec{u} = (u_1(t), \dots , u_N(t))^T \qquad (3)$$

$$u_j(t) = u(x_j, t) \; ; \; j = 1, \dots , N .$$

Thus, we have a system of ordinary differential equations with the vector function $\vec{u}(t): R^1 \to R^N$; $\vec{F}: R^N \times R \to R^N$. The simplest method for estimation of system (3) is the Euler - PC method. If we introduce the notation

$$\vec{y}^{\,n} \cong \vec{u}(t_n) \; ; \quad y_j^n \cong u_j(t_n) = u(x_j, t_n) \; ; \; j = 1, \dots, N , \; n \geq 1,$$

then instead of (3) we can formulate the system of difference equations

$$P: \; y_j^{n+1,I} = y_j^n + \tau F_j^n = y_j^n + (\lambda/2)\left\{f_{j+1}^n - f_{j-1}^n\right\} \qquad (4.1)$$

$$C: \; y_j^{n+1,II} = y_j + \tau F_j^{n,I} = y_j + (\lambda/2)\left\{f_{j+1}^{n,I} - f_{j-1}^{n,I}\right\} \qquad (4.2)$$

$$j = 1, \dots , N.$$

Again, the linearized method with $f(u) = a\,u$, $a = \text{const}$, is stable for $\lambda a \leq 1$.

We have the following local error estimations:

$$u_j^{n+1} - y_j^{n+1,I} = \frac{\tau^2}{2}\frac{\partial^2 u_j}{\partial t^2}\bigg|_{t=\Theta_1, x_j} - \tau\frac{h^2}{6}\frac{\partial^3 f_j^n}{\partial x^3}\bigg|_{x=\xi_1, t_n}$$

$$u_j^{n+1} - y_j^{n+1,II} = -\frac{\tau^2}{2}\frac{\partial^2 u_j}{\partial t^2}\bigg|_{t=\Theta_2, x_j} - \tau\frac{h^2}{6}\frac{\partial^3 f_j^n}{\partial x^3}\bigg|_{x=\xi_2, t_n}$$

$$F_j^{n,I} - F_j^n = \frac{h^2}{6}\frac{\partial^3 f_j^n}{\partial x^3}\bigg|_{x=\xi_3, t_n}$$

$$\Theta_i \in U(t_n) , \quad \xi_i \in U(x_j) , \quad i = 1, 2, 3;$$

U - neighbourhood of x_j and t_n, respectively.

We assume that the corresponding derivatives do not differ very much. We can find, for the difference of the space components of functions f_j,

$$\tilde{F}_j = \frac{1}{6h}\left\{-6 f_j + 9 f_{j+1} - f_{j-1} - 2 f_{j+2}\right\}$$

and moreover

$$\tilde{F}_j - F_j = -\frac{h^2}{6}\frac{\partial^3 f_j}{\partial x^3}\bigg|_{x=\xi_4} .$$

---

1) The Paragraphs 2.2 and 2.3 have arisen from a cooperation with C. Bernert (cf. / 3, 8/).

Therefore, we can get for the local error estimation

$$u_j^{n+1} - y_j^{n+1,II} = - \frac{y_j^{n+1,II} - y_j^{n+1,I}}{2} - \frac{\tau'}{2} (F_j^{n,I} - \tilde{F}_j^n).$$

For $u_j^{n+1} - y_j^{n+1,I}$ an analogous estimation is possible.

2.3 Fehlberg's imbedding principle

We argue that the classical imbedding principle, developed by Fehlberg for parabolic differential equations, is unstable in the case of hyperbolic equations, because the Courant condition is violated. Therefore, we construct a simple method using an approximately chosen free parameter. From system (3) we derive (analogous of the technique used in chapter 2.2) the formulas

$$\begin{cases} \vec{y}^{\,n+1,(K)} = \vec{y}^{\,n} + \tau \sum_{j=1}^{3} \gamma_j^{(K)} \vec{k}_j^{\,n} & ; \ K = I, II \\ \vec{k}_j^{\,n} = \vec{F}(\vec{y}^{\,n} + \tau \sum_{i<j} \beta_{ji} \vec{k}_i^{\,n}) & , \ j = 1,2,3 . \end{cases} \qquad (5)$$

By means of a Taylor expansion we can calculate the coefficients of the formula, which are given in Table 3.

Table 3: Coefficients for the parameterized imbedding method of the order of approximation 1 (K = I) and 2 (K = II)

j	β_{j1}	β_{j2}	$\gamma_j^{(I)}$	$\gamma_j^{(II)}$
1	-	-	$1-\delta$	$(1-\delta)/2$
2	0.5	-	δ	δ
3	$1-\delta$	δ	0	$(1-\delta)/2$

The comparison of the terms $\vec{y}^{\,n+1,K}$ for K = I, II yields the local error estimation

$$\vec{u}^{\,n+1} - \vec{y}^{\,n+1,I} = \vec{y}^{\,n+1,II} - \vec{y}^{\,n+1,I} = C_1 \tau (\vec{k}_3 - \vec{k}_1).$$

For stability (Courant condition), we need $\lambda a \leq C_2$ in the linear case. Here, $C_1 = (1-\delta)/2$, $C_2 = 2\sqrt{\delta - 1}/\delta$. On account of the constant C_2 we must select $\delta > 1$. On the other hand, we also want to ensure that the requirements

$$C_1 \longrightarrow \min , \quad C_2 \longrightarrow \max$$

are considered. Numerical experience shows that $\delta = 1.08$ seems to be a suitable value. Then, $C_1 = - 0.04$, $C_2 = 0.5$.

We note that the amount of computation of the method described in this chapter is rather small.

3. Numerical results

All the methods described here may be transferred to the two-dimensional case in space without great difficulty. In addition to the example in chapter 2.1, we have applied the methods to various test examples. We present a few special results (for other tests see /4, 8/; the numerical calculations were carried out by C. Bernert).

(a) $u_t - u_x = 0$, i. c.: $u(x,0) = x^2$

Exact solution: $u(x,t) = (x + t)^2$

(b) $u_t + u\,u_x = 0$, i. c.: $u(x,0) = \sqrt{2x}$

Exact solution: $u(x,t) = \sqrt{t^2 + 2x} - t$

(c) $u_t - 0.5\,u_x - 0.5\,u_y = 0$, i. c.: $u(x,y,0) = (x + y)^2$

Exact solution: $u(x,y,t) = (x + y + t)^2$

(d) $u_t + 0.5\,u\,u_x + 0.5\,u\,u_y = 0$, i. c.: $u(x,y,0) = \sqrt{2x + 2y}$

Exact solution: $u(x,y,t) = \sqrt{t^2 + 2x + 2y} - t$.

The computational results are shown in Table 4.

Table 4: Numerical results for examples (a),...,(d) with $h_x = h_y = 0.05$; $\tau = 0.025$

time step	error	(a), 1D	(b), 1D	(c), 2D	(d), 2D
10	exact truncat.	2.718 E-5	3.505 E-10	2.397 E-4	2.711 E-5
	estimation	2.696 E-5	3.755 E-9	2.472 E-4	3.655 E-4
100	exact truncat.	1.675 E-4	-	8.774 E-3	-
	estimation	1.647 E-4	-	8.884 E-3	-

Additionally, we have considered a system of isentropic gas dynamics, for which we know the exact solution for suitable initial conditions (see /11/):

$$\begin{cases} \rho_t + (\rho u)_x = 0 \\ u_t + (\rho u^2 + p)_x = 0, \quad p = A\rho^{\varkappa}, \quad \varkappa = 3. \end{cases}$$

Notice that for small times the solution is smooth, but for $t \geq 1$ a shock front is produced. The approximate solution obtained by the imbedding method of the Section 2.3 ($\delta = 1.08$) has been plotted in the diagram (see Fig. 2). It is well-known there exist other methods, which better approximate non-steady solutions (e.g. methods of Godunov-type, van Leer-method etc., see /15/). However our main goal is the derivation of error estimations. Really, we could give some results, which allow formulating an acceptable error approximation in this case.

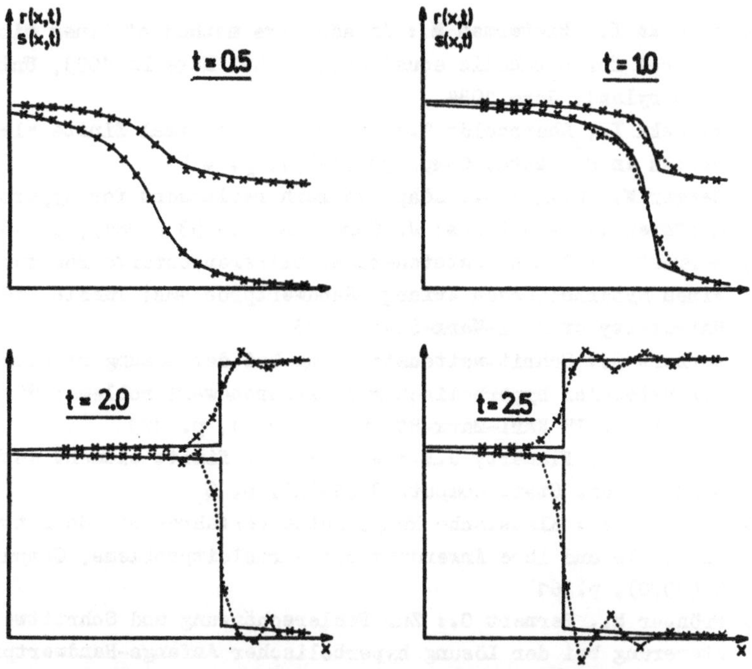

Fig. 2a: Comparison of analytical (——) and numerical (×–×–×) solution at different values of time for the components r = u + a, s = u − a, u - velocity, a - local sound speed

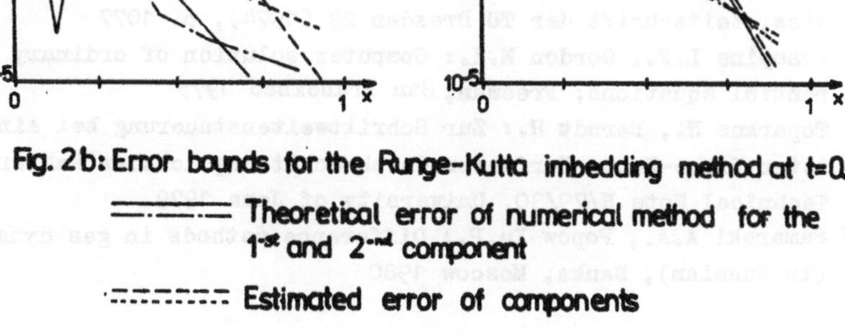

Fig. 2b: Error bounds for the Runge-Kutta imbedding method at t=0.01

Theoretical error of numerical method for the 1-st and 2-nd component

Estimated error of components

4. References

/1/ Babuska I., Bieterman M.: An adaptive method of lines with error control for parabolic equations; Techn. Note BN-1023, University of Maryland, June 1984

/2/ Babuska I., Rheinboldt W.: Analysis of optimal finite element meshes in R^1, Math. Comp. 33 (1979), p. 435

/3/ Berger M., Oliger J.: Adaptive mesh refinement for hyperbolic differential equations; J. Comp. Physics 53 (1984), p. 484

/4/ Bernert C.: Ein selbststeuerndes Differenzenverfahren zur Lösung eines hyperbolischen Anfangs-Randwertproblems; Thesis, Technical University of Karl-Marx-Stadt 1985

/5/ Bernert C.: Schrittweitensteuerung bei der Lösung räumlich mehrdimensionaler hyperbolischer Anfangsrandwertprobleme; Wiss. Zeitschrift d. TH Karl-Marx-Stadt 26 (1984), p. 173

/6/ Davis S.F., Flaherty J.E.: An adaptive finite element method; SIAM J. Sci. Stat. Comput. 3 (1982), p. 6

/7/ Fehlberg E.: Klassische Runge-Kutta-Verfahren mit Schrittweitenkontrolle und ihre Anwendung auf Wärmeleitprobleme, Computing 6 (1970), p. 61

/8/ Fröhner M., Bernert C.: Zur Fehlerschätzung und Schrittweitensteuerung bei der Lösung hyperbolischer Anfangs-Randwertprobleme; Wiss. Beiträge der Martin-Luther-Universität Halle-Wittenberg 24 (M 33), (1984), p. 22

/9/ Hedstrom G.W.: Models of difference schemes for $u_t + u_x = 0$ by partial differential equations, Math. Comp. 29 (1975), p. 964

/10/ Reiher T.: Ein adaptives Verfahren zur Lösung einer linearen parabolischen Differentialgleichung in einer Raumdimension; Thesis, Humboldt-University of Berlin 1985

/11/ Rusanov V.: A test case for checking computational methods for gas flow with discontinuities; in K. Förster (ed.), Vieweg-Verlag Braunschweig - Wiesbaden 1978, p. 104

/12/ Schnerr F., Scholz S.: Fehlerschätzungen für das charakteristische Anfangswertproblem bei der hyperbolischen Differentialgleichung; Wiss. Zeitschrift der TU Dresden 23 (1974), p. 1077

/13/ Shampine L.F., Gordon M.K.: Computer solution of ordinary differential equations; Freeman, San Francisco 1975

/14/ Toparkus H., Berndt H.: Zur Schrittweitensteuerung bei eingebetteten Runge-Kutta-Verfahren durch Schätzung lokaler Fehler; Technical Note N/79/30, University of Jena 1979

/15/ Samarski A.A., Popow Yu.P.: Difference methods in gas dynamics (in Russian), Nauka, Moscow 1980

MONOTONE DISCRETIZATION OF TWO-POINT BOUNDARY VALUE PROBLEMS AND RELATED NUMERICAL METHODS

Ch.Großmann [1)]

1. Introduction

In the numerical solution of differential equations for several purposes it is useful to generate lower and upper bounds for the desired solution. In the present paper we investigate some special discretization techniques generating such bounds by means of monotonicity principles and appropriate modifications of the operators. The problems under consideration are weakly nonlinear two point boundary value problems. The given approach can be applied to other types of problems in a similar way (see [8],[13] e.g.). First, we summarizes the main convergence results of the monotone discretization schemes. Contrary to [5],[6] the technique of variational inequalities for monotone operators is used in existence theory and another approach better adapted to further considerations is applied to show convergence. Then we deal with important schemes to determe bounding operators and investigate their main properties. In the next chapter we prepare the numerical solution by considering appropriate bases in the finite-dimensional spaces under consideration. The main attemptof the present paper is concentrated on iteration schemes for solving the auxiliary problems generated by the monotone discretization principle. These methods we investigate in the following two chapters. Finally, we report on numerical experiments showing the usefulness of monotone discretizations. The generation of lower and upper bounds for the solution of two-point boundary value problems as considered here take advantage of weak maximum principles leading to operators of monotone kind. Also based on these principles in [1],[2],[10],[14],[16] discretization techniques are proposed delivering enclosing lower and upper solutions. Contrary to the approach considered here, in these papers the enclosure is constructed by means of correction terms or by using semiinfinite programming technique. Another interesting method based upon monotonicity principles was proposed in [11]. There the enclosure is constructively guaranteed without the concept of bounding operator by direct applying maximum principles to select appropriate base functions.

---

1) Technische Universität Dresden, Sektion Mathematik,
Wissenschaftsbereich Numerische Mathematik
DDR - 8027 Dresden, Mommsenstraße 13

2. The monotone discretization scheme

Let $\Omega = (a,b)$ be some given interval and let U be the Sobolev space $H_0^1(\Omega)$, i.e. the space of functions possessing quadratically integrable generalized first order derivatives and vanishing at the endpoints. Furthermore U^* and $\langle.,.\rangle$ denote the dual space $H^{-1}(\Omega)$ and the dual pairing, respectively. The norms in U and U^* are denoted by $\|\cdot\|$ and $\|\cdot\|_*$.

Let continuous operators $L,G : U \to U^*$ be given which satisfy the following properties:

(i) L is linear and coercive, i.e.

$$\langle Lu,u\rangle \geq \gamma\|u\|^2 \quad \forall\, u \in U \tag{2.1}$$

holds with some $\gamma > 0$.

(ii) G is monotone according to

$$\langle Gu-Gv,u-v\rangle \geq 0 \quad \forall\, u,v \in U. \tag{2.2}$$

(iii) $L+G$ is of monotone kind, i.e.

$$(L+G)u \leq (L+G)v, \quad u,v \in U \text{ implies } u \leq v. \tag{2.3}$$

Here the partial orderings in U,U^* are defined via the cones

$$\mathfrak{K} = \{v \in U : v(x) \geq 0 \ \text{ a.e. in } \Omega\}$$

and

$$\mathfrak{K}^* = \{w \in U^* \ \ w,v \ \geq 0 \ \ \forall\, v \in \mathfrak{K}\} .$$

With the operators L,U the following abstract problem is under consideration:
Find some $u \in U$ such that

$$Lu + Gu = 0 . \tag{2.4}$$

The supposed continuity of $L+G$ and properties (2.1),(2.2) guarantee a unique solution u of (2.4) to exist (compare [4],[12], e.g.).

Corresponding to G we introduce continuous operators $\underline{G}_h,\overline{G}_h : U \to U^*$ satisfying

$$\underline{G}_h v \geq Gv \geq \overline{G}_h v \quad \forall\, v \in U. \tag{2.5}$$

Operators $\underline{G}_h,\overline{G}_h$ according to (2.5) are called <u>bounding operators</u>. Here $h > 0$ indicates some discretization parameter. Now, by means of $\underline{G}_h,\overline{G}_h$ we define the following approximate problems:
Find $\underline{u}_h,\overline{u}_h \in U$ such that

$$L\underline{u}_h + \underline{G}_h\underline{u}_h = 0, \quad L\overline{u}_h + \overline{G}_h\overline{u}_h = 0 . \tag{2.6}$$

For appropriate relaxation of (2.6) similarly to [5],[6] we obtain easily

<u>Theorem 2.1</u>: Let $\underline{u}_h, \bar{u}_h \in U$ satisfy the inequalities

$$L\underline{u}_h + \underline{G}_h\underline{u}_h \le 0 \le L\bar{u}_h + \bar{G}_h\bar{u}_h. \tag{2.7}$$

Then the solution u of problem (2.4) can be estimated by

$$\underline{u}_h(x) \le u(x) \le \bar{u}_h(x) \quad \forall\, x \in \Omega. \tag{2.8}$$

Furthermore the inequalities

$$\|u-\underline{u}_h\|_{C(\bar{\Omega})} \le \frac{c}{\gamma}\left(\|G\underline{u}_h-\underline{G}_h\underline{u}_h\|_* + \|L\underline{u}_h+\underline{G}_h\underline{u}_h\|_*\right),$$

and (2.9)

$$\|u-\bar{u}_h\|_{C(\bar{\Omega})} \le \frac{c}{\gamma}\left(\|G\bar{u}_h-\bar{G}_h\bar{u}_h\|_* + \|L\bar{u}_h+\bar{G}_h\bar{u}_h\|_*\right)$$

hold, where $c>0$ denotes the constant of continuous embedding $U \hookrightarrow C(\bar{\Omega})$.

We remark that theorem 2.1 also remains valid if the coerciveness condition (2.1) is replaced by $\|L^{-1}\| \le 1/\gamma$. This can be used to investigate a wider range of problems as initial value problems or parabolic problems e.g. (compare [8],[13]).

To shorten the notation in the sequel we use G_h instead of $\underline{G}_h, \bar{G}_h$ if the properties for both types of bounding operators hold in the same way and no confusion is possible.

To investigate the existence of solutions of the auxiliary problem (2.6) we assume the bounding operators to satisfy the following conditions:

$$\|G_hv-Gv\|_* \le s_1(h,R) \quad \forall\, v \in B_R \tag{2.10}$$

with

$$B_R = \{v \in U : \|v\| \le R\}$$

and

$$\|(G_h-G)u-(G_h-G)v\|_* \le s_2(h,R)\|u-v\| \quad \forall\, u,v \in B_R. \tag{2.11}$$

Here $s_1, s_2 : R_+^2 \to R_+$ denote some functions with

$$\lim_{h\to+0} s_k(h,R) = 0, \; k=1,2 \quad \text{for any } R > 0. \tag{2.12}$$

<u>Theorem 2.2</u>: Let $R > \|u\|$ denote an arbitrary constant, where u is the solution of (2.4). Then some $\bar{h} > 0$ exists such that for any $h \in (0,\bar{h}]$ the problems (2.6) have unique solutions $\underline{u}_h, \bar{u}_h \in B_R$. Furthermore there holds

$$\lim_{h\to+0} \|\bar{u}_h-\underline{u}_h\| = 0\,. \tag{2.13}$$

<u>Proof</u>: Because of (2.12) some $\hat{h} > 0$ exists with $s_2(h,R) < \gamma \quad \forall\, h \in (0,\hat{h}]$. By (2.1),(2.2),(2.10) we obtain

$$\langle (L+G_h)u-(L+G_h)v, u-v\rangle = \langle L(u-v), u-v\rangle + \langle G_hu-G_hv, u-v\rangle$$

$$\ge (\gamma - s_2(h,R))\|u-v\|^2 \quad \forall\, u,v \in B_R.$$

With the continuity of $L+G_h$ it can be shown (see [4],[12], e.g.) that a unique $u_h^R \in B_R$ exists satisfying the variational inequality

$$\langle (L+G_h)u_h^R, v-u_h^R \rangle \geq 0 \quad \forall v \in B_R. \tag{2.14}$$

Now, we estimate $\|u-u_h^R\|$.

Using (2.14) with $v = u$ and taking into account $(L+G)u = 0$ we obtain

$$\langle L(u_h^R-u), u-u_h^R \rangle + \langle Gu_h^R-Gu, u-u_h^R \rangle + \langle G_h u_h^R - Gu_h^R, u-u_h^R \rangle \geq 0.$$

With (2.1),(2.2) and (2.10) this leads to

$$\gamma \|u-u_h^R\| \leq \|G_h u_h^R - Gu_h^R\|_* \leq s_1(h,R).$$

Because of (2.12) this results in

$$\lim_{h\to+0} \|u-u_h^R\| = 0 . \tag{2.15}$$

Thus some $\bar{h} \in (0,\hat{h}]$ exists such that

$$u_h^R \in \text{int } B_R \quad \forall h \in (0,\bar{h}].$$

Now, the variational inequality (2.14) is equivalent to

$$(L+G_h)u_h^R = 0 \quad \forall h \in (0,\bar{h}].$$

With (2.15) this proves the theorem. ∎

Another approach to existence theory for the auxiliary problems (2.6) was given in [5],[6]. There the operator was supposed to be defined by

$$\langle Gu,v \rangle = \int_a^b g(x,u(x))v(x)dx \quad \forall u,v \in U \tag{2.16}$$

with some locally Lipschitz-continuous function $g(\cdot,\cdot): \bar{\Omega} \times R \to R$ being monotone in the second argument, i.e.

$$g(x,s) \leq g(x,t) \quad \forall x \in \bar{\Omega},\ s \leq t. \tag{2.17}$$

3. Basic construction of bounding operators and their properties

Let some grid $Z(\bar{\Omega}) = \{x_i\}_{i=0}^N$ be given on the interval $\bar{\Omega} = [a,b]$, i.e.

$$a = x_0 < x_1 < \dots < x_{N-1} < x_N = b .$$

The corresponding step sizes and subintervals we denote by

$$h_i := x_i - x_{i-1} \quad \text{and} \quad \Omega_i = (x_{i-1},x_i),\ i=1(1)N,$$

respectively. Furthermore we denote by $C_h(\bar{\Omega})$ the subspace of piecewise continuous functions on the fixed grid defined by

$$C_h(\bar{\Omega}) = \{v \in L_2(\Omega) : v|_{\Omega_i} \in C(\bar{\Omega}_i)\}$$

Now we choose some finite-dimensional subspace $V_h \subset C_h(\bar{\Omega}) \subset U^*$ and some projection

$$S_h : C_h(\bar{\Omega}) \to V_h.$$

Furthermore we define a staircase mapping $\pi_h : C_h(\bar{\Omega}) \to V_h$ by

$$[\pi_h w](x) := \sup_{\xi \in \bar{\Omega}_i} w(\xi) \quad \forall\, x \in \Omega_i,\ i=1(1)N. \tag{3.1}$$

Obviously this results in

$$\pi_h w \geq w \quad \forall\, w \in C_h(\bar{\Omega}).$$

In the sequel we suppose $Gu \in C(\bar{\Omega}_i)$ $\forall\, u \in U$. Based upon the property of π_h mentioned bounding operators $\underline{G}_h, \bar{G}_h$ can be defined by

$$\underline{G}_h := (S_h + \pi_h(I-S_h))G$$

and (3.2)

$$\bar{G}_h := (S_h - \pi_h(S_h-I))G.$$

With the abbreviation

$$\underline{P}_h := S_h + \pi_h(I-S_h),\ \bar{P}_h := S_h - \pi_h(S_h-I) \tag{3.3}$$

we have $\underline{G}_h = \underline{P}_h G$, $\bar{G}_h = \bar{P}_h G$. The following important relations

$$G - \underline{G}_h = (I-\pi_h)(I-S_h)G,$$

(3.4)

$$\bar{G}_h - G = (I-\pi_h)(S_h-I)G$$

result directly from (3.2).
To investigate the approximation properties (2.10)-(2.12) we specify as well the subspace V_h as the projection operator S_h. We define

$$\mathcal{P}_m := \{v \in C_h(\bar{\Omega}) : v|_{\Omega_i} \text{ polynomial of degree} \leq m\}$$

and choose $V_h = P_m$. For $m \geq 1$ we use an equidistant auxiliary grid on every $\bar{\Omega}_i$ given by

$$\sigma_j^i := x_{i-1} + \frac{j}{m} h_i,\quad j=0(1)m.$$

We define the projection $S_h : C(\bar{\Omega}) \to V_h$ by piecewise interpolation due to

$$[S_h w](\sigma_j^i) = w(\sigma_j^i),\ j=0(1)m,\ i=1(1)N.$$

Let $\mu_j^i(.)$ denote the Lagrange interpolation polynomials corresponding to the knots σ_j^i, $j=0(1)m$ for the interpolation on the subinterval $\bar{\Omega}_i$. In the case $m=0$ we define $\sigma_0^i = x_{i-1}$ and $\mu_0^i(x) \equiv 1$ for any $i=1(1)N$. Then we have the representation

$$[S_h w](x) = \sum_{j=0}^{m} w(\sigma_j^i)\mu_j^i(x) \quad \forall\, x \in \bar{\Omega}_i. \tag{3.5}$$

Now, we assume $Gu \in C^{0,\alpha}(\bar{\Omega})$ $\forall\, u \in U$ with some fixed $\alpha \in (0,1/2]$. Here $C^{0,\alpha}(\bar{\Omega})$ denotes the space of Hölder continuous functions equipped with the norm

$$\|v\|_{0,\alpha} = \|v\|_{C(\bar{\Omega})} + |v|_\alpha$$

with

$$|v|_\alpha = \sup_{\substack{x,y\in\bar{\Omega}\\ x\neq y}} \frac{|v(x)-v(y)|}{|x-y|^\alpha}.$$

In the sequel α denotes a constant contained in $(0,1/2]$. To qualify the approximation properties of G_h with respect to G, furthermore, we suppose G to satisfy the following conditions:

$$|Gv|_\alpha \le \sigma_1(R) \quad \forall\, v \in B_R\,, \tag{3.6}$$

$$|Gu-Gv|_\alpha \le \sigma_2(R)\|u-v\| \quad \forall\, u,v \in B_R \tag{3.7}$$

with some nondecreasing functions $\sigma_1,\sigma_2 : R_+ \to R_+$. The following lemmas state sufficient conditions for (3.6),(3.7) and (2.10)-(2.12), respectively.

<u>Lemma 3.1</u>: Let the operator G be defined via some function $g(.,.)$ by (2.16). If g has a continuous derivative g_u w.r.t. the second argument and the Hölder conditions

$$|g(x,s)-g(y,t)| \le \ell_1(r)(|x-y|^\alpha+|s-t|) \tag{3.8}$$

$$\forall\, x,y \in \bar{\Omega}, |s|,|t| \le r$$

$$|g_u(x,s)-g_u(y,t)| \le \ell_2(r)(|x-y|^\alpha+|s-t|)$$

hold then (3.6),(3.7) are satisfied with

$$\sigma_1(R) = \ell_1(cR)(1+cR),$$

$$\sigma_2(R) = c_2 + c(1+cR)\ell_2(cR)$$

with some constants $c, c_2 > 0$.

<u>Proof</u>: Because of the continuous embedding $U \hookrightarrow C^{0,\alpha}(\bar{\Omega})$ we have some $c > 0$ with

$$\|v\|_{0,\alpha} \le cR \quad \forall\, v \in B_R.$$

Due to (2.16) we identify G with the Nemyckii operator defined by

$$[Gv](x) = g(x,v(x)) \quad \forall\, x \in \bar{\Omega}.$$

With (3.8) this results in

$$|[Gv](x)-[Gv](y)| = |g(x,v(x))-g(y,v(y))| \le$$

$$\le \ell_1(\|v\|_{C(\bar{\Omega})})(|x-y|^\alpha + |v(x)-v(y)|) \le$$

$$\le \ell_1(\|v\|_{C(\bar{\Omega})})(1+|v|_\alpha)|x-y|^\alpha .$$

Thus we have

$$|Gv|_\alpha \le \ell_1(cR)(1+cR).$$

This proves the given representation for $\sigma_1(.)$.
Using (3.9) we obtain

$$|g(x,u(x))-g(x,v(x))-g(y,u(y))+g(y,v(y))| \le$$

$$\le |w(x)-w(y)| \int_0^1 |g_u(x,\tau v(x)+(1-\tau)u(x))| d\tau +$$

$$+ |w(y)| \int_0^1 |g_u(x,\tau v(x)+(1-\tau)u(x))-g_u(y,\tau v(y)+(1-\tau)u(y))| d\tau$$

$$\le c_1|w|_\alpha |x-y|^\alpha + \|w\|_{C(\bar{\Omega})}\ell_2(cR)(1+\tfrac{1}{2}|u|_\alpha + \tfrac{1}{2}|v|_\alpha)|x-y|^\alpha$$

with $w = u-v$. Now $u,v \in B_R$ leads to

$$|Gu-Gv|_\alpha \le [c_1c + c\ell_2(cR)(1+cR)]\|u-v\|. \quad \blacksquare$$

Lemma 3.2: Let $V_h = P_m$ and P_h be chosen according to (3.3) with π_h, S_h given by (3.1) and (3.5), respectively. Then constants $c_3, c_4 > 0$ exist such that the following estimations hold:

$$\|(I-P_h)w\|_* \le c_3 h^\alpha |w|_\alpha \tag{3.10}$$

and

$$\|(I-P_h)v-(I-P_h)w\|_* \le c_4 h^\alpha |v-w|_\alpha \quad \forall\, v,w \in C^{0,\alpha}(\bar{\Omega}). \tag{3.11}$$

Proof: With $\sum_{j=0}^{k} \mu_j^i(x) \equiv 1 \quad \forall\, x \in \bar{\Omega}_i$ we obtain

$$[(I-S_h)w](x) = \sum_{j=0}^{m} \mu_j^i(x)(w(x)-w(\sigma_j^i)) \quad \forall\, x \in \bar{\Omega}_i . \tag{3.12}$$

By definition of the Lagrange polynomials $\mu_j^i(.)$ some $\hat{c} > 0$ exists with

$$|\mu_j^i(x)| \le \hat{c} \quad \forall\, x \in \bar{\Omega}_i,\ j=0(1)k.$$

We remark that $\hat{c}$ is independent of the parameter h. Due to (3.12) and $|x-\sigma_j^i| \le h, \ \forall\, x \in \Omega_i, \ j=0(1)m$, this results in

$$|[(I-S_h)w](x)| \le (m+1)\hat{c}|w|_\alpha h^\alpha \quad \forall\, x \in \bar{\Omega}_i,\ i=1(1)N. \tag{3.13}$$

Furthermore from the definition (3.1) of the operator π_h we obtain

$$\|(I-\pi_h)v\|_{L_\infty(\Omega)} \le 2\|v\|_{C(\bar{\Omega})}.$$

Combining with (3.13) this results in (3.10) with $c_3 = 2(m+1)\hat{c}$.
Assume $u,z \in C(\bar{\Omega})$ and we denote $d := u-z$. Then by definition of π_h there holds the estimation

$$|[(I-\pi_h)u-(I-\pi_h)z](x)| \le \sup_{y \in \Omega_i} |d(x)-d(y)| \quad \forall\, x \in \bar{\Omega}_i .$$

This results in

$$\|(I-\pi_h)u-(I-\pi_h)z\|_* \le \tilde{c}\|d\|_{C(\bar{\Omega})} \tag{3.14}$$

with some $\tilde{c} > 0$. Now we set $u = (I-S_h)v$, $z = (I-S_h)w$. With (3.3) and the linearity of the operator $I-S_h$ from (3.14) we obtain

$$\|(I-P_h)v-(I-P_h)w\|_* \le \tilde{c}\|(I-S_h)(v-w)\|_{C(\bar{\Omega})} .$$

Using (3.13) this proves (3.11) with

$$c_4 = (m+1)\tilde{c}\ \hat{c} . \qquad \blacksquare$$

Remark: The Lemma 3.2 remains valid also for elements $v,w \in L_2(\Omega)$ with $v|_{\Omega_i}, w|_{\Omega_i} \in C^{0,\alpha}(\bar{\Omega}_i)$ $i=1(1)N$ if we replace $|\cdot|_\alpha$ by $|||\cdot|||_\alpha$ with

$$|||v|||_\alpha := \max_{1\le i\le N} \sup_{\substack{x,y\in\Omega_i \\ x\ne y}} \frac{|v(x)-v(y)|}{|x-y|^\alpha} . \tag{3.15}$$

Lemma 3.3: Let the properties (3.8),(3.9) hold and let be $V_h = \mathcal{P}_m$. Then the bounding operators G_h defined by (3.1),(3.2),(3.5) satisfy (2.10)-(2.12) with

$$s_1(h,R) = c_3(1+cR)\ell_1(cR)h^\alpha, \quad s_2(h,R) = c_4(c_2+c(1+cR)\ell_2(cR))h^\alpha .$$

Proof: The statements above immediately result from lemmas 3.1, 3.2 and (3.2): $\blacksquare$

By lemma 3.3 and theorems 2.1, 2.2 we established the convergence of the monotone discretization technique. From these investigations we obtain a minimal order $h^{1/2}$ for the convergence. By more detailed estimations of $\|G\underline{u}_h-\underline{G}_h\underline{u}_h\|_*$ higher order of convergence can be shown. In the present paper we concentrate our attention to effective numerical methods for solving the auxiliary problems (2.6). The reader interested in further results in higher order approximations is refered to [6]. Especially the following refined convergence result has been proved there by recursively using the smoothing property $L^{-1}: H^\ell(\Omega) \to H^{\ell+2}(\Omega)$.

Theorem 3.1: Let the operators G,L be defined by (2.16) with a sufficiently smooth function $g(.,.)$ and by

$$\langle Lu,v\rangle = \int_a^b u'v'dx \quad \forall\, u,v \in U, \tag{3.16}$$

respectively.

Then for bounding operators $\underline{G}_h, \bar{G}_h$ according to (3.2),(3.5) the estimation

$$\|\underline{u}_h - \bar{u}_h\| \le c\, h^{m+1} \quad \forall\, h \in (0,\bar{h}]$$

holds with some c, $\bar{h} > 0$. Here $\underline{u}_h, \bar{u}_h$ denote solutions of the auxiliary problems (2.6).

Remark: We note that problem (2.4) with operators L, G due to (3.16), (2.16) is just the weak formulation of the two point boundary value problem

$$\begin{aligned} -u''(x) + g(x,u(x)) &= 0 \text{ in } \Omega \\ u(a) = u(b) &= 0 \,. \end{aligned} \tag{3.17}$$

4. Representation of the finite-dimensional problems

In this chapter we deal with the finite-dimensional realization of the general monotone discretization scheme described in the previous sections. An essential problem consists in the appropriate representation of the solution $u_h \in U$ of the operator equation

$$L\, u_h = g_h \tag{4.1}$$

for fixed $g_h \in V_h \subset U$. Due to the linearity of the operator L we can use the superposition principle to separate the influence of boundary conditions and of the right hand sides related to the subdomains Ω_j.

We define functions $\varphi_i \in H^1(\Omega)$, i=0(1)N by

$$\begin{aligned} \langle L\varphi_i, \varphi \rangle &= 0 \quad \forall\, \varphi \in D(\Omega_j) \\ \varphi_i(x_k) &= \delta_{ik} \qquad i,k=0(1)N;\ j=1(1)N. \end{aligned} \tag{4.2}$$

Furthermore we suppose a base $\{\zeta_\ell\}_{\ell=1}^M$ of V_h to be known, i.e. $V_h = \mathrm{lin}\{\zeta_\ell\}_{\ell=1}^M$. To this base $\{\zeta_\ell\}_{\ell=1}^M$ we relate function $\psi_\ell \in U, \ell=1(1)M$ via the linear operator by

$$\begin{aligned} \langle L\psi_\ell, \varphi \rangle &= \langle \zeta_\ell, \varphi \rangle \quad \forall\, \varphi \in D(\Omega_j) \\ \psi_\ell|_{\Omega_j} &\in H_0^1(\Omega_j). \end{aligned} \tag{4.3}$$

Now, we have the representation

$$u_h(x) = \sum_{i=1}^{N-1} u_i \varphi_i(x) + \sum_{\ell=1}^{M} w_\ell \psi_\ell(x) \tag{4.4}$$

with real coefficients u_i, i=1(1)N-1; w_ℓ, ℓ=1(1)M. We denote this subspace by U_h, i.e.

$$U_h = \mathrm{lin}\,\{\varphi_i\}_{i=1}^{N-1} + \mathrm{lin}\{\psi_\ell\}_{\ell=1}^{M} \,. \tag{4.5}$$

Because of $g_h \in V_h \subset L_2(\Omega)$ and of the smoothing property of L^{-1}, we know $u_h \in H^2(\Omega)$ for the solution of (4.1). Using the continuous embedding $H^2(\Omega) \hookrightarrow C'(\bar{\Omega})$ we obtain the further conditions

$$u_h'(x_i-0) = u_h'(x_i+0), \quad i=1(1)N-1. \tag{4.6}$$

With (4.4) this is equivalent to

$$\begin{aligned} &u_{i-1}\varphi_{i-1}'(x_i-0) + u_i\varphi_i'(x_i-0) + \sum_{\ell=1}^{M} w_\ell\psi_\ell'(x_i-0) = \\ &= u_i\varphi_i'(x_i+0) + u_{i+1}\varphi_{i+1}'(x_i+0) + \sum_{\ell=1}^{M} w_\ell\psi_\ell'(x_i+0) \ , \quad i=1(1)N-1. \end{aligned} \tag{4.7}$$

On the other hand, if g_h has a representation

$$g_h(x) = \sum_{\ell=1}^{M} z_\ell\zeta_\ell(x), \tag{4.8}$$

then (4.2),(4.3) result in

$$w_\ell = z_\ell, \quad \ell=1(1)M \tag{4.9}$$

Thus the remaining coefficients u_i, $i=1(1)N-1$ in (4.4) can be computed by (4.7). Here the system (4.7) is a tridiagonal system of linear equations. Now we investigate the selection of appropriate bases in V_h and U_h for $V_h = \mathcal{P}_m$. We choose

$$V_h = \text{lin}\,\{\zeta_{ij}\}_{i=1,\,j=o}^{N\,,\,m} \tag{4.10}$$

with

$$\zeta_{ij}(x) = \begin{cases} (x-x_{i-1})^j, & x \in \Omega_i \\ 0 \ , & \text{otherwise} \end{cases} \qquad i=1(1)N, j=0(1)m \tag{4.11}$$

and consider two types of the linear operator L.

<u>Type a)</u>: The operator L is defined by (3.16), i.e.

$$\langle Lu,v \rangle = \int_a^b u'v'dx.$$

In this case (4.2) leads to

$$\varphi_i(x) = \begin{cases} \frac{1}{h_i}(x-x_{i-1}), & x \in \bar{\Omega}_i \\ \frac{1}{h_{i+1}}(x_{i+1}-x), & x \in \bar{\Omega}_{i+1} \\ 0 \ , & \text{otherwise.} \end{cases} \quad , \ i=0(1)N \tag{4.12}$$

For i=0 and i=N the undefined part in (4.12) has to be neglected.

Now (4.3),(4.11) result in

$$\phi_{ij}(x) = \begin{cases} \frac{1}{(j+1)(j+2)} (h_i^{j+2}\varphi_i(x)-(x-x_{i-1})^{j+2}), & x \in \Omega_i \\ 0, & \text{otherwise.} \end{cases} \tag{4.13}$$

If G is defined by (2.16) and m = 0 then (4.9) leads to

$$w_{io} = -\sup_{x\in\Omega_i} g(x,u_{i-1}\varphi_{i-1}(x) + u_i\varphi_i(x) + w_{io}\phi_{io}(x)), \quad i=1(1)N \tag{4.14}$$

with $u_o = u_N = 0$. Using (4.12),(4.13) the differentiability condition (4.7) results in

$$\frac{u_i-u_{i-1}}{h_i} - \frac{u_{i+1}-u_i}{h_{i+1}} = \frac{1}{2} (h_i w_{io} + h_{i+1}w_{i+1,o}), \quad i=1(1)N-1. \tag{4.15}$$

The special case of the nonlinear system (4.14),(4.15) to determine u_i, i=1(1)N-1; w_{io}, i=1(1)N was considered in [5],[6].

<u>Type b)</u>: The operator L is defined by

$$\langle Lu,v\rangle = \int_a^b (u'v'+p(x)uv)dx \quad \forall\, u,v \in U \tag{4.16}$$

with a piecewise constant function

$$p(x) = p_i > 0 \quad \forall\, x \in \Omega_i, \; i=1(1)N. \tag{4.17}$$

In this case (4.2),(4.3) and (4.11) lead to

$$\varphi_i(x) = \begin{cases} \dfrac{\sinh\sqrt{p_i}\,(x-x_{i-1})}{\sinh\sqrt{p_i}\,h_i}, & x \in \bar{\Omega}_i \\ \dfrac{\sinh\sqrt{p_{i+1}}(x_{i+1}-x)}{\sinh\sqrt{p_{i+1}}\,h_{i+1}}, & x \in \Omega_{i+1} \\ 0, & \text{otherwise}; \quad i=0(1)N \end{cases} \tag{4.18}$$

and

$$\phi_{io}(x) = \begin{cases} \frac{1}{p_i} (1-\varphi_{i-1}(x)-\varphi_i(x)), & x \in \Omega_i \\ 0, & \text{otherwise} \end{cases} \quad i=1(1)N \tag{4.19}$$

$$\phi_{i1}(x) = \begin{cases} \frac{1}{p_i} ((x-x_{i-1})-h_i\varphi_i(x)), & x \in \Omega_i \\ 0, & \text{otherwise.} \end{cases} \quad i=1(1)N \tag{4.20}$$

Similarly the remaining functions ϕ_{ij}, j=2(1)k can be determined.

5. Iterative solution of the discrete problems

In this chapter we consider various methods for the iterative solution of the auxiliary problems

$$Lu_h + G_h u_h = 0 \,. \tag{5.1}$$

By construction (3.2) the operator G_h is not differentiable. On the other hand it is not useful to base an iteration scheme for (5.1) upon the derivative of G because we want to realize the iterations in the fixed finite-dimensional subspace $U_h \subset U$. A mapping $DG(.): U \to \mathcal{L}(U,U^*)$ is called an <u>almost derivative</u> of G if the following estimation holds

$$\|Gv-Gu-(DG)(u)(v-u)\|_* \le (s_3(h,R)+s_4(R)\|v-u\|)\|v-u\| \quad \forall\, u,v \in B_R \tag{5.2}$$

with some nondecreasing functions $s_3: R_+^2 \to R_+$ and $s_4: R_+ \to R_+$ satisfying

$$\lim_{h\to+0} s_3(h,R) = 0 \text{ for any } R > 0. \tag{5.3}$$

In the sequel we consider the following two types of the operator DG related to G:

i) $$[(DG)(u)w](x) := q_i(u)\,\frac{w(x_{i-1})+w(x_i)}{2}, \quad \forall\, x \in \Omega_i \tag{5.4}$$

and

ii) $$[(DG)(u)w](x) := q_i(u)w(x), \quad \forall\, x \in \Omega_i, \quad i=1(1)N \tag{5.5}$$

with

$$q_i(u) := g_u\left(\frac{x_{i-1}+x_i}{2}, \frac{u(x_{i-1})+u(x_i)}{2}\right). \tag{5.6}$$

We remark that DG according to (5.5),(5.6) approximates the derivative

$$[G'(u)v](x) = g_u(x,u(x))v(x)$$

by replacing $g_u(x,u(x))$ by a staircase function due to (5.6). Operator DG defined by (5.4) forms a further approximation using a staircase function instead of the whole function $G'(u)v$. This simplifies some numerical realizations resting uppon almost derivatives.

<u>Lemma 5.1</u>: Let the operator G satisfy the conditions of Lemmas 3.1,3.2. Then the operators DG defined by (5.4),(5.6) or (5.5),(5.6) are almost derivatives of G.

<u>Proof</u>: Definition (5.4) leads to

$$[Gv-Gu-(DG)(u)(v-u)](x) = w(x)\int_0^1 [g_u(x,\tau u(x)+(1-\tau)v(x))-q_i]d\tau + $$
$$+ q_i\left(w(x) - \frac{w(x_{i-1})+w(x_i)}{2}\right) \tag{5.7}$$
$$\forall\, x \in \Omega_i, \; i=1(1)N.$$

with $w := v-u$.

By (3.9),(5.4) the integrand can be estimated in the following way

$$
\begin{aligned}
&|g_u(x,\tau u(x)+(1-\tau)v(x))-g_u(\frac{x_{i-1}+x_i}{2}, \frac{u(x_{i-1})+u(x_i)}{2})| \le \\
&\le \ell_2(cR)((\frac{h_i}{2})^\alpha + \|w\|_{C(\bar{\Omega})} + |u(x) - \frac{u(x_{i-1})+u(x_i)}{2}|) \\
&\le \ell_2(cR)((\frac{h_i}{2})^\alpha + \|w\|_{C(\bar{\Omega})} + 2|u|_\alpha(\frac{h_i}{2})^\alpha) \\
&\qquad\qquad\qquad\qquad \forall\, x \in \Omega_i,\ i=1(1)N.
\end{aligned}
\tag{5.8}
$$

With the continuous embedding $U \to C^{0,\alpha}(\bar{\Omega}) \to C(\bar{\Omega})$ (5.7),(5.8) result in (5.6) for DG(.) according to (5.4).

In the case (5.5) the proof is simplified because the last term in (5.7) does not appear. ∎

Lemma 5.2: If the almost derivative DG is defined by (5.4) then for any $R > 0$ some $\hat{h} = \hat{h}(R) > 0$ exists with

$$
\langle (L+(DG)(u))v,v\rangle \ge \frac{\gamma}{2}\|v\|^2 \quad \forall\, v \in U,\ u \in B_R,\ h \in (0,\hat{h}]. \tag{5.9}
$$

If DG is defined by (5.5), then there holds

$$
\langle (L+(DG)(u))v,v\rangle \ge \gamma\|v\|^2 \quad \forall\, u,v \in U,\ h > 0.
$$

Proof: Property (2.17) guarantees $q_i \ge 0$, $i=1(1)N$. Now, the definition (5.4) results in

$$
\begin{aligned}
\langle (DG)(u)v,v\rangle &= \sum_{i=1}^{N} (q_i \int_{\Omega_i} v^2(x)dx) + \\
&+ \sum_{i=1}^{N} (q_i \int_{\Omega_i} (\frac{v(x_{i-1})+v(x_i)}{2} - v(x))v(x)dx) \\
&\ge - c_1\, h^\alpha \|v\|^2 \text{ with some } c_1 > 0.
\end{aligned}
$$

Using (2.1) this proves the first assertion.
By $q_i \ge 0$, $i=1(1)N$ and (2.1) the second assertion holds trivially. ∎

Based upon the estimations given above some iteration process for solving problem (5.1) can be defined by

$$
(L+(DG)(u^k))(u^{k+1}-u^k)+(L+G_h)u^k = 0\ ,\quad k=0,1,\ldots \tag{5.10}
$$

with an appropriate initial guess $u^0 \in U$ for u_h.

Theorem 5.1: Let the assumptions of lemma 5.1 be fulfilled. Then constants $\bar{h} > 0$, $r > 0$ exist such that for any $h \in (0,\bar{h}]$ and any $u^0 \in U$ with $\|u^0-u_h\| \le r$ the sequence $\{u^k\}$ generated by (5.10) is well defined and converges to u_h. Furthermore we have

$$\lim_{h\to+0} \overline{\lim_{k\to\infty}} \frac{\|u^{k+1}-u_h\|}{\|u^k-u_h\|} = 0 \tag{5.11}$$

Proof: The iteration (5.10) is equivalent to

$$(L+(DG)(u^k)u^{k+1} + G_h u^k - (DG)(u^k)u^k = 0.$$

Taking into account (5.1) this results in

$$\begin{aligned}(L+(DG)(u^k))(u^{k+1}-u_h) + (G-G_h)u_h-(G-G_h)u^k \\ -G\,u_h + Gu^k+(DG)(u^k)(u_h-u^k) = 0.\end{aligned} \tag{5.12}$$

Because of (5.2) and lemmas 5.1.,5.2. we can select $r > 0$, $\bar{h} \in (0,\hat{h}]$, with $\hat{h} > 0$ from lemma 5.2. such that

$$\rho := \frac{2}{\gamma}\,(s_2(\bar{h},R) + s_3(\bar{h},R) + s_4(R)r) < 1 \tag{5.13}$$

with $R := \|u_h\| + r$. Let $B_r(u_h)$ denote the neighbourhood

$$B_r(u_h) = \{u \in U : \|u-u_h\| \le r\}.$$

Then $B_r(u_h) \subset B_R$ holds. Inductively we show $\{u^k\} \subset B_r(u_h)$. The starting element u^0 was just selected due to $u^0 \in B_r(u_h)$. Now, we suppose $u^k \in B_r(u_h) \subset B_R$. Then by $h \in (0,\hat{h}]$ and lemma 5.2, the linear operator $L+Q(u^k)$ is coercive. Thus equation (5.10) has a unique solution u^{k+1}. Lemmas 5.1, 5.2 with (5.12) result in the estimation

$$\begin{aligned}\|u^{k+1}-u_h\| &\le \frac{2}{\gamma}(\|(G-G_h)u^k-(G-G_h)u_h\|_* + \|Gu_h-Gu^k-(DG)(u^k)(u_h-u^k)\|_*) \le \\ &\le \frac{2}{\gamma}(s_2(h,R)+s_3(h,R)+s_4(R)\|u^k-u_h\|)\|u^k-u_h\|.\end{aligned} \tag{5.14}$$

With (5.13) we obtain

$$\|u^{k+1}-u_h\| \le \rho\|u^k-u_h\| < r.$$

Thus $u^{k+1} \in B_r(u_h)$ and $\lim_{h\to\infty} u^k = u_h$ hold. The asymptotically superlinear convergence (5.11) is a direct consequence of (5.14). ∎

Now we investigate the finite-dimensional performance of the algorithm (5.10). Let (DG)(.) be defined by (5.4) and let $V_h = \mathcal{P}_m$ hold. Furthermore we specify L according to (3.16). To use the advantage of an appropriate base in U_h we replace L by $L^k := L+(DG)(u^k)$ and define $\{\psi_{ij}^k\}$,

$\{\varphi_i^k\}$ similarly to (4.2),(4.3). With the operator L under consideration we obtain

$$-(\psi_{ij}^k)'' + \frac{1}{2} q_i^k(\psi_{ij}^k(x_{i-1}) + \psi_{ij}^k(x_i)) = \zeta_{ij}(x) \text{ in } \Omega_i$$
$$\psi_{ij}|_{\Omega_i} \in H_0^1(\Omega_i). \tag{5.15}$$

Because of the homogeneous boundary conditions the influence of $(DG)(u^k)$ vanishes in (5.15). Thus we have $\psi_{ij}^k = \psi_{ij}$ with ψ_{ij} according to (4.13), i.e. the base $\{\psi_{ij}^k\}$ is independent of the iteration index k.

For $\{\varphi_i^k\}$ we obtain from (4.2), with L_k instead of L, the boundary value problems

$$-(\varphi_i^k)'' + \frac{1}{2} q_j^k(\varphi_i^k(x_{j-1}) + \varphi_i^k(x_j)) = 0 \text{ in } \Omega_j \qquad i=1(1)N-1, j=1(1)N,$$
$$\varphi_i^k(x_\ell) = \delta_{i\ell}. \qquad \ell=0(1)N \tag{5.16}$$

This results in

$$\varphi_i^k(x) = \varphi_i(x) - \frac{1}{2}(q_i^k\psi_{io}(x) + q_{i+1}^k\psi_{i+1,o}(x)), \quad i=1(1)N-1 \tag{5.17}$$

with φ_i, ψ_{io} according to (4.12),(4.13).

Now, we deal with the realization of method (5.10). Let Q^k denote the operators $Q^k := (DG)(u^k)$. With $L^k := L+Q^k$, there holds

$$L^k u^{k+1} = Q^k u^k - G_h u^k. \tag{5.18}$$

Because of (5.4) we have $Q^k u^k - G_h u^k \in V_h$. Thus the representation

$$Q^k u^k - G_h u^k = \sum_{i=1}^{N} \sum_{j=o}^{m} z_{ij}^k \zeta_{ij} \tag{5.19}$$

in the base $\{\zeta_{ij}\}$ can be computed. By definition of $\{\psi_{ij}^k\}$ and $\psi_{ij}^k \equiv \psi_{ij}$ in the case (5.4) we obtain

$$w_{ij}^{k+1} = -z_{ij}^k, \quad i=1(1)N, \quad j=0(1)m \tag{5.20}$$

for the second set of the coefficients in the representation

$$u^{k+1} = \sum_{i=1}^{N-1} u_i^{k+1}\varphi_i^k(x) + \sum_{i=1}^{N} \sum_{j=o}^{m} w_{ij}^{k+1}\psi_{ij}^k(x). \tag{5.21}$$

The remaining coefficients u_i^{k+1}, $i=1(1)N-1$ result from $u^{k+1} \in H^2(\Omega) \hookrightarrow C^1(\bar{\Omega})$. We have $(u^{k+1})'(x_i-0) = (u^{k+1})'(x_i+0)$, $i=1(1)N$. Taking into account the supports of the base functions this is equivalent to

$$u_{i-1}^{k+1}(\varphi_{i-1}^{k})'(x_i-0)+u_i^{k+1}(\varphi_i^{k})'(x_i-0)+\sum_{j=o}^{m} w_{ij}^{k+1}(\psi_{ij}^{k})'(x_i-0) =$$

$$= u_i^{k+1}(\varphi_i^{k})'(x_i+0)+u_{i+1}^{k+1}(\varphi_{i+1}^{k})'(x_i+0) + \sum_{j=o}^{m} w_{i+1j}^{k+1}(\psi_{i+1j}^{k})'(x_i+0) \tag{5.22}$$

$$i=1(1)N-1,\ u_o^{k+1} = u_N^{k+1} = 0 .$$

With (4.12),(4.13),(5.17) from (5.22) we obtain the symmetric tridiagonal system

$$a_i^k u_{i-1}^{k+1} + b_i^k u_i^{k+1} + a_{i+1}^k u_{i+1}^{k+1} = \rho_i^k,\ i=1(1)N-1$$

$$u_o^{k+1} = u_N^{k+1} = 0 \tag{5.23}$$

with

$$a_i^k = -\frac{1}{h_i} + \frac{1}{2} h_i q_i^k ,$$

$$b_i^k = \frac{1}{h_i} + \frac{1}{h_{i+1}} + \frac{1}{2}(h_i q_i^k + h_{i+1} q_{i+1}^k)$$

and

$$s_i^k = \sum_{j=o}^{m} (w_{i+1j}^{k+1}(\psi_{i+1j}^{k})'(x_i+0) - w_{ij}^{k+1}(\psi_{ij}^{k})'(x_i-0)). \tag{5.24}$$

For sufficiently small $h_i > 0$ system (5.23) can be solved by fast factorization (without pivoting).

Now we turn our attention to algorithm (5.10) with analmost derivative (DQ)(u) defined by (5.5). In this case, however, $Q^k u^k - G_h u^k \in V_h$ does not hold. To guarantee this relation, which is essential for a finite dimensional performance of the method we apply the concept of bounding operators to $G-Q^k : U \to U^*$ in the same way as earlier to G. We remember the representation $G_h = P_h G$ and replace the original problem (2.4) by a modified auxiliary problem

$$(L+(DG)(u_h))u_h + P_h(G-(DG)(u_h))u_h = 0 . \tag{5.25}$$

The lower and upper estimation of the solution u of problem (2.4) by solutions u_h of (5.25) directly result from the properties

$$\underline{P}_h w \ge w,\ \overline{P}_h w \le w \qquad \forall\, w \in C(\overline{\Omega})$$

and (2.3). Because of

$$(D(G-(DG)(w)))(x) = 0$$

the idea of method (5.10) leads to the iteration scheme

$$(L+Q^k)u^{k+1} + P_h(G-Q^k)u^k = 0 ,\quad k=0,1,\dots . \tag{5.26}$$

The approximation property of (5.25) for (2.4) can be proved similarly to the approximation property of (2.6) for (2.4). Thus theorem 2.2 can be carried over to the auxiliary problems (5.25).

Now we investigate the convergence of method (5.26).

Theorem 5.2: The propositions of theorem 5.1 hold in the same way for method (5.26).

Proof: From (5.25),(5.26) we obtain

$$(L+Q^k)(u^{k+1}-u_h)+(Q^k-Q_h)u_h + P_h(Gu^k-Q^ku^k) - P_h(Gu_h-Q_hu_h) = 0$$

with $Q_h := (DG)(u_h)$. This results in

$$(L+Q^k)(u^{k+1}-u_h) - Gu_h + Gu^k + Q^k(u_h-u^k) + \\ +(I-P_h)(G-Q_h)u_h-(I-P_h)(G-Q^k)u^k. \tag{5.27}$$

Now we estimate

$$|||(G-Q_h)u_h - (G-Q^k)u^k|||_\alpha$$

where $|||\cdot|||_\alpha$ is defined by (3.15). By the triangle inequality we have

$$|||(G-Q_h)u_h-(G-Q^k)u^k|||_\alpha \le |Gu_h-Gu^k|_\alpha + |||Q^k(u_h-u^k)|||_\alpha + \\ |||(Q_h-Q^k)u_h|||_\alpha . \tag{5.28}$$

Due to lemma 3.1 the estimation

$$|Gu_h-Gu^k|_\alpha \le \sigma_2(R)\|u^k-u_h\|$$

holds.

By definition (5.5),(5.6) of the almost derivative DG and the supposed Hölder continuity some nondecreasing function $\sigma_5 : R_+ \to R_+$ exists such that

$$|||(DG)(u^k)(u_h-u^k)|||_\alpha \le \sigma_5(R)\|u_h-u^k\| \tag{5.29}$$

provided $u^k \in B_R$. Finally from (3.9),(5.5),(5.6) we have

$$|||(Q_h-Q^k)u_h|||_\alpha \le \ell_2(cR)\|u_h-u^k\|_{C(\bar\Omega)}|u_h|_\alpha \tag{5.30}$$

for $u_h, u^k \in B_R$. With the remark concerning lemma 3.2 we obtain

$$\|(I-P_h)(G-Q_h)u_h-(I-P_h)(G-Q^k)u^k\|_* \le s_5(h,R)\|u_h-u^k\|$$

if $u_h, u^k \in B_R$

with some nondecreasing function $s_5 : R_+^2 \to R_+$. The remaining part of the proof can be carried over from the proof of theorem 5.1 because of (5.27).∎

Now we deal with the finite-dimensional performance of method (5.26). Like for method (5.10) we construct adapted bases $\{\psi_{ij}^k\}$, $\{\varphi_i^k\}$ related to $L^k := L+(DG)(u^k)$. Let us consider the case of the operator L defined by (4.16),(4.17). Relation (4.2) with L^k instead of L results in

$$
\begin{aligned}
&-(\varphi_i^k)'' + (p_j+q_j^k)\varphi_i^k = 0 \text{ in } \Omega_j \\
&\qquad\qquad \varphi_i^k(x_\ell) = \delta_{i\ell}, \quad i,\ell=0(1)N, \; j=1(1)N.
\end{aligned}
\tag{5.31}
$$

With the abbreviation $r_i^k := \sqrt{p_i+q_i^k}$ this leads to (compare (4.18))

$$
\varphi_i^k(x) = \begin{cases} \dfrac{\sinh r_i^k(x-x_{i-1})}{\sinh r_i^k h_i}, & x \in \bar{\Omega}_i \\[2ex] \dfrac{\sinh r_{i+1}^k(x_{i+1}-x)}{\sinh r_{i+1}^k\, h_{i+1}}, & x \in \Omega_{i+1} \\[2ex] 0, & \text{otherwise}; \quad i=0(1)N. \end{cases}
\tag{5.32}
$$

Similarly to (4.19),(4.20) we obtain ψ_{ij}^k by replacing p_i, φ_i by $p_i+q_i^k$ and φ_i^k, respectively.

Now, algorithm (5.26) can be executed in the same way as method (5.10). We suppose the new iterative u^{k+1} to be described by the parameters u_i^{k+1}, w_{ij}^{k+1} due to (5.21).

First, we represent $P_h(Gu^k-Q^ku^k)$ in the base $\{\zeta_{ij}\}$. According to (4.8), (4.9) this defines w_{ij}^{k+1}, namely we have

$$
P_h(Gu^k-Q^ku^k) = -\sum_{i=1}^{N}\sum_{j=o}^{m} w_{ij}^{k+1}\zeta_{ij}.
$$

The smoothness condition $u^{k+1} \in C^1(\bar{\Omega})$ results in (5.22). In the case considered here we obtain the symmetric tridiagonal system (5.23) for determining u_i^{k+1}, $i=1(1)N-1$, but with the coefficients

$$
\begin{aligned}
a_i^k &= -\frac{r_i^k}{\sinh r_i^k h_i}, \quad i=1(1)N; \\
b_i^k &= \frac{r_i^k \cosh r_i^k h_i}{\sinh r_i^k h_i} + \frac{r_{i+1}^k \cosh r_{i+1}^k h_{i+1}}{\sinh r_{i+1}^k h_{i+1}}, \quad i=1(1)N-1.
\end{aligned}
\tag{5.33}
$$

The right hand side of (5.23) is given by (5.24). We remark that (5.33) results in a positive definite system (5.23).

6. Monotone iterations for the auxiliary problems

The approximation of the original problem (2.4) by auxiliary problems (2.6) via the concept of bounding operators guarantees the inclusion

$$\underline{u}(x) \le u(x) \le \bar{u}(x). \tag{6.1}$$

The methods (5.10),(5.26) generate locally convergent sequences $\{\underline{u}^k\}$, $\{\bar{u}^k\}$ with

$$\underline{u} = \lim_{k\to\infty} \underline{u}^k, \quad \bar{u} = \lim_{k\to\infty} \bar{u}^k.$$

To extend the estimation (6.1) to the iterates in the form

$$\underline{u}^k(x) \le u(x) \le \bar{u}^k(x), \quad \forall\, x \in \bar{\Omega},\ k=1,2,\dots \tag{6.2}$$

we combine the monotone discretization with appropriate monotone iteration methods ([15],[16]e.g.). In [9] for a special finite-dimensional technique monotone convergence was considered. Here we propose a general operator scheme being a modification of algorithm (5.26). Here we suppose the operators $\underline{P}_h, \bar{P}_h : C_h(\bar{\Omega}) \to V_h$ to satisfy the monotonicity condition

$$v,w \in C_h(\bar{\Omega}), v \le w \Rightarrow \underline{P}_h v \le \underline{P}_h w;\ \bar{P}_h v \le \bar{P}_h w. \tag{6.3}$$

Property (6.3) holds for P_h according to (3.3) in the case $V_h = \mathcal{P}_0$. However, this is not valid for $V_h = \mathcal{P}_m$, $m \ge 1$.

We denote by $D : C(\bar{\Omega}) \to C_h(\bar{\Omega})$ the operator

$$[Dv](x) := d_i v(x) \quad \forall\, x \in \Omega_i,\ i=1(1)N \tag{6.4}$$

with $d_i > 0$, $i=1(1)N$ and suppose $(L+D)$ to be of monotone kind.

The monotone iteration technique for solving the auxiliary problems

$$(L+D)\underline{u}_h + \underline{P}_h(G-D)\underline{u}_h = 0, \quad (L+D)\bar{u}_h + \bar{P}_h(G-D)\bar{u}_h = 0 \tag{6.5}$$

can be described by the following scheme:

<u>Step 1</u>: Select $\underline{u}^0, \bar{u}^0 \in U$ and the parameters $d_i > 0$, $i=1(1)N$ such that

$$\begin{aligned} &(L+D)\underline{u}^0 + \underline{P}_h(G-D)\underline{u}^0 \le 0 \\ &(L+D)\bar{u}^0 + \bar{P}_h(G-D)\bar{u}^0 \ge 0 \end{aligned} \tag{6.6}$$

and

$$\underline{u}^0 \le v \le w \le \bar{u}^0 \Rightarrow (G-D)v \ge (G-D)w. \tag{6.7}$$

Set $k := 0$.

<u>Step 2</u>: Determine $\underline{u}^{k+1}, \bar{u}^{k+1} \in U$ by

$$(L+D)\underline{u}^{k+1} + \underline{P}_h(G-D)\underline{u}^k = 0\ , \quad (L+D)\bar{u}^{k+1} + \bar{P}_h(G-D)\bar{u}^k = 0 \tag{6.8}$$

successively for $k = 0,1,\dots$.

Theorem 6.1: Let the operator $G : C(\bar\Omega) \to C_h(\bar\Omega)$ be continuous. Then method (6.6) - (6.8) generates sequences $\{\underline{u}^k\},\{\bar{u}^k\} \subset U$ with the properties:

i) $\underline{u}^0(x) \le \underline{u}^1(x) \le \ldots \le \underline{u}^k(x) \le \underline{u}^{k+1}(x) \le \ldots \le \underline{u}_h(x) \le u(x) \le \bar{u}_h(x) \le$

$$\le \ldots \le \bar{u}^{k+1}(x) \le \bar{u}^k(x) \le \ldots \le \bar{u}^1(x) \le \bar{u}^0 \quad \forall\, x \in \bar\Omega \,; \tag{6.9}$$

ii) $\lim\limits_{k\to\infty} \underline{u}^k = \underline{u}_h;\ \lim\limits_{k\to\infty} \bar{u}^k = \bar{u}_h$.

There $\underline{u}_h, \bar{u}_h$ denote solutions of the auxiliary problems

$$(L+D)\underline{u}_h + \underline{P}_h(G-D)\underline{u}_h = 0,\quad (L+D)\bar{u}_h + \bar{P}_h(G-D)\bar{u}_h = 0 \tag{6.10}$$

and u denotes the solution of the original problem (2.4).

Proof: Inductively we show $\underline{u}^k \le \underline{u}^{k+1} \le \bar{u}^{k+1} \le \bar{u}^k$. (6.11)

We suppose

$$\begin{aligned} (L+D)\underline{u}^k + \underline{P}_h(G-D)\underline{u}^k &\le 0 \,,\\ (L+D)\bar{u}^k + \bar{P}_h(G-D)\bar{u}^k &\ge 0 \,,\\ \underline{u}^k &\le \bar{u}^k. \end{aligned} \tag{6.12}$$

By (6.6) this is guaranteed for $k = 0$.
Now, (6.8),(6.12) result in

$$(L+D)\underline{u}^{k+1} + \underline{P}_h(G-D)\underline{u}^k = 0 \ge (L+D)\underline{u}^k + \underline{P}_h(G-D)\underline{u}^k.$$

Because L+D is of monotone kind we obtain $\underline{u}^{k+1} \ge \underline{u}^k$. (6.13)
Similarly $\bar{u}^{k+1} \le \bar{u}^k$ can be shown. Using (6.3),(6.7),(6.8),(6.12) and $\underline{P}_h w \ge \bar{P}_h w \quad \forall\, w \in C_h(\bar\Omega)$ we have

$$\begin{aligned} 0 &= (L+D)\underline{u}^{k+1} + \underline{P}_h(G-D)\underline{u}^k \ge (L+D)\underline{u}^{k+1} + \bar{P}_h(G-D)\underline{u}^k \ge \\ &\ge (L+D)\underline{u}^{k+1} + \bar{P}_h(G-D)\bar{u}^k = (L+D)(\underline{u}^{k+1} - \bar{u}^{k+1}). \end{aligned}$$

The operator L+D is of monotone kind. Thus, we obtain

$$\underline{u}^{k+1} \le \bar{u}^{k+1}.$$

The relations (6.3),(6.7),(6.8),(6.13) result in

$$(L+D)\underline{u}^{k+1} + \underline{P}_h(G-D)\underline{u}^k = 0 \ge (L+D)\underline{u}^{k+1} + \underline{P}_h(G-D)\underline{u}^{k+1}.$$

Thus the first inequality in (6.12) holds also for k+1 . This completes the induction.
Now, we investigate the convergence. The boundedness of the sequence $\{\underline{u}^k\} \subset C(\bar\Omega)$ leads to a bounded sequence $\{\underline{P}_h(G-D)\underline{u}^k\}$ in $L_2(\Omega)$.

By the smoothing property

$$(L+D)^{-1}: L_2(\Omega) \to H^2(\Omega) \qquad (6.14)$$

and (6.8) the sequence $\{\underline{u}^k\}$ is bounded in $H^2(\Omega)$. With the compact embedding $H^2(\Omega) \hookrightarrow C(\bar{\Omega})$ and the monotonicity

$$\underline{u}^k(x) \le \underline{u}^{k+1}(x), \quad \forall\, x \in \bar{\Omega},\ k=0,1,\ldots$$

we obtain $\{\underline{u}^k\}$ to be convergent in $C(\bar{\Omega})$, i.e.

$$\lim_{k\to\infty} \|\underline{u}^k-\underline{u}\|_{C(\Omega)} = 0 \quad \text{with some } \underline{u} \in C(\bar{\Omega})$$

By the smoothing property (6.14) this results in

$$\lim_{k\to\infty} \|\underline{u}^k-\underline{u}\| = 0 \ .$$

With continuity of the operators $(L+D)$ and $P_h(G-D): U \to U_*$ and (6.8) we finally obtain the complete proof of the theorem. ∎

Remarks: - Method (6.6) - (6.8) can be improved by an iterative refinement of D in the sense

$$[D^k v](x) = d_i^k v(x) \quad \forall\, x \in \Omega_i \ ,$$

$d_i^k > 0$, $i=1(1)N$. Here D^k has to satisfy

$$\underline{u}^k \le v \le w \le \bar{u}^k \Rightarrow (G-D^k)v \ge (G-D^k)w.$$

- Taking into account that the boundary conditions can be relaxed in an appropriate way in monotonicity principles we can use in method (6.6) - (6.8) also starting iterates $\underline{u}^0, \bar{u}^0 \in H^1(\Omega)$ with $\underline{u}^0(a) \le 0$, $\underline{u}^0(b) \le 0$ and $\bar{u}^0(a) \ge 0$, $\bar{u}^0(b) \ge 0$. This simplifies significantly the selection of the starting iterates.

7. Numerical examples

Finally we want to illustrate the efficiency of the proposed methods by means of results obtained for the following two test problems (compare [6]):

Example 1 $\quad -u''+u^3-(9\pi^2+\sin^2 3\pi x)\sin 3\pi x = 0$ in $\Omega = (0,1)$

$$u(0) = u(1) = 0 \ .$$

Example 2 $\quad$ Find $u \in H_0^1(\Omega)$ with

$$\int_0^1 [u'v'+\alpha(\sinh u - \mathrm{sign}((x-0.6)(x-0.8)))v]dx = 0$$

$$\forall\, v \in H_0^1(\Omega),\ \Omega = (0,1).$$

Method (5.10) using an equidistant grid and starting with $\underline{u}^o \equiv \bar{u}^o \equiv 0$, $\alpha = 100$ (in example 2) generated the results

$$\delta_h := \max_{0 \le i \le N} (\bar{u}_h(x_i) - \underline{u}_h(x_i))$$

reported in table 1 and table 2, respectively. To economize the computations, for both the lower iterates $\underline{u}^k$ and the upper iterates $\bar{u}^k$, we applied the same linearization due to (5.4) but with

$$q_i(u^k) := g_h \left(\frac{x_{i-1}+x_i}{2}, \frac{\underline{u}_{i-1}^k + \underline{u}_i^k + \bar{u}_{i-1}^k + \bar{u}_i^k}{4}\right)$$

instead of (5.6).

Table 1 (example 1)

	m = 0	m = 1	m = 2
N = 5	5.967 685 (13)	2.367 227 (6)	0.599 570 (6)
N = 10	4.186 350 (9)	0.701 473 (6)	0.084 508 (6)
N = 20	2.493 946 (7)	0.175 436 (5)	0.009 893 (4)
N = 40	1.346 794 (6)	0.043 601 (4)	0.001 214 (3)

Table 2 (example 2)

	m = 0	m = 1	m = 2
N = 5	1.181 196 (50)	0.779 038 (45)	no convergence
N = 10	0.710 983 (15)	0.149 880 (8)	0.016 216 (11)
N = 20	0.323 309 (8)	0.035 133 (5)	0.002 203 (5)
N = 40	0.158 828 (6)	0.008 845 (4)	0.000 304 (4)

The values in the brackets show the number of iterations needed to obtain the accuracy 10^{-6}.

Figures 1, 2 illustrate the iterates obtained by method (6.6) - (6.8) with starting iterates $\underline{u}^o \equiv -2$, $\bar{u}^o \equiv 2$. In figure 2 the methods with updating D_k is used.

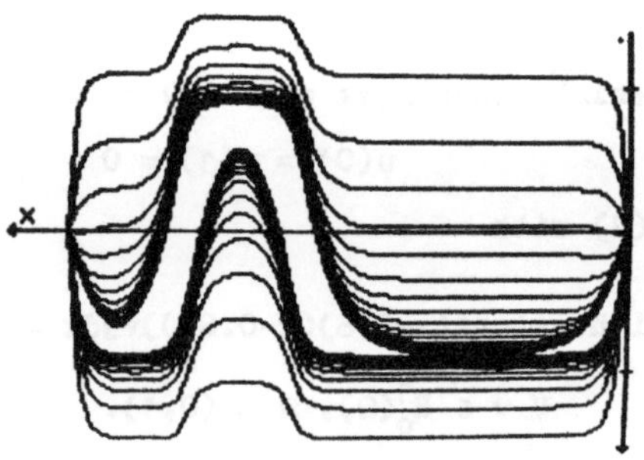

Fig. 1

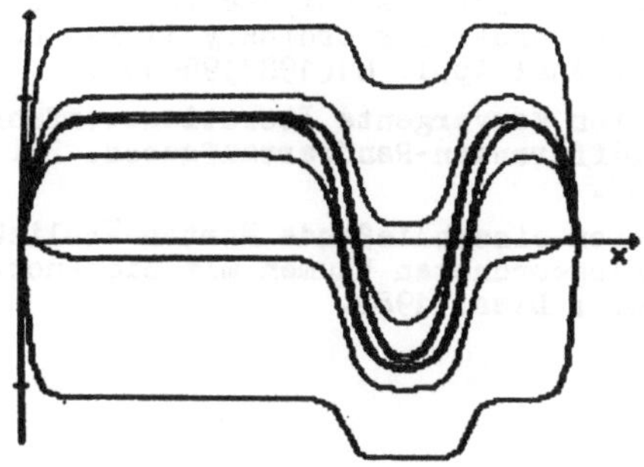

Fig. 2

References

[1] Adams, E.: Invers-Monotonie, direkte und indirekte Intervallmethoden. Bericht Nr.185(1982), Forschungszentrum Graz.

[2] Adams, E., H.Spreuer: Konvergente numerische Schrankenkonstruktionen mit Spline-Funktionen für nichtlineare gewöhnliche bzw. lineare parabolische Randwertaufgaben. "Int.Math." (ed.: K.Nickel), Berlin: Springer Verlag 1975.

[3] Faaß, E.: Beliebig genaue numerische Schranken für die Lösung parabolischer Randwertaufgaben. Karlsruhe: Diss. 1975.

[4] Glowinski, R.: Numerical methods for nonlinear variational problems. Springer Verlag New York-Berlin, 1984.

[5] Großmann, Ch., M.Krätzschmar and H.-G.Roos: Gleichmäßig einschließende Diskretisierungsverfahren für schwach nichtlineare Randwertaufgaben (to appear in Numer.Math.).

[6] Großmann, Ch., M.Krätzschmar and H.-G.Roos: Uniformly enclosing discretization methods of high order for boundary value problems (submitted to Math.Comp.).

[7] Großmann, Ch. and H.-G.Roos: Feedback grid generation via monotone discretization for two-point boundary value problems (to appear in IMA J.Numer.Anal.).

[8] Großmann, Ch., S.Scholz and H.Schönheinz: Generating upper and lower solutions of initial value problems via monotone discretization. (submitted to Computing).

[9] Krätzschmar, M.: Iterationsverfahren zur Lösung schwach nichtlinearer elliptischer Randwertaufgaben mit monotoner Lösungseinschließung. Dresden: Diss. TU Dresden 1983.

[10] Meyn, K.H.: Monotonieaussagen für elliptische und parabolische Randwertaufgaben und Anwendungen auf Finite-Element-Funktionen. Hamburg: Diss. 1979.

[11] Nickel, K.: The construction of a priori bounds for the solution of a two point boundary value problem with finite elements I. Computing 23(1979), 247-265.

[12] Pascali, D. and S.Sburlan: Nonlinear mappings of monotone type. Editure Academiei, Bukarest, 1978.

[13] Roos, H.-G.: The Rothe method and monotone discretization for 1D-parabolic equations. (This volume p. ...)

[14] Spreuer,H.: A method for the computation of bounds with convergence of arbitrary order for ordinary linear boundary value problems. J.Math.Anal.Appl. 81(1981)99-133.

[15] Törnig,W.: Monoton konvergente Iterationsverfahren zur Lösung nichtlinearer Differenzen-Randwertaufgaben. Beitr. Numer.Math. 4(1975), 245-257.

[16] Voller,R.: Monoton einschließende Newton-ähnliche Iterationsverfahren in halbgeordneten Räumen mit nichtnotwendig regulärem Kegel. Düsseldorf: Diss. 1982.

MULTIGRID AND IMPROVED ACCURACY FOR THE EULER EQUATIONS

P.W. Hemker

CWI, Centre for Mathematics and Computer Science
P.O.Box 4079, 1009 AB Amsterdam, The Netherlands

ABSTRACT

In this paper we describe high resolution discretisations for the 2-dimensional steady Euler equations. Two new schemes (superbox and starbox) are presented. The second order discrete equations are solved with the help of a specially efficient first order multigrid solver. Moreover, a tau-extrapolation technique is applied to further improve the accuracy. In this way an iterative procedure is developed to find, during the solution process, an increasingly more accurate approximation to the solution (higher than 2nd order if an asymptotic expansion exists). A few basic multigrid cycles are usually sufficient to obtain the final approximation up to truncation error accuracy.

1. DISCRETISATION PROCEDURES

Neglecting heat conduction, the flow of an inviscid gas is described by the Euler equations. In two dimensions these equations are

$$\frac{\partial q}{\partial t} + \frac{\partial}{\partial x} f(q) + \frac{\partial}{\partial y} g(q) = 0 , \tag{1.1}$$

with

$$q = \begin{bmatrix} \rho \\ \rho u \\ \rho v \\ \rho e \end{bmatrix}, \quad f = \begin{bmatrix} \rho u \\ \rho u^2 + p \\ \rho uv \\ \rho ue + up \end{bmatrix}, \quad g = \begin{bmatrix} \rho v \\ \rho vu \\ \rho v^2 + p \\ \rho ve + vp \end{bmatrix}, \tag{1.2}$$

where ρ , u , v , e and p represent density, velocity in x- and y- direction, specific energy and pressure respectively. The pressure is obtained from the equation of state, which - for a perfect gas - is given by

$$p=(\gamma-1)\rho(e-\frac{1}{2}(u^2+v^2)),$$

γ is the ratio of specific heats. $q(t,x,y)$ describes the state of the gas as a function of time and space and f and g are the fluxes in the x- and y- direction. We denote the open domain of definition of (1.1) by Ω^* .

To allow discontinuous solutions, (1.1) can be rewritten in its integral form

$$\frac{\partial}{\partial t}\iint_{\Omega} q\,dx\,dy + \int_{\partial\Omega}(f.n_x + g.n_y)\,ds = 0, \tag{1.3}$$

for all $\Omega\subset\Omega^*$;

$\partial\Omega$ is the boundary of Ω and (n_x,n_y) is the unit outward normal vector at the boundary $\partial\Omega$.

The form (1.3) of equation (1.1) shows the character of the system of conservation laws: the increase of q in Ω is caused only by the inflow of q over $\partial\Omega$. In symbolic form we write (1.3) as

$$q_t + N(q) = 0. \tag{1.4}$$

In this paper we are only interested in the solution of the steady state equation

$$N(q) = 0. \tag{1.5}$$

The solution of the weak form (1.3) of (1.1) is known to be non-unique and a physically realistic solution (which is the limit of a flow with vanishing viscosity) is known to satisfy the additional entropy condition (cf. [1,2]). Further, the equation (1.1) is hyperbolic, i.e. written in the form

$$\frac{\partial q}{\partial t} + \frac{\partial f}{\partial q}.\frac{\partial q}{\partial x} + \frac{\partial g}{\partial q}.\frac{\partial q}{\partial y} = 0$$

the matrix

$$k_1\frac{\partial f}{\partial q} + k_2\frac{\partial g}{\partial q}$$

has real eigenvalues for all (k_1,k_2) and the 4 eigenvectors span the 4-D state space.

These eigenvalues are $(k_1u+k_2v)\pm c$ and (k_1u+k_2v) (a double eigenvalue), where $c = \sqrt{\gamma p/\rho}$ is the local speed of sound. The sign of these eigenvalues determines the direction in which the information about the solution is carried along the line (k_1,k_2) as time develops. It locates the domain of dependence. The entropy condition implies that characteristics do not emerge at a discontinuity in the flow.

In order to discretise (1.1) or (1.3) on a domain with an irregular grid, we use a finite volume technique. The domain of definition, Ω^*, is divided into a number of disjunct cells $\{\Omega_\alpha\}$ and equation (1.3) is required to hold on each $\bigcup\Omega_\alpha$. In this way the conservation character is preserved, provided that we take the same approximation for the flow quantities $\int_{\Gamma_{\alpha\beta}} f.n_x + g.n_y\,ds$, both for the computation of the outflow of Ω_α and for the inflow of Ω_β for any two neighbouring cells Ω_α and Ω_β with $\Gamma_{\alpha\beta} = \partial\bar{\Omega}_\alpha \cap \partial\bar{\Omega}_\beta$. In that case (1.3) holds for any Ω which is the union of an arbitrary subset of $\{\Omega_\alpha\}$. In this method there is no need to transform the equations (1.1) or the boundary conditions with respect to the independent variables (x,y).

In order to generate a nested sequence of discretisations with decreasing meshwidth for our multigrid solution procedure, and to minimise the administrative overhead of the computer code, we found it most convenient to to divide the domain Ω^* in quadrilateral cells Ω_{ij}, in a way topologically equivalent with a regular division in squares. Then $\Omega_{i\pm1,j\pm1}$ are the only possible neighbours of Ω_{ij} . In this way a mapping is conceived from a rectangular and regular "computational domain" to the irregular physical domain. This mapping plays a role only for the generation of the sequence of refining irregular grids, and in the proofs of the accuracy of the resulting schemes. We take this mapping non-singular (i.e. with non-vanishing Jacobian J on $\bar{\Omega}^*$) and sufficiently smooth (with bounded partial derivatives of J). The mapping determines the vertices of the quadrilateral cells in the physical space, where all cell-edges are straight line segments. As a consequence any domain Ω^* with a curved boundary is approximated by a polygonal domain, but in subsequent refinements the curved lines are better approximated.

The discrete approximation q_h of $q(x,y)$ is represented by a (vector-) value q_{ij} for each Ω_{ij}. This q_{ij} is associated with the mean value of q over Ω_{ij}. Using (1.3), the space discretisation method is now completely determined by the method of approximation of

$$\int_{\Gamma_{ijk}} (f.n_x + g.n_y)\, ds\,, \qquad k = N,E,S,W, \tag{1.6}$$

at the four walls Γ_{ijk} of the cell Ω_{ij}. The wall Γ_{ijk} may be either a common boundary with another cell Ω_{ijk} or a part of the boundary $\partial\Omega^*$. In both cases the approximation of (1.6) is computed as

$$f^k(q^k_{ij}, q^k_{ijk}) \cdot \mathrm{meas}(\Gamma_{ijk}), \tag{1.7}$$

i.e. at each Γ_{ijk} we approximate $fn_x + gn_y$ by a constant value , which depends only on q^k_{ij}, an approximation to $q(x,y)$ in Ω_{ij} at the edge Γ_{ijk}, and also on q^k_{ijk}, an approximation to $q(x,y)$ in the neighbouring cell Ω_{ijk} at Γ_{ijk}. (Notice that we allow the approximation of q to be discontinuous over Γ_{ijk}.)

The semi-discretisation of the equation (1.5) is now the set of non-linear equations

$$N_h(q_h)\,|_{i,j} := \sum_{k=N,E,S,W} f^k(q^k_{ij}, q^k_{ijk})\, \mathrm{meas}(\Gamma_{ijk}) = 0\,, \tag{1.8}$$

for all (i,j) with $\Omega_{ij} \subset \Omega^*$.

The approximate flux $f^k(q^k_{ij}, q^k_{ijk})$ depends on q_h, the approximation of q, near Γ_{ijk} and also on the direction (n^k_x, n^k_y) of the edge Γ_{ijk}. However, by the rotation invariance of the Euler equations, we may relate $f^k(.,.)$ to a local coordinate system, rotated such that it is aligned with Γ_{ijk}. Then this computation comes down to

$$f^k(q_{ij}^k,q_{ijk}^k) = T_k^{-1} f(T_k q_{ij}^k, T_k q_{ijk}^k)$$

Here $f(.,.)$ is a *numerical flux function,* independent of the orientation of Γ_{ijk}. The operator T_k takes care of the local rotation of the coordinate system at Γ_{ijk}. If $(n_x^k,n_y^k) = (1,0)$, then T_k is the identity operator. In this way only a single function $f(.,.)$ is needed to approximate the flux between two cells.

The description of the discretisation is completed by the choice of the numerical flux function and by the determination of q_{ij}^k and q_{ijk}^k from $\{q_{ij} \mid \Omega_{ij} \subset \Omega^*\}$.

For consistency of the resulting scheme, $f(.,.)$ should satisfy $f(q,q) = f(q)$, cf. [3]. A usual representation of $f(.,.)$ is given by

$$f(q_0,q_1) = \frac{1}{2} f(q_0) + \frac{1}{2} f(q_1) - \frac{1}{2}\, d(q_0,q_1). \tag{1.9}$$

A central difference flux is defined by $d(q_0,q_1) = 0$. For reasons explained in [4], in our multigrid procedure we use a slight modification of an upwind numerical flux function that was proposed by Osher [5,6]; viz. (1.9) with $d(.,.)$ defined by

$$d(q_0,q_1) = \int_{q_0}^{q_1} \left|\frac{\partial f}{\partial q}\right|(w)\, dw\,, \tag{1.10}$$

where the integration path in the state space follows three sub-paths along the eigenspaces of $\partial f/\partial q$. These sub-paths correspond to the eigenvalues $\lambda_1 = u-c$, $\lambda_2 = \lambda_3 = u$, and $\lambda_4 = u+c$ respectively. This implies that, except at a location of a shock, we can write $f(q_0,q_1) = f(q^*)$, where q^* is connected with q_0 by the Riemann invariants corresponding to $\lambda \geqslant 0$, and q^* is connected with q_1 by the Riemann invariants for $\lambda \leqslant 0$. In the case of a shock we find a numerical flux of the form $f(q_0,q_1) = f(q_1^*) - f(q_2^*) + f(q_3^*)$, for some proper q_k^*, $k=1,2,3$.

At the boundary of the domain Ω^*, interpolation of q_h from the interior of Ω^* yields a value q_{ijk}^{IN} at that boundary, which corresponds to a mean value of q at Γ_{ijk} in Ω_{ij}. For well posed boundary conditions $B(q) = 0$ at Γ_{ijk}, a value q_{ijk}^{OUT} can be determined such that it satisfies the boundary conditions and is connected to q_{ijk}^{IN} by the Riemann invariants for $\lambda<0$ at the left (or $\lambda<0$ at the right) boundary. This implies that

$$f^k(q_{ijk}^{OUT},q_{ijk}^{IN}) = T_k^{-1} f(T_k\, q_{ijk}^{OUT}), \tag{1.11a}$$

and

$$B(q_{ijk}^{OUT}) = 0\,, \tag{1.11b}$$

is satisfied at a point on Γ_{ijk}. For details see [4].

The basic first order scheme.

For our first order scheme we use a piecewise constant numerical approximation for q for each quadrilateral cell:

$$q_h(x,y) = q_{ij} \qquad \text{for} \quad (x,y)\in\Omega_{ij}. \tag{1.12a}$$

This uniform state in Ω_{ij} is assumed for all (i,j), and hence

$$f^k(q^k_{ij},q^k_{ijk}) = f^k(q_{ij},q_{ijk})\,. \tag{1.12b}$$

The flux at Γ_{ijk} now corresponds with the flux at a discontinuity between two uniform states. Such a flux can be computed by solving the Riemann problem of gasdynamics (i.e. Godunov's method). However, this is a nontrivial non-linear computation, and we approximate it by (1.9) - (1.10), which is essentially Osher's "approximate Riemann solver".

The order of accuracy of the resulting schemes on the nonuniform mesh is not immediate. It can be proved that, in this way, at most second order accuracy can be obtained for irregular grids, when q^k_{ij},q^k_{ijk} are computed properly. With the approximation (1.12.b) the scheme is first order accurate.

For a function q, the truncation error is defined as

$$\tau_h(q) = N_h(R_h q) - \bar{R}_h N(q)\,, \tag{1.13}$$

where the restriction operators R_h and $\bar{R}_h$ take mean values over cells. For (1.12.b) and a smooth q it can be shown that $\tau_h(q) = \mathcal{O}(h)$ for $h \to 0$. We denote the first order discrete equations (1.8) - (1.12) in symbolic form by

$$N^1_h(q_h) = 0\,. \tag{1.14}$$

This first order discretisation is conservative, satisfies an entropy condition, is monotonous and gives a sharp representation of discontinuities (shocks and contact discontinuities), as long as these are aligned with the mesh. Further it allows an efficient solution of the discrete equations by a multigrid method [7] . Disadvantages are the low order of accuracy and the fact that it is highly diffusive for oblique discontinuities. For a first order (upwind) scheme these are well known facts and it leads to the search for other grids and higher order methods.

The superbox scheme.

A property that we want to maintain in a second order scheme, is the conservation of q, because it allows discontinuities to be captured as weak solutions of (1.1). Therefore, we consider only schemes that are still based on (1.8), and we select $f^k(q^k_{ij},q^k_{ijk})$ to yield a better approximation to (1.6) than (1.12b).

A straightforward way to form a more accurate approximation is to replace the 1st order approximation (1.12) by a 2nd order one. Instead of the piecewise constant approximation for $q(x,y)$, we now use a piecewise bilinear function $q_h(x,y)$ on a set of 2×2 cells (a "superbox"). Such a superbox on the h-level corresponds with a single cell at the 2h-level. Over the boundaries of the superbox, the trial function $q_h(x,y)$ can be discontinuous again. Inside the superbox it is determined by q_{ij}, $q_{i+1,j}$, $q_{i,j+1}$ and $q_{i+1,j+1}$, the mean values over the sub-cells of the superbox (see figure 1). Using the bilinear function, we see that a central difference approximation is used for the four flux computations inside the superboxes; at superbox boundaries interpolation is made from the inside of the superbox by the bilinear interpolation. In this way the values $\{q_{ij}^k\}$ are computed and the approximate Riemann solver is used to compute the numerical flux. We denote the corresponding discrete operator by N_h^{2S}. It is easily shown that this *superbox scheme* is 2nd order accurate in the sense that

$$\overline{R}_{2h,h}(N_h^{2S}(R_h q) - \overline{R}_h N(q)) = \mathcal{O}(h^2).$$

We elucidate this superbox discretisation by its 1-dimensional analogue. Let $\{q_j\}_{j=1,\dots,2N}$ denote states in the cells $\{\Omega_j\}$ with barycentres at x_j and walls between cells at $x_{j+1/2}$. Superboxes are $\Omega_i^B = \Omega_{2i-1} \bigcup \Omega_{2i}$, $i = 1,2,\dots,N$. For the computed fluxes at $x_{j+1/2}$ we have to distinguish between odd and even j. At an internal wall $x_{2i-1/2}$ we find the interpolated value $q_{2i-1/2} = (q_{2i-1} + q_{2i})/2$ and at a superbox wall $x_{2i+1/2}$ we have interpolated values from the left, $q_{2i+1/2}^L$, and from the right, $q_{2i+1/2}^R$. We define $\Delta q_{j+1/2} = q_{j+1} - q_j$ and find the 2nd order one-sided interpolated values

$$q_{2i+1/2}^L = q_{2i} + \frac{1}{2}\Delta q_{2i-1/2}, \tag{1.15}$$

$$q_{2i+1/2}^R = q_{2i+1} - \frac{1}{2}\Delta q_{2i+3/2},$$

The computed fluxes at $\{x_{j+1/2}\}$ now are $f_{2i-1/2} = f(q_{2i-1/2} + \Delta q_{2i-1/2})$ and $f_{2i+1/2} = f(q_{2i+1/2}^L, q_{2i+1/2}^R)$. We notice that the states $q_{2i+1/2}^L$ and $q_{2i+1/2}^R$ are well determined, even near the boundary $\partial\Omega*$, so that the procedure to compute the boundary fluxes is also competely described by (1.11).

Star schemes

A disadvantage of the quadrilateral cells in a more complex flow configuration can be the existence of (only) two special directions in which possible discontinuities are allowed. Discontinuities aligned with the gridlines are easily represented by q_h, whereas discontinuities skew to the meshlines are hard to resolve. Although higher order approximations may relieve this situation [8] , they can lead to too large spurious extrema (a Gibbs-like effect). Some improvement of this situation may be expected if more mesh-line directions are introduced. For this purpose we constructed a mesh with triangular cells. In the "computational domain" these triangles are formed by refinement of the quadrilaterals as previously described, in such a way that each quadrangle is divided into 4 triangles by connecting its barycentre with the vertices (see figure 1). For the new first order scheme based on these triangles (the *1st-order star scheme*), uniform states are assumed in all triangular cells (i.e. $q_h(x,y)$ is a piecewise constant function over the triangles), and the discrete operator is defined by

$$N_h^{1*}(q_h)\,|_\alpha := \sum_{k=1,2,3} f^k(q_\alpha, q_{\alpha,k})\,\mathrm{meas}(\Gamma_{\alpha,k}) = 0\,, \tag{1.16}$$

where the summation is over the 3 sides of the triangular cell Ω_α; $q_{\alpha,k}$ is the state in cell $\Omega_{\alpha,k}$, the k-th neighbour triangle of Ω_α.

For the second order scheme based on these triangles (the *2nd order star-scheme*), we introduce a linear representation of $q_h(x,y)$ in each set of 4 triangles that were all formed in the same quadrangle. This interpolation is made such that

$$q_h(x_{2i},y_{2j}) = (q_{2i+2/3,2j} + q_{2i-2/3,2j} + q_{2i,2j+2/3} + q_{2i,2j-2/3})/4,$$

$$\frac{\partial q_h}{\partial x} = \frac{q_{2i+2/3,2j} - q_{2i-2/3,2j}}{x_{2i+2/3,2j} - x_{2i-2/3,2j}}; \quad \frac{\partial q_h}{\partial y} = \frac{q_{2i,2j+2/3} - q_{2i,2j-2/3}}{x_{2i,2j+2/3} - x_{2i,2j-2/3}}.$$

The indices correspond with the barycentres of the triangles in the computational domain. By this interpolation all the values at the cell walls, $\{q_\alpha^k \mid \Omega_\alpha \subset \Omega; k=1,2,3\}$, are determined, and the discrete scheme reads

$$N_h^{2*}(q_h)\,|_\alpha := \sum_{k=1,2,3} f^k(q_\alpha^k, q_{\alpha,k}^k)\,\text{meas}(\Gamma_{\alpha,k}) \tag{1.17}$$

$$= \sum_{k=1,2,3} T_k^{-1} f(T_k q_\alpha^k, T_k q_{\alpha,k}^k)\,\text{meas}(\Gamma_{\alpha,k}) = 0.$$

2. MULTIGRID ALGORITHMS AND TAU-EXTRAPOLATION

The usual way to find the solution of the steady state equations

$$N_h(q_h) = 0, \tag{2.1}$$

is to integrate for $t \to \infty$, the semi-discrete system of equations

$$(q_h)_t + N_h(q_h) = 0, \tag{2.2}$$

i.e. to take an initial guess and to compute $q_h(t)$ until initial disturbances have sufficiently died out. The advantage is that, starting with a physically meaningful situation, we expect that a meaningful steady state will be reached, even when unicity of the steady equations is not guaranteed. The drawback is that many timesteps may be necessary before the solution has sufficiently converged. For the acceleration of the convergence, many devices have been developed such as local time stepping, residual smoothing, implicit residual averaging or enthalpy damping [9].

Multigrid is also used as an acceleration device for (2.2) [10, 11, 9]. In that approach space-discretisations (2.1) are given on a sequence of grids. The coarse grids are used to move low frequency disturbances rapidly out of the domain Ω^* by large timesteps, whereas high frequency disturbances are locally damped on the fine grids by sufficiently dissipative timestepping. This accelerated procedure still may require a reasonably large number of steps (>100).

For the multigrid solution of (2.1), we take another approach [12, 13, 4]. We consider directly the steady state equations and we first construct a sequence of nested 1st order discretisations as described in the previous section. By the stability of the first order discretisation, a relatively simple relaxation method (Collective Symmetric Gauss Seidel (CSGS) iteration, i.e. a SGS relaxation where the 4 variables corresponding to a single cell Ω_{ij} are relaxed collectively) is able to reduce the high frequency error components efficiently, and -therefore- a FAS multigrid algorithm with this relaxation is well suited to solve the discrete first order equations [7] .

Although no explicit artificial viscosity is added to the scheme, a suitable amount of "numerical diffusivity" is automatically introduced by the upwind discretisation. As $h \to 0$, this "artificial diffusion" vanishes and the sequence of discretisations converges to the Euler equations as the limit of an equation with vanishing viscosity. This also motivates us to apply a simple mesh-continuation procedure (FMG) for $h \to 0$, to find the initial estimates for the FAS multigrid iteration. It appears that, under quite general circumstances, the convergence factor of a FAS iteration cycle (with CSGS-relaxation and $p = q = s = 1$ as described in [7]) ranges between 0.4 and 0.85, so that from n=1 to n=8 of these FAS cycles suffice to compute the 1st order accurate solution up to truncation error accuracy (see e.g. Chapter 5 in [14]). The complete Full Multigrid (FMG) method to compute the 1st order approximation is given in the following ALGOL68 procedures, cf. [15] .

```
proc FMG = ( int n,L, ref [] field q, r) void:
begin
  for m to L
  do
    to n do FAS ( m,q,r) od;
    if m<L
    then q[m+1]:= interpolation q[m]
    fi
  od
end;

proc FAS = ( int m, ref [] field q,r) void:
begin
  to p do relax (q[m],r[m]) od;
  if m>1
  then
    field qcoarse = restriction q[m];
    q[m-1]:= qcoarse;
    r[m-1]:= euler q[m-1]+ restrict (r[m] - euler q[m]);
    to s do FAS ( m-1, q, r) od;
    q[m]:= q[m] + prolongation (q[m-1] - qcoarse)
  fi;
  to q do relax (q[m], r[m] ) od
end;
```

In these programs q[1],...,q[L], and r[1],...,r[L] are sequences of grid functions with increasingly finer meshes; q for the solution and r for the right-hand-side. A call to the procedure relax(q_h, r_h) has the effect of a relaxation sweep for the improvement of the approximate solution q_h of $N_h^1(q_h) = r_h$. The operators

euler , **restrict** , **restriction and prolongation** correspond with the operators N_h. $\bar{R}_{2h,h}$ or $R_{2h,h}$ and $P_{h,2h}$ in the mathematical notation.

The procedure FMG assumes an initial estimate q[1] of the solution on the coarsest grid. In r[1], ..., r[L] it takes as input the right-hand side of (2.1), i.c. it should be initialized at zero. After a call it delivers in q[L] an approximate solution on the finest grid. The procedure FAS updates the approximation at the level m in q[m] by a FAS-cycle. By a single call of FMG and -depending on the problem- a few subsequent calls of FAS at the level L we obtain the 1st order discrete solution up to truncation error accuracy.

If we try to solve the 2nd order discretisation (2.1) in the same manner as we do the first order equations, we may expect difficulties for two reasons. First, the construction of the set of 4 equations to be solved at each cell Ω_{ij} in the CSGS relaxation is much more complex . Secondly, the nonlinear equations (2.1) are less stable. The 2nd order discretisations are less diffusive, (in the case of central differences clearly "anti-diffusive" [16]). This may lead not only to non-monotonous solutions, but it can also cause a Gauss Seidel relaxation not to reduce sufficiently the rapidly varying error components.

To obtain 2nd order accurate solutions, we do not try to solve the system

$$N_h^2(q_h) = 0 \tag{2.3}$$

as such. We use the first order operator N_h^1 to find the higher order accurate approximation in a defect correction iteration:

$$N_h^1(q_h^{(1)}) = 0\,, \tag{2.4.a}$$

$$N_h^1(q_h^{(i+1)}) = N_h^1(q_h^{(i)}) - N_h^2(q_h^{(i)}). \tag{2.4.b}$$

For an introduction to the defect correction principle see [17] . By a well-known technique [18] , it can be proved that -if the problem is smooth enough- the accuracy of $q_h^{(i)}$ is of order 2 for $i \geqslant 2$.

In fact we may use $q_h^{(i+1)} - q_h^{(i)}$ as an error indicator. In the smooth parts of the solution $q_h^{(1)} - q_h^{(1+i)} = \mathcal{O}(h)$, $q_h^{(2)} - q_h^{(2+i)} = \mathcal{O}(h^2)$; where these differences are larger, e.g. $\mathcal{O}(1)$, the solution is not smooth (relative to the the grid used). Then grid adaptation is to be considered rather than the choice of a higher order method, if a more accurate solution is wanted.

In a multigrid environment, where solutions on more grids are available, we can consider other approaches to compute more accurate solutions, such as (1) Richardson extrapolation or (2) τ-extrapolation. Both extrapolation methods can be well used to find a more accurate solution if the solution is smooth indeed. A drawback is that these methods rely on the existence of an asymptotic expansion of the (truncation) error for $h \rightarrow 0$, and -globally- no a-priori information about the validity of this assumption is available.

Since the evaluation of $N_h^2(q_h)$ is hardly more expensive than the evaluation of $N_h^1(q_h)$, the costs to compute the defect in (2.4.b) is of the same order as the evaluation of the relative truncation error $\tau_{2h,h}(q_h) = N_{2h}^1(R_{2h,h}q_h) - \bar{R}_{2h,h}N_h^1(q_h)$. This makes us to prefer (2.4.b) to τ-extrapolation for the computation of a 2nd order discrete system.

Another disadvantage of extrapolation is that the accurate solution (for Richardson) or the estimate for the truncation error (τ-extrapolation) is obtained at the one-but-finest level and no high resolution of local phenomena is obtained. Whereas we want not only a high order of accuracy, but also an accurate representation of possible discontinuities, in [8] we used Richardson extrapolation (only) as a possibility to find a higher order initial estimate for the iteration process (2.4.b). In the present paper we concentrate for a while on τ-extrapolation to improve the accuracy of the smooth components in the solution as they are obtained from the 2nd order scheme.

Tau extrapolation

Let the nonlinear equation

$$N_h(q_h) = r_h\,, \tag{2.5}$$

with $q_h \in X_h$, $r_h \in Y_h$, be a discretisation of

$$N(q) = r\,, \tag{2.6}$$

where $q \in X$, $r \in Y$, and let the discretisation be such that $r_h = \bar{R}_h r$ and let the operator $N_h : X_h \rightarrow Y_h$ satisfy

$$N_h(R_h q) = r_h + \tau_h(q)\,, \tag{2.7}$$

$$\tau_h(q) = h^p \bar{R}_h \tau(q) + \mathcal{O}(h^{\bar{p}}), \quad \text{for } h \rightarrow 0\,, \tag{2.8}$$

where $R_h : X \rightarrow X_h$ and $\bar{R}_h : Y \rightarrow Y_h$ are restrictions (linear surjections) and $\tau(q)$ is independent of h. The latter requirement with $\bar{p} > p$ means that $\tau_h(q)$, the local truncation error for the solution q, satisfies an asymptotic expansion.

Further, assume that we have a sequence of nested discretisations for $h = 2^{-k} h_0$, $k = 1,2,\ldots$; this means that, for h and $2h$ from this sequence, restrictions $R_{2h,h} : X_h \rightarrow X_{2h}$ and $\bar{R}_{2h,h} : Y_h \rightarrow Y_{2h}$ exist such that $R_{2h,h} R_h = R_{2h}$ and $\bar{R}_{2h,h} \bar{R}_h = \bar{R}_{2h}$.

Then for the *relative truncation error*

$$\tau_{2h,h} := N_{2h}\, R_{2h,h} - \bar{R}_{2h,h}\, N_h\,, \tag{2.9}$$

we easily derive

$$\tau_{2h,h}\, R_h = \tau_{2h} - \bar{R}_{2h,h}\, \tau_h\,. \tag{2.10}$$

With the procedure as described in section 1 we construct a sequence of nested discretisations for the Euler equations. Under the assumption (2.8), with $\bar{p} > p$, we can compute $\tau_{2h,h}(q_h)$ and apply τ-extrapolation [19, 14] to improve the accuracy of our solution.

Combining (2.8) and (2.10) we see

$$h^p \bar{R}_{2h} \tau = \frac{1}{2^p - 1}\, \tau_{2h,h}\, R_h + \mathcal{O}(h^{\bar{p}}) \tag{2.11}$$

$$= \bar{R}_{2h,h}\, \tau_h + \mathcal{O}(h^{\bar{p}})$$

$$= 2^{-p}\, \tau_{2h} + \mathcal{O}(h^{\bar{p}})\,.$$

and hence

$$N_h(R_h q) = r_h + h^p\, \bar{R}_h\, \tau(q) + \mathcal{O}(h^{\bar{p}}) \tag{2.12}$$

$$= r_h + h^p\, \bar{P}_{h,2h} \bar{R}_{2h}\, \tau(q)$$

$$+ h^p\,(I_h - \bar{P}_{h,2h}\bar{R}_{2h,h})\bar{R}_h\, \tau(q) + \mathcal{O}(h^{\bar{p}})$$

$$= r_h + \frac{1}{2^p - 1}\bar{P}_{h,2h}\, \tau_{2h,h}(R_h q)$$

$$+ (I_h - \bar{P}_{h,2h}\bar{R}_{2h,h})\, \tau_h(q) + \mathcal{O}(h^{\bar{p}}).$$

We conclude that for $\bar{P}_{h,2h}$ and $\tau(q)$ such that $(I_h - \bar{P}_{h,2h}\bar{R}_{2h,h})\bar{R}_h\tau(q) = \mathcal{O}(h^{\bar{p}-p})$ we obtain a discretisation scheme consistent of $\mathcal{O}(h^{\bar{p}})$ if we solve for q_h^* the equation

$$N_h(q_h^*) = r_h + \frac{1}{2^p - 1}\bar{P}_{h,2h}\, \tau_{2h,h}(q_h^*)\,. \tag{2.13}$$

We see that, with the accuracy restriction, we still have some freedom in the choice of $\bar{P}_{2h,h}$.

Since for a given $\tilde{q}_h \in X_h$ the $\tau_{2h,h}(\tilde{q}_h)$ is easily evaluated from (2.9), the solution of (2.13) is readily found by a defect correction iteration. For the accurate approximation of the Euler equations, we apply the τ-extrapolation to the 2nd order discretisation (superbox or starbox)

$$N_h^2(q_h) = 0\,. \tag{2.14}$$

However, since the efficient solution procedure FMG is available only for the first order discrete system

$$N_h^1(q_h) = r_h\,,$$

we solve in an iterative procedurte the equations

$$N_h^1(q_h^{(n+1)}) = N_h^1(q_h^{(n)}) - N_h^2(q_h^{(n)}) + \frac{1}{3}\bar{P}_{h,2h}\, \tau_{2h,h}(q_h^{(n)}). \tag{2.15}$$

In the initial step the right-hand-side is taken equal to zero, so that approximately the first order accurate discrete system is solved. In the *next few* steps the last term in the right-hand-side of (2.15) is replaced by zero. At the end of the iteration the formula (2.15) is applied in full.

As (2.15) converges, we find an approximate solution $q_h = R_h\tilde{q}$, that satisfies

$$N_h^2(R_h\tilde{q}) = \bar{P}_{h,2h}\, \bar{R}_{2h}\, \tau_h(\tilde{q}) + \mathcal{O}(h^{\bar{p}})\,. \tag{2.16}$$

An algorithm with increasing accuracy

We aim at the approximate solution q_h of the discrete Euler equations (2.16) for a given fine mesh and we assume that also $L-1$ coarser meshes exist. We denote the level of refinement by m and the approximate solution at level m by q[m] = $q_{2^{L-m}h}$. The coarser grids, $m<L$, are not only used for the realisation of FAS-iteration steps, but also for the construction of the initial estimate for the iteration process. The algorithm used to obtain the initial estimate and further iterands in the defect correction process reads as follows:

```
proc IDEC TAU = ( int L,few,maxit, ref [] field q,r) void:
begin
  initialise at zero (r);
  initialise at zero (q);
  initialise (q[1]);
  FMG2 (2, L, q, r );
  r[L]:= euler1 q[L] - euler2 q[L];
  FMG2 (2, L, q, r );

  for f to maxit
  do
    r[L]:= euler1 q[L] - euler2 q[L];
    if f>few then
      field tau= euler2 restrict q[L] - restrict euler2 q[L];
      r[L]:= r[L] + prolon tau / 3
    fi;
    FAS (L, q, r)
  od
end
```

Here, q and r have the same meaning as in the procedure FMG. The operators **euler1** and **euler2** are respectively the 1st and the 2nd order accurate discrete operators N_h. The procedure FMG2 is a generalisation of FMG for the case that non-trivial initial estimates for q_h and r_h are known on all levels m=1,...,L.

```
proc FMG2 = ( int n,L, ref [] field q,r) void:
begin
  for m from L-1 by -1 to 1
  do r[m] := restrict r[m+1] od;

  for m to L
  do
    to n do FAS (m,q,r) od;
    if m<L
    then
      q[m+1]:= q[m+1] +
        + interpolation (q[m] - restriction q[m+1])
    fi
  od
end;
```

REFERENCES

1. P.D. LAX (1971.). Shock waves and entropy E.H. ZARANTONELLO (eds.). *Contributions to Nonlinear Functional Analysis*, Acad. Press, New York.
2. P.D. LAX (1973). *Hyperbolic systems of conservation laws and the mathematical theory of shock waves*, Regional conference series in applied mathematics, 11, SIAM Publication, Philadelphia.
3. A. HARTEN, P.D. LAX, and B.VAN LEER (1983). On upstream differencing and Godunov-type schemes for hyperbolic conservation laws. *SIAM Review25*, 35-61.
4. P.W. HEMKER and S.P. SPEKREIJSE (1985). *Multiple Grid and Osher's Scheme for the Efficient Solution of the Steady Euler Equations*, CWI Report NM-R8507, To appear in Appl. Num. Math. 1986.
5. S. OSHER (1981). Numerical solution of singular perturbation problems and hyperbolic systems of conservation laws. O. AXELSSON, L.S. FRANK AND A. VAN DER SLUIS (eds.). *Analytical and Numerical Approaches to Asymptotic problems in Analysis*, Regional conference series in applied mathematics, North Holland Publ. Comp..
6. S. OSHER and F. SOLOMON (1982). Upwind difference schemes for hyperbolic systems of conservation laws. *Math. Comp.38*, 339-374.
7. P.W. HEMKER and S.P. SPEKREIJSE (1985). Solution of the Stationary Euler equations by Multigrid, Finite Volumes and Flux Difference Splitting. *Proceedings Multigrid Conference, Oberwolfach, dec. 1984,*, Regional conference series in applied mathematics, Vieweg Publ. Comp., To appear.
8. P.W. HEMKER (1985). *Defect correction and higher order schemes for the multigrid solution of the steady Euler equations*, CWI Report NM-R8523, To appear in Proceedings Multigrid Conference, Cologne, Oct. 1985.
9. W. SCHMIDT and A. JAMESON. Euler solvers as an analysis tool for aircraft aerodynamics. W.G.HABASHI (eds.). *Advances in Computational Transonics*, Regional conference series in applied mathematics, Pineridge Press, Swansea.
10. RON-HO NI (1982). A multiple grid scheme for solving the Euler equations. *AIAA Journal20*, 1565-1571.
11. D.C. JESPERSEN (1985). *A time-accurate multiple-grid algorithm.*, AIAA-85-1493-CP..
12. D.C. JESPERSEN (1984). *Recent developments in multigrid methods for the steady Euler equations*, Lecture Notes, March 12-16, 1984, von Karman Inst., Rhode-St.Genese, Belgium.
13. W.A. MULDER (1985). Multigrid relaxation for the Euler equations. *J. Comput. Phys.*, W.A. Mulder, thesis Leiden Univ..
14. W. HACKBUSCH (1985). *Multigrid Methods and Applications*, Springer Series Comp. Mathematics, 4, Springer Verlag, Berlin, Heidelberg.
15. P.W. HEMKER (1980). On the structure of an adaptive multi-level algorithm. *BIT20*, 289-301.

16. A.N. BROOKS and T.J.R. HUGHES (1982). Streamline-Upwind Petrov-Galerkin formulations for convective dominated flows with particular emphasis on the incompressible Navier-Stokes equations. *Comp. Meth. Appl. Mech. Engng.*.
17. K. BÖHMER and P. HEMKER, and H.J. STETTER (1984). The Defect Correction Approach. *Computing Suppl.5*, 1-32.
18. W. HACKBUSCH (1981). Bemerkungen zur iterierten Defektkorrektur und zu ihrer Kombination mit Mehrgitterverfahren. *Rev. Roum. Math. Pures Appl.26*, 1319-1329.
19. A. BRANDT (1982). Guide to Multigrid Development. W. HACKBUSCH AND U. TROTTENBERG (eds.). *Multigrid Methods*, Lect. Notes in Mathematics, 220-312, Springer Verlag.

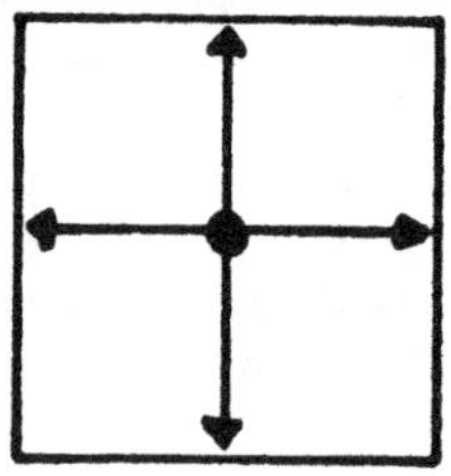

fig. 1a

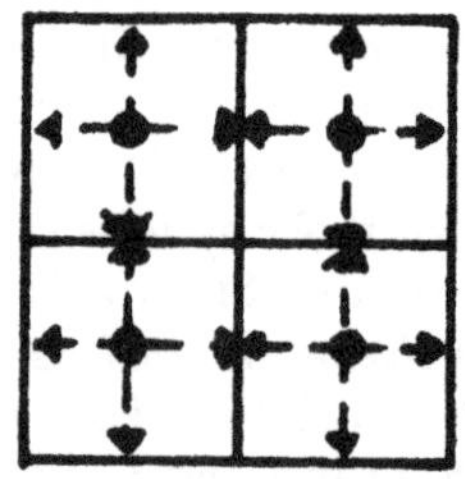

fig. 1b

fig. 1c

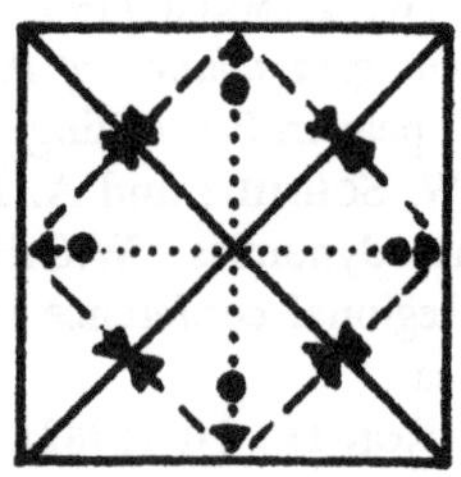

fig. 1d

Figure 1. The computation of q_{ij}^k from q_{ij} for the different schemes.
• : q_{ij} values,
← : q_{ij}^k values,
- - - - : lines of linear interpolation,
—— : lines of piecewise constant interpolation.
fig. 1a: the basic 1st order scheme,
fig. 1b: the superbox scheme,
fig. 1c: the 1st order starbox scheme,
fig. 1d: the 2st order starbox scheme.

A COMPUTABLE FINITE ELEMENT INTERPOLATION ERROR BOUND IN THE UNIFORM METRIC

Reinhard Lehmann [1)]

Using a multi-dimensional Taylor formula with remainder in integral form, a numerical value for the Sobolev imbedding constant of $H^2(\Omega)$ in $C(\bar{\Omega})$ is obtained when Ω is either the unit triangle or the unit square. This result is applied to derive a computable finite element interpolation error bound in the uniform metric.

0. Numerical methods which make use of a partial ordering monotonicity, frequently need a priori enclosing intervals for the solution. Here, a priori error estimates are of interest provided that they are computable.

The last ten years have brought reasonable results for such estimates in the finite element method. Convergence analysis in the FEM leads finally to the problem of estimating the interpolation error

$$||u - \Pi u||_1 \leq C\ ||u||_2$$

over some reference element. Here, $||\cdot||_1$ is some norm (or semi-norm) measuring the distance between an arbitrary element u and its interpolate Πu, and $||u||_2$ is some other norm (or semi-norm) essentially determined by the assumed smoothness of the function u under consideration.

A first theoretical approach is due to BIRKHOFF, SCHULTZ & VARGA (1968) who gave a representation of the error functional

$$F_\alpha(u) := \partial^\alpha u(x_0) - \partial^\alpha \Pi u(x_0)$$

using Peano and Sard kernel theorem. This result was used by BARNHILL & GREGORY (1976) and BARNHILL, BROWN & MITCHELL (1981) who obtained bounds for the constant $C(\Omega)$ in

$$u \in H^2(\Omega), \quad |u - \Pi u|_{1,\Omega} \leq C(\Omega)\ |u|_{2,\Omega}\ ,$$

namely $C(\Delta) = 1.207$ (Δ the unit triangle) and $C(\square) = 0.625$ ($\square$ the unit square).

A Taylor series expansion technique was applied in the contributions by CIARLET & WAGSCHAL (1971), ARCANGELI & GOUT (1976), MEINGUET & DESCLOUX (1976), and GOUT (1977). These results are reported in CIARLET's book (1978), see also MEINGUET (1984).

A general formula for the constant C in the finite element method for an elliptic boundary value problem of order 2s,

1) Martin-Luther-Universität Halle-Wittenberg, Sektion Mathematik, Universitätsplatz 6, Halle (Saale), DDR-4010

$$|u - u_h|_A \leq C\, |u|_{2s,\Omega} \quad , \quad u \in H^{2s}(\Omega),$$

was given by MIKHLIN (1986).

A relatively simple estimation technique leads NATTERER (1975) to the estimate

$$0.65 \leq C(\Delta) \leq 0.81$$

where in the lower bound the square is missing, as it turned out later.

Another approach was developped by ATTEIA (1977). He formulated a general eigenvalue problem the smallest positive eigenvalue of which is the reciprocal to the squared constant under consideration. In his thesis ARBENZ (1982) solved the corresponding eigenvalue problems numerically and obtained $C(\Delta) = 0.4888$, $C(\square) = 0.3184$. Using these results, we derived (1986) estimates for the constant $\tilde{C}_{2,s}(\Delta)$ in the inequality

$$||u - \Pi u||_{s,\Delta} \leq \tilde{C}_{2,s} |u|_{2,\Delta} \quad , \quad u \in H^2(\Delta), \; s = 0,\, 1, \tag{0.1}$$

namely

$$\tilde{C}_{2,s}(\Delta) \leq \begin{cases} 0.6952, & s = 0, \\ 0.8498, & s = 1. \end{cases} \tag{0.2}$$

Solving the corresponding eigenvalue problem numerically, WEISS (1985) obtained

$$\tilde{C}_{2,1}(\Delta) = 0.5086 \; . \tag{0.3}$$

In the present note, we shall give a pointwise interpolation error estimate over the triangle and the square. After globalizing it, we obtain a conditionally computable interpolation error estimate for plane convex and polygonally bounded domains. It is conditionally computable in that it needs an a priori estimate of the second derivative semi-norm of the function under consideration.

1. By Ω we denote either the unit triangle Δ or the unit square $\square$ in the plane. Let u be a function belonging to $C^s(\bar{\Omega})$, then the Taylor formula for $x,\, y \in \Omega$ reads

$$u(x) = \sum_{|\alpha|<s} \frac{1}{\alpha!} \partial^\alpha u(y)(x-y)^\alpha + \sum_{|\alpha|=s} \frac{s}{\alpha!} \int_0^1 (1-t)^{s-1} \partial^\alpha u(y+t(x-y))(x-y)^\alpha dt \; .$$

If $u \in H^2(\Omega)$ then, by the Sobolev Lemma, the function u belongs to $C(\bar{\Omega})$, and there is a constant $C(\Omega)$ such that

$$||u||_{C(\bar{\Omega})} \leq C(\Omega)\; ||u||_{2,\Omega} \; .$$

We want to estimate that constant. As an aside, there will be another proof of the Sobolev Lemma.

Without loss of generality we may assume the function u to be sufficiently smooth: The set $H^2(\Omega) \cap C^\infty(\Omega)$ is dense in $H^2(\Omega)$.

For such a function u we start from the Taylor formula with $s = 2$, integrate it over Ω with respect to y and fixed x. Then, squaring it, we obtain

$$|\Omega|^2|u(x)|^2 = \{ \int_\Omega u(y)dy + \sum_{|\alpha|=1} \int_\Omega u^{(\alpha)}(y)\ (x-y)^\alpha dy +$$

$$+ \sum_{|\alpha|=2} \frac{2}{\alpha!} \int_\Omega \int_0^1 (1-t)\ u^{(\alpha)}(y+t(x-y))(x-y)^\alpha dt\ dy\}^2$$

$$=: \{ \int_\Omega u\ dy + \sum_{|\alpha|=1} K_\alpha(x) + \sum_{|\alpha|=2} \frac{2}{\alpha!} J_\alpha(x) \}^2 \tag{1.1}$$

First, we square the integrals:

$$a^2 := \{ \int_\Omega u(y)\ dy \}^2 \leq |\Omega| \int_\Omega |u(y)|^2 dy = |\Omega|\ |u|^2_{0,\Omega}\ , \tag{1.2}$$

$$|\alpha| = 1,\ K_\alpha^2(x) := \{ \int_\Omega u^{(\alpha)}(y)\ (x-y)^\alpha dy\}^2 \leq \int_\Omega (x-y)^{2\alpha} dy\ |u^{(\alpha)}|^2_{0,\Omega} =$$

$$= I_\Omega^\alpha(x)\ |u^{(\alpha)}|^2_{0,\Omega}\ , \tag{1.3}$$

$$|\alpha| = 2,\ J_\alpha^2(x) := \{ \int_\Omega \int_0^1 (1-t)\ u^{(\alpha)}(y+t(x-y))dt\ (x-y)^\alpha dy\}^2$$

$$\leq I_\Omega^\alpha(x) \int_\Omega \int_0^1 (1-t)^2\ |u^{(\alpha)}(y+t(x-y))|^2 dt\ dy$$

$$= I_\Omega^\alpha(x) \int_0^1 \int_\Omega (1-t)^2\ |u^{(\alpha)}(y+t(x-y))|^2 dy\ dt\ .$$

In the last integral we substitute $z = z_{t,x}(y) := y + t(x - y) = tx + (1 - t)y$, $0 \leq t \leq 1$, such that $dz = (1 - t)^2 dy$, and $y \in \Omega$ is transformed into $z \in G_t(x) := \{ z \in R^2 : z = tx + (1-t)y,\ y \in \Omega \}$, $0 \leq t \leq 1$. Since Ω is convex, $G_t(x)$ is completely contained in Ω. Hence, the last equality can be continued by

$$J_\alpha^2(x) \leq I_\Omega^\alpha(x) \int_0^1 \int_{G_t(x)} |u^{(\alpha)}(z)|^2 dz\ dt \leq I_\Omega^\alpha(x)\ |u^{(\alpha)}|^2_{0,\Omega}\ . \tag{1.4}$$

It remains to estimate the integrals

$$I_\Omega^\alpha(x) := \int_\Omega (x-y)^{2\alpha} dy, \quad |\alpha| = 1,\ 2,$$

over $\Omega = \Delta$ or $\square$.

2. In case of the unit square, $\Omega = \square := \{ (y_1,y_2) \in R^2 : 0 \leq y_i \leq 1\}$, we find $(x = (x_1,x_2))$

$$I_\square^\alpha(x) = \int_0^1 (x_1-y_1)^{2\alpha_1}\ dy_1 \int_0^1 (x_2-y_2)^{2\alpha_2} dy_2\ ,\ x \in \square\ .$$

For the integral

$$I_k(\xi) := \int_0^1 (\xi-\eta)^{2k}\ d\eta\ ,\ k \geq 0,$$

we obtain

$$I_k(\xi) = \frac{1}{2k+1} [\xi^{2k+1} - (\xi-1)^{2k+1}]\ .$$

Obviously, $I_k(\xi)$ is a polynomial in ξ of degree $2k$ with leading power ξ^{2k}, its maximum over $0 \leq \xi \leq 1$ is attained at the end points of that interval and equals $(2k+1)^{-1}$. Therefore,

$$I^{\alpha}_{\square}(x) \le \frac{1}{2\alpha_1+1}\,\frac{1}{2\alpha_2+1}\,, \quad \alpha = (\alpha_1,\alpha_2), \quad x \in \square .$$

Hence, $I^{\alpha}_{\square}(x) \le \frac{1}{3}$, $|\alpha| = 1$, and for $|\alpha| = 2$ we obtain

$$I^{\alpha}_{\square}(x) \le \begin{cases} \frac{1}{5}\,, & \alpha = (2,0) \text{ or } (0,2), \\ \frac{1}{9}\,, & \alpha = (1,1). \end{cases}$$

Thus, inequality (1.3) becomes

$$|\alpha| = 1, \quad K^2_{\alpha}(x) \le \frac{1}{3}\, |u^{(\alpha)}|^2_{0,\square}$$

and in (1.4) we obtain

$$|\alpha| = 2, \quad J^2_{\alpha}(x) \le \left\{ \begin{matrix} \frac{1}{5}\,, & \alpha = (2,0) \text{ or } (0,2) \\ \frac{1}{9}\,, & \alpha = (1,1) \end{matrix} \right\} |u^{(\alpha)}|^2_{0,\square}$$

Next, we have in (1.1)

$$b^2 := \{\sum_{|\alpha|=1} K_{\alpha}(x)\}^2 \le 2 \sum_{|\alpha|=1} K^2_{\alpha}(x) \le \frac{2}{3}\, |u|^2_{1,\square}\,, \tag{2.1}$$

and

$$c^2 := \{\sum_{|\alpha|=2} \frac{2}{\alpha!}\, J_{\alpha}(x)\}^2 = (J_{20} + J_{02} + 2J_{11})^2 .$$

To the last relation we apply the ε-inequality, namely

$$c^2 \le (1 + \frac{1}{\varepsilon})\, (J_{20} + J_{02})^2 + 4(1+\varepsilon)\, J^2_{11}\,, \quad \varepsilon > 0.$$

Equilibrating the parameter ε and using the estimates above, we find

$$c^2 \le \frac{28}{45}\, |u|^2_{2,\square}\,. \tag{2.2}$$

Finally, eq. (1.1) is of the form (here, $|\Omega| = 1$)

$$|u(x)|^2 = (a + b + c)^2$$

with the estimates (1.2), (2.1) and (2.2). Equilibrating with another two ε-inequalities, we obtain

$$|u(x)|^2 \le \frac{103}{45}\, ||u||^2_{2,\square}\,, \quad x \in \square\,,$$

or

$$||u||_{C(\square)} \le 1.513\, ||u||_{2,\square}\,. \tag{2.3}$$

3. In case of $\Omega = \Delta = \{ (y_1,y_2) \colon 0 \le y_i \le 1,\ y_1 + y_2 \le 1 \}$ we find for the integrals

$$I^{\alpha}_{\Delta}(x) = \int_{\Delta} (x-y)^{2\alpha} dy = \int_{y_1=0}^{1} \int_{y_2=0}^{1-y_1} (x_1 - y_1)^{2\alpha_1} (x_2-y_2)^{2\alpha_2} dy_2\, dy_1 \tag{3.1}$$

the following estimates. For $\alpha = (1,0)$ or $(0,1)$ this integral leads to the function

$$I^{\alpha}_{\Delta}(x) = \frac{1}{2}\xi^2 - \frac{1}{3}\xi + \frac{1}{12}\,, \quad \xi = x_1 \text{ or } x_2\,, \quad (x_1,x_2) \in \Delta\,,$$

over $0 \le \xi \le 1$. That polynomial attains its maximum at the end points of the interval [0,1] such that

$$|\alpha| = 1, \; I_\Delta^\alpha(x) \leq \frac{1}{4}, \; x \in \Delta . \tag{3.2}$$

Similarly, we find for $\alpha = (2,0)$ or $(0,2)$ the expression

$$I_\Delta^\alpha(x) = \frac{1}{30}(15\,\xi^4 - 20\,\xi^3 + 15\,\xi^2 - 6\,\xi + 1), \; \xi = x_1 \text{ or } x_2, \; (x_1,x_2)\in\Delta,$$

such that

$$I_\Delta^\alpha(x) \leq \frac{1}{6}, \; x \in \Delta, \; \alpha = (2,0) \text{ or } (0,2). \tag{3.3}$$

For $\alpha = (1,1)$, after tedious calculations the integral (3.1) turns out to be an incomplete polynomial of degree 2, and then it is not difficult to estimate

$$I_\Delta^\alpha(x) \leq \frac{1}{18}, \; x \in \Delta, \; \alpha = (1,1). \tag{3.4}$$

The estimates (3.2) - (3.4) lead via (1.3) and (1.4) to the following inequalities:

$$b^2 := \{\sum_{|\alpha|=1} K_\alpha(x)\}^2 \leq 2 \sum_{|\alpha|=1} K_\alpha^2(x) \leq \frac{1}{2}\,|u|_{1,\Delta}^2 , \tag{3.5}$$

and applying the ε-inequality to

$$c^2 := (J_{20} + J_{02} + 2J_{11})^2$$

we obtain

$$c^2 \leq \frac{4}{9}\,|u|_{2,\Delta}^2 . \tag{3.6}$$

Finally, eq. (1.1) is of the form

$$|\Delta|^2\,|u(x)|^2 = (a + b + c)^2 \tag{3.7}$$

with the estimates (1.2), (3.5), and (3.6). The ε-inequality for (3.7) yields

$$|u(x)|^2 \leq \frac{52}{9}\,||u||_{2,\Delta}^2 .$$

The final result is

$$||u||_{C(\Delta)} \leq 2.404\,||u||_{2,\Delta} . \tag{3.8}$$

4. From the results obtained in Sections 2 and 3 we can derive an estimate for the linear interpolation error $e := u - \Pi u$, $u \in H^2(\Omega)$, where Πu denotes the linear Lagrange interpolate of u at the vertices of $\Omega = \Delta$ or $\square$. As an example, we shall demonstrate it for $\Omega = \Delta$.

Eq. (3.7) can be estimated as

$$|\Delta|^2\,|e(x)|^2 \leq 3(a^2 + b^2 + c^2) \tag{4.1}$$

with corresponding estimates of a^2, b^2, c^2 with respect to the error semi-norms:

$$a^2 \leq \frac{1}{2}\,|e|_{0,\Delta}^2 , \qquad b^2 \leq \frac{1}{2}\,|e|_{1,\Delta}^2 ,$$

$$c^2 \leq \frac{4}{9}\,|e|_{2,\Delta}^2 = \frac{4}{9}\,|u|_{2,\Delta}^2 ,$$

since $\partial^\alpha e = \partial^\alpha u$, $|\alpha| \geq 2$. Therefore, eq. (4.1) yields

$$|e(x)|^2 \leq 6\ ||e||^2_{1,\Delta} + \frac{16}{3}\ |u|^2_{2,\Delta}\ .$$

To the H^1-norm of the error e we apply the estimate (0.1) with the constant (0.3),

$$|e(x)|^2 \leq 6.8854\ |u|^2_{2,\Delta}$$

such that

$$||e||_{C(\Delta)} \leq 2.624\ |u|_{2,\Delta}\ ,\ u \in H^2(\Delta). \tag{4.2}$$

5. Now the estimate (4.2) will be transformed onto an arbitrary triangle T. In order to distinguish local and global coordinates, we shall mark quantities related to Δ by a hat and use Greek letters for local coordinates. Then, eq. (4.2) reads

$$\max_{(\xi,\eta)\in\Delta} |\hat{e}(\xi,\eta)|^2 \leq 6.8854\ |\hat{u}|^2_{2,\Delta}\ . \tag{5.1}$$

Let T be a triangle with nodes $P_i = P_i(x_i,y_i)$, $i = 1, 2, 3$, numbered in positive direction. Then the mapping $\Delta \to T$ is given by

$$\begin{aligned} x = x(\xi,\eta) &= x_1 + (x_2-x_1)\xi + (x_3-x_1)\eta\ , \\ y = y(\xi,\eta) &= y_1 + (y_2-y_1)\xi + (y_3-y_1)\eta\ , \end{aligned} \qquad (\xi,\eta) \in \Delta, \tag{5.2}$$

its inverse will be denoted by $\xi = \xi(x,y)$, $\eta = \eta(x,y)$.

We showed (1986) that

$$|\hat{u}|^2_{2,\Delta} \leq C(T)\ |u|^2_{2,T} \tag{5.3}$$

where $u = u(x,y) = \hat{u}(\xi(x,y), \eta(x,y))$ and

$$C(T) = \frac{1}{2|T|}\ \max\ \{\ (\bar{x}_2^2 + \bar{x}_3^2 + \theta^2)^2,\ (\bar{y}_2^2 + \bar{y}_3^2 + \theta^2)^2\ \}$$

and the abbreviations $\bar{x}_i = x_i - x_1$, $\bar{y}_i = y_i - y_1 (i = 2, 3)$, $\theta^2 = |\bar{x}_2\bar{y}_2 + \bar{x}_3\bar{y}_3|$.

Since the mapping (5.2) and its inverse are affine linear, we have

$$\max_{(\xi,\eta)\in\Delta} |\hat{e}(\xi,\eta)| = \max_{(x,y)\in T} |e(x,y)|.$$

Thus, we obtain from (5.1) and (5.3) the estimate

$$||e||_{C(T)} \leq 2.624\ \sqrt{C(T)}\ |u|_{2,T}\ ,\ u \in H^2(T), \tag{5.4}$$

which is true for an arbitrary triangle T of positive measure.

Finally, let T_h denote a family of triangles which forms a regular triangularization of a plane convex and polygonally bounded domain G such that $\sum_{T\in T_h} T = G$. Then, from (5.4),

$$\begin{aligned} ||e||_{C(\bar{G})} &= \max_{T\in T_h} ||e||_{C(T)} \leq 2.624 \max_{T\in T_h} \sqrt{C(T)}\ \sqrt{\sum_{T\in T_h} |u|^2_{2,T}} \\ &= 2.624 \max_{T\in T_h} \sqrt{C(T)}\ |u|_{2,G}\ ,\ u \in H^2(G), \end{aligned} \tag{5.5}$$

and that is a global interpolation error estimate.

References

Arbenz, P. (1982): Computable finite element error bounds for Poisson's equation. IMA J. Numer. Anal. 2 (1982), 4, 475-479.

Arcangeli, R. & Gout, J. L. (1976): Sur l'évaluation de l'erreur d'interpolation de Lagrange dans un ouvert de R^n. RAIRO, Sér. Anal. Numer. 10 (1976), 5-27.

Atteia, M. (1977): Evaluation de l'erreur dans la méthode des éléments finis. Numer. Math. 28 (1977), 295-306.

Barnhill, R. E., Brown, J. M. & Mitchell, A. R. (1981): A comparison of finite element error bounds for Poisson's equation. IMA J. Numer. Anal. 1 (1981), 95-103.

Barnhill, R. E. & Gregory, J. A. (1976): Interpolation remainder theory from Taylor expansions on triangles. Numer. Math. 25 (1976), 401-408.

Birkhoff, G., Schultz, M. H. & Varga, R. S. (1968): Piecewise Hermite interpolation in one and two variables with applications to PDE. Numer. Math. 11 (1968), 232-256.

Ciarlet, Ph.-G. & Wagschal, C. (1971): Multipoint Taylor formulas and applications to the finite element method. Numer. Math. 17 (1971), 84-100.

Ciarlet, Ph.-G. (1978): *The finite element method for elliptic problems*. North Holland 1978.

Gout, J. L. (1977): Estimation de l'erreur d'interpolation d'Hermite dans R^n. Numer. Math. 28 (1977), 407-429.

Lehmann, R. (1986): Computable error bounds in the finite element method. IMA J. Numer. Anal. 6 (1986) (to appear).

Meinguet, J. & Descloux, J. (1977): An operator-theoretical approach to error estimation. Numer. Math. 27 (1977), 307-326.

Meinguet, J. (1984): A practical method for obtaining a-priori error bounds in pointwise and mean square approximation problems. In: *Approximation Theory and Spline Functions*. D. Reidel Publ. Company, 1984, pp. 97-125.

Mikhlin, S. G. (1986): *Constants in some inequalities of analysis*. John Wiley & Sons Ltd. 1986.

Natterer, F. (1975): Berechenbare Fehlerschranken für die Methode der Finiten Elemente. ISNM vol. 28, pp. 109-121. Birkhäuser Basel 1975.

Weiss, R. (1985): Diploma Thesis. Martin-Luther-Universität Halle-Wittenberg, Sektion Mathematik, 1985.

AGGREGATION AND HOMOGENIZATION IN REACTOR DIFFUSION

Ivo M a r e k [1)]

1. I n t r o d u c t i o n

Recently much attention has been paid to multi-level iterative processes ([13]). The main concern naturally has been given to the finite-dimensional (discretized) models (see e.g. [4], [6]). In this contribution we would like to show how some of the ideas of multi-level approach can be applied to a nondiscretized model. We present an analysis giving a rigorous mathematical justification of a very powerful method of solving a class of problems typical for reactor physics computations ([7], [3]). We show that the well known method of homogenization of the data and the aggregation of the velocity variables can be viewed as a particular case of an abstract two (multi)-level iterative scheme. It follows from our analysis that the two-level approach very naturally explains the most important features of the homogenization-aggregation techniques, in particular, the algorithms and formulas of the homogenization-aggregation method.

2. A n a b s t r a c t t w o - l e v e l s c h e m e

Let Y and Z be Banach spaces over the field of real numbers. Let Y' denote the dual space of Y and $B(Y,Z)$ the Banach space of bounded linear maps of Y into Z ; we let $B(Y) = B(Y,Y)$

Let us consider the equation

$$u - Au = b , \tag{2.1}$$

where $b \in Y$ and $A \in B(Y)$

We assume that we are given two maps R and $S(u)$ such that $R \in B(Y,Z)$ and $S(u) \in B(Z,Y)$ for $u \in \mathcal{D}$, where $\mathcal{D}$ is a suitable nonempty set in Y . We also require the following relations to be valid for every $u \in \mathcal{D}$

$$RS(u) = I_Z \tag{2.2}$$

and

$$P(u)u = S(u)Ru = u . \tag{2.3}$$

In (2.2) I_Z denotes the identity map of Z .

As a consequence of (2.2) we deduce that $P(u) = [P(u)]^2$. We let

$$A_Z = R\,A\,S(u) \tag{2.4}$$

---

1) Katedra numerické matematiky na matematicko-fyzikální fakultě Univerzity Karlovy, Malostranské nám. 25, 118 00 Praha 1, Czechoslovakia

and introduce the aggregated equation

$$z - A_Z z = Rb \ . \tag{2.5}$$

Our task is to propose a two-level iterative algoritm of solving (2.1) by means of (2.5) in a way that the "expensive operations" are allowed only for equation (2.5) whilst the "cheap operations" are allowed for both equations (2.1) and (2.5).

REMARK. It is easy to identify the expensive and the cheap operations are in our abstract model. As an example one can think of a two grid model where the exact solving of an algebraic system is considered as an expensive operation and one single matrix multiplication with a sparse matrix as a cheap operation. In fact, any two or multi-level algorithm is defined by having determined the "value" of operations and the allowance of each of the operations under considerations in the process of computation.

Now, we define a general two-level algorithm as follows:

(i) Choose $u^{(p)} \in Y$ an approximation of the desired exact solution of (2.1).

(ii) Solve the equation (2.5) with $u = u^{(p)}$; denote the solution by $z^{(p)}$;

(iii) Disaggregate according to the formula

$$u^{(p+1)} = AS(u^{(p)})z^{(p)} + b \ ;$$

(iv) Iterate for $p = 0,1,\ldots$.

Let us note that the cheap operations of the algorithm just introduced appear in the step (iii) that is, they are determined by the operators $S(u)$ whilst the expensive operations appearing in (ii) mean the solving of (2.5) determined by the $A_Z(u^{(p)})$ and Rb .

In the terms of the elements of Y our iterative process can be expressed as

$$u^{(k+1)} = AT(u^{(k)}) + b \ , \tag{2.6}$$

where

$$T(u) = S(u)\left[I_Z - A_Z(u)\right]^{-1} Rb \ . \tag{2.7}$$

The following hypotheses H1 - H11 guarantee first that the algorithm introduced in (i) - (iv) is well defined that is, each its step is realizable and, second that the process is locally convergent.

Let W be a subspace of Y .

H1. The spaces Y and Z are generated by closed normal cones K and K_Z respectively ([5] see also [4]).

Let K' be the dual cone, i.e. let $K' = \{u' \in Y' : u'(u) \geq 0$ for all $u \in K\}$,
$K^d = \{u \in K : u'(u) > 0$ for all $u' \in K'$, $u' \neq 0\}$ and

$K_{u_0} = \{u \in K : \text{There is an } \alpha = \alpha(u) > 0 \text{ such that } u - \alpha(u)u_0 \in K\}$.

H2. There is a set $\mathcal{D} \subset W \subset Y$ such that $\mathcal{D}$ is W-relatively open, $\emptyset \neq \mathcal{D} \subset K$ and $u^* \in \mathcal{D}$, where $u^* = Au^* + b$.

H3. The operators $A \in B(Y)$ and $A_Z \in B(Z)$ leave invariant the cones K and K_Z respectively i.e. $AK \subset K$ and $A_Z K_Z \subset K$.

H4. The vector $b \in Y$ and the operator T defined in (2.7) are such that $b + T(u) \in \mathcal{D}$ whenever $u \in \mathcal{D}$.

H5. The operator $A_Z(u)$ is positively homogeneous with respect to $u \in \mathcal{D}$, i.e. $A_Z(\gamma u) = A_Z(u)$ for $u \in \mathcal{D}$ and $\gamma > 0$, $\gamma \in R^1$.

H6. The operator A is convergent, i.e. its spectral radius $r(A) < 1$.

H7. There is an $\alpha \in R^1$, $0 < \alpha < 1$ such that $r(P(u)A) \leq \alpha$, $u \in \mathcal{D}$.

H8. There is a β , $\beta \in R^1$, $0 < \beta < 1$, such that $r(J(u^*)) \leq \beta < 1$ where $J(u) = A[I - P(u)A]^{-1}[I - P(u)]$, and where I denotes the identity operator on Y .

H9. The operator $S(u)$ is continuous with respect to $u \in W$ in the norm $||\cdot||_W$ of W .

H10. The norms of Y and W fulfil $||u|| \leq ||u||_W$, $u \in W$.

H11. Let $P(u)u = u$, $u \in \mathcal{D}$.

Our convergence result can be formulated as follows:

THEOREM 2.1. Besides the relations (2.2) and (2.3) let the hypotheses H1 - H11 be fulfilled. Then the iteration process (2.6) is locally convergent with respect to W , i.e. there exists an W-open neigh-borhood U of $u^* \in \mathcal{D} \subset W$ such that

$$\lim_{k\to\infty} ||u^{(k)} - u^*|| = 0 \quad \text{for } u^{(0)} \in U .$$

The speed of convergence is characterized by the estimates

$$||u^{(k)} - u^*|| \leq ||u^{(k)} - u^*||_W \leq \kappa\rho^k , \tag{2.8}$$

where

$$\rho = r(J(u^*)) + \eta < 1 ,$$

with some $\eta > 0$ and κ independent of k .

Proof The following relations are easily obtained

$$[A_Z(u)]^k = R[P(u)A]^k S(u) ,$$

$$S(u)[I_Z - A_Z(u)]^{-1}R = [I - P(u)A]^{-1}P(u) ,$$

$$T(u) = u - S(u)y = u - S(u)[I_Z - A_Z(u)]^{-1}R[(I - A)u - b] ,$$

where $y = Ru - z$, z is the exact solution of (2.5),

$$T(u) = u - [I - P(u)A]^{-1} P(u)[(I - A)u - b] . \qquad (2.9)$$

From these formulae it follows that

$$AT(u) + b = J(u)u' + N(u)b ,$$

where

$$N(u)b = b + A[I - P(u)A]^{-1} P(u)b .$$

Furthermore ,

$$J(u)u^* = u^* - N(u)b$$

and thus

$$AT(u) + b - u^* = J(u)(u - u^*) .$$

Since

$$Ru^* - A_Z(u^*)Ru^* = Rb ,$$

we derive that

$$AT(u) - AT(u^*) = J(u)(u - u^*) . \qquad (2.10)$$

Let $\zeta > 0$ be arbitrary. It is well known (e.g. [12, p. 55]), that there is a norm $\|\cdot\|_\zeta$ equivalent with the norm $||\cdot||_W$ such that

$$||J(u^*)||_\zeta \leq r(J(u^*)) + \zeta . \qquad (2.11)$$

The continuity of $S(u)$ with respect to the W-norm implies the W--continuity of $J(u)$. Thus, for a given $\varepsilon > 0$ there is a $\delta > 0$ such that $||J(u) - J(u^*)||_\zeta < \varepsilon$ as soon as $||u - u^*||_\zeta < \delta$.

According to (2.10) and (2.11) this implies that

$$||AT(u) - AT(u^*)||_\zeta \leq [r(J(u^*)) + \varepsilon + \zeta] \; ||u - u^*||_\zeta . \qquad (2.12)$$

The map $H(u) = AT(u)$ is thus contractive on a neighborhood U of u^* . Since all the iterants $u^{(k)}$ belong to $\mathcal{D} \subset W$ we may conclude that the process (2.6) is W-convergent. The equivalence of the norms $||\cdot||_W$ and $||\cdot||_\zeta$ together with (2.12) then imply the desired result (2.8). The proof is complete.

3. A modified process

So far the operator A in (2.1) was assumed to be bounded. In order to be able to apply the homogenization-aggregation method to equations with unbounded operators we have to modify our considerations slightly.

Let us investigate the following operator equation

$$Lu = Bu + b , \qquad (3.1)$$

where B is a bounded linear map of the space Y into Y , $b \in Y$ and L is a linear densely defined operator mapping $\mathcal{D}(L) \subset Y$ into Y , assume that L is boundedly invertible and that both L^{-1} and B leave invariant a generating normal cone $K \subset Y$. Furthermore, let R and $S(u)$ fulfil the hypotheses of the previous section we can form the appropriate aggregated analogue to (3.1):

$$L_Z z = B_Z z + Rb , \tag{3.2}$$

where $L_Z = RLS(u)$ and $B_Z = RBS(u)$.

Now we are able to define a modified two-level algorithm:

1° Find an approximation $u^{(k)}$ to the required solution u^* to (3.1);

2° Compute a correction $z^{(k)}$ using (3.2);

3° Disaggregate by setting

$$Lu^{(k+1)} = BT(u^{(k)}) + b \tag{3.3}$$

where

$$T(u) = S(u)z$$

and where

$$L_Z z = B_Z z + Rb .$$

4° Iterate for $k = 0,1,2,\dots$.

The convergence of the above process is described in the following theorem.

THEOREM 3.1. Assume the validity of the hypotheses H1 - H11 hold true with the following alterations: $A = L^{-1}B$, $A_Z(u) = L_Z^{-1}B_Z^{-1}$. Moreover, let

H12. For every $u \in \mathcal{D}$ and $v \in \mathcal{D}$ we have that

$$|| L^{-1}[S(u)RB - B]S(u)v||_W < \frac{1}{2}\tau||v||_W ,$$

and

$$|| L^{-1}S(u)RLv - v||_W < \frac{1}{2}\tau||v||_W$$

where τ is sufficiently small (i.e. at least $1 - r(L^{-1}B) > \tau > 0$).

Then there exists a W-open neighborhood of $u^* \in \mathcal{D} \subset W$ such that

$$\lim_{k\to\infty} ||u^{(k)} - u^*|| < \hat{\tau}, \quad u^0 \in U ,$$

where $\hat{\tau} \leq \dfrac{\kappa\tau}{1 - r(L^{-1}B) - \tau}$, and where κ is such that

$$||(I - L^{-1}B)^{-1}||_W \leq \frac{\kappa}{1 - r(L^{-1}B)} .$$

4. Multigroup systems

Let $d = 3$ and $G \subset R^d$ be a region with a lipschitzian boundary ∂G . It is assumed that $\bar{G} = \bigcup^{p} G_m$, where G_m is an open domain (the bar denotes the closure in the standard topology) and $G_j \cap G_k = \emptyset$ if

$j \neq k$.

Let D_j , σ_j , σ^e_{jk} , σ^f_{jk} , q_j be nonnegative functions in G analytic in $G^{(m)}_t$, $m = 1,\dots,p$, $t = 1,\dots,t_m$ where

$$\overline{G}_m = \bigcup_{t=1}^{t_m} \overline{G}^{(m)}_t , \quad G^{(m)}_t \cap G^{(m)}_s = \emptyset$$

if $s \neq t$.

Let us consider the diffusion system [1,2,7]

$$\begin{aligned} L_j u_j &\equiv - \sum_{\ell=1}^{d} \frac{\partial}{\partial x_\ell}\left(D_j \frac{\partial u_j}{\partial x_\ell}\right) + \sigma_j u_j \\ &= \sum_{k=1}^{N} \left(\sigma^e_{jk} + \frac{1}{\lambda} \sigma^f_{jk}\right) u_k + q_j \quad \text{in } G \end{aligned} \tag{4.1}$$

with the boundary conditions

$$u_j + \alpha_j \frac{\partial u_j}{\partial \overline{n}} = 0 \quad \text{on } \partial G \tag{4.2}$$

where $\alpha_j \geq 0$ and $\overline{n}$ is the direction of the outer normal of ∂G and with the interface conditions

$$u_j \text{ and } D_j \frac{\partial u_j}{\partial n} \text{ continuous in } G , \tag{4.3}$$

where $\frac{\partial}{\partial n}$ denotes the derivative in the direction of the outer normal with respect to $\partial G^{(m)}_t$.

The problem of finding a solution to the problem (4.1) - (4.3) is rather complex because of the nonsmoothness of the data. Another difficulty of numerical solving this problem appears if N - the order of the diffusion system - is large. It should be noted that in the early works concerning the reactors diffusion systems the order N was rather small say $N = 2$ or $N = 3$.

A problem of particular importance is the so called basic problem of reactor physics, i.e. the finding the principal eigenelements of (4.1) - (4.3); in this context, an eigenpair (λ_0, u_0) is principal, if u_0 is nonnegative within G . In such case the problems (4.1) - (4.3) is solved by using the source iteration method [7, p. 46]

$$Lu^{(k+1)} = Bu^{(k+1)} + \frac{1}{\lambda_{(k)}} Cu^{(k)} , \tag{4.4}$$

where $u^{(0)} \geq 0$ in G , $u \neq 0$
and

$$\lambda_{(k)} = \frac{\sum_{j=1}^{N} \int_G u_j^{(k)} u_j^{(k)}}{\sum_{j=1}^{N} \int_G u_j^{(k)} u_j^{(k-1)}} \tag{4.5}$$

and where

$$L = \mathrm{dig}\ \{L_1, \dots, L_N\} , \tag{4.6}$$

$$B = (\sigma^e_{jk}) \ , \tag{4.7}$$

$$C = (\sigma^f_{jk}) \ . \tag{4.8}$$

We see that the most common problem is to solve equation (4.4) for different right hand sides $b = \frac{1}{\lambda_{(k)}} Cu^{(k)}$

To avoid the complications mentioned above, a very effective method of numerical treatment of the problem (4.1) - (4.3) has been invented in the past: the method of data homogenization and energy aggregation. This method is still used as a standard tool of numerical computations.

In this paper we show that the aggregation in the energy domain can be considered as a particular case of the modifice two-level scheme presented in Section 3.

A similar analysis can be elaborated for the homogenization for the space dependent variables of the neutron flux. In contrary to the older computation schemes our algorithms give ways of constructing approximations of arbitrary high degree of accuracy. We are able to show that the abstract and modified schemes can also be applied to numerical solving multigroup Boltzmann neutron transport systems [15, 16, 10].

In order to apply some of the results of sections 2 and 3 we let

$$Y = L^2(G) \times \ldots \times L^2(G) \ , \quad W = C(\overline{G}) \times \ldots \times C(\overline{G})$$

with the standard inner products and norms, so that

$$||u||_Y \leq ||u||_W \ , \quad u \in W \ .$$

Further, let $K_1 = \{u \in L^2(G): u \geq 0 \text{ a.e. in } G\}$ and $K = K_1 \times \ldots \times K_1$ and $Z = R^n = R^n_+ - R^n_+$, where $R^n_+ = \{x \in R^n: (x)_j \geq 0 \ , \ j = 1,\ldots,n\}$.

Let $N = \sum_{j=1}^{n} n_j$, $n_0 = 0 < n_1 < \ldots < n_n$ and define $(\ell) = \{\ell :$ $n_{j-1} + 1 \leq \ell \leq n_j\}$, $j = 1,\ldots,n$.

Further, we let $\mathcal{D} = K_0 \cap W$, where

$$K_0 = \{v \in K: v \geq \alpha(v)u_0 \ , \ \alpha(v) > 0\}$$

and u_0 is a vector-function which has the following property:

For every $v \in K$, $v \neq 0$, there are quantities $\alpha(v) > 0$ and $\beta(v)$ such that

$$L^{-1}(B + C)v - \alpha(v)u_0 \in K \quad \text{and} \quad \beta(v)u_0 - L^{-1}(B + C)v \in K \ .$$

The existence of such u_0 is guaranteed by the properties of L , B and C (see [14]).

Further, let $w = (w_1,\ldots,w_N)$ be such that $w_j(x) > 0$ in $G_t^{(m)}$, $t = 1,\ldots,t_m$, $m = 1,\ldots,p$, $w \in W$.

To define the operators R and $S(u)$, we let

$$Ru = z = (z_1^{(1)},\dots,z_1^{(p)},z_2^{(1)},\dots,z_2^{(p)},\dots,z_n^{(1)},\dots,z_n^{(p)}) ,$$

where

$$z_j^{(m)} = \sum_{\ell=n_{j-1}+1}^{n_j} \int_{G_m} w_\ell u_\ell$$

and

$$S(u)z = v , \quad z \in Z = R^{np} \times \dots \times R^{np} , \quad u \in \mathcal{D} ,$$

where

$$v_j(x) = \frac{u_j(x)}{(Ru)_{(j)}^{(m)}} z_j^{(m)} , \quad j = 1,\dots,N , \quad x \in G_m .$$

It can be shown, that all the hypotheses of Section 2 are fulfilled and hence the corresponding aggregation process is convergent to the exact solution of the problem $u = L^{-1}Bu + b$.

Assuming further that $\mathcal{D}^* \subset \mathcal{D}$ is a "small" neighborhood of the exact solution u^* and that $\hat{L}$, $\hat{B}$ and $\hat{C}$ are good approximations of the original operators, we are able to prove the following result.

THEOREM 4.1. Let u be a solution to the diffusion system

$$- \sum_{\ell=1}^{d} \frac{\partial}{\partial x_\ell}\left(\hat{D}_{(j)} \frac{\partial \hat{u}_{(j)}}{\partial x_\ell}\right) + \hat{\sigma}_{(j)}\hat{u}_{(j)} = \sum_{k=1}^{n} \hat{\sigma}^e_{(j)(k)}\hat{u}_{(k)} + \hat{q}_{(j)} \tag{4.9}$$

with the boundary and interface conditions analogous to those (4.2) and (4.3), where (the summation is taken over $n_{j-1} + 1 \le \ell \le n_j$)

$$\hat{D}_{(j)} = \frac{\Sigma D_\ell \phi_\ell \psi_\ell}{\Sigma \phi_\ell \psi_\ell} , \quad \hat{\sigma}_{(j)} = \frac{\Sigma \sigma_j \phi_\ell \psi_\ell}{\Sigma \phi_\ell \psi_\ell} ,$$

$$\hat{\sigma}^e_{(j)(k)} = \frac{\sum\limits_{\ell=n_{j-1}+1}^{n_j} \sum\limits_{n_{k-1}+1}^{n_k} \sigma^e_{\ell t}\phi_\ell\psi_\ell\phi_t\psi_t}{\sum\limits_{\ell=n_{j-1}+1}^{n_j} \phi_\ell\psi_\ell \sum\limits_{n_{k-1}+1}^{n_k} \phi_t\psi_t} ,$$

$$\hat{q}_{(j)} = \frac{\Sigma q_\ell \phi_\ell \psi_\ell}{\Sigma \phi_\ell \psi_\ell} , \quad \hat{\alpha}_{(j)} = \frac{\Sigma \alpha_\ell \phi_\ell \psi_\ell}{\Sigma \phi_\ell \psi_\ell}$$

and where $\phi_1,\dots,\phi_n$, $\psi_1,\dots,\psi_n$ are components of vectors Φ and Ψ both in $\overset{o}{R}{}^n_+ = \{x \in R^n_+ : (x)_j > 0 , \ j = 1,\dots,n\}$.

If the operators $L^{-1}B$ and $\tilde{L}^{-1}\tilde{B}$ where

$$\tilde{L} = \text{diag}\ \{\tilde{L}_1,\dots,\tilde{L}_N\}$$

and $\tilde{B} = (\tilde{\sigma}_{jk})$ and where

$$\tilde{L}_j u_j \equiv - \sum_{\ell=1}^{d} \frac{\partial}{\partial x_\ell}\left(\tilde{D}_{(j)} \frac{\partial u_j}{\partial x_j} \phi_j\right) + \tilde{\sigma}_{(j)} u_j \phi_j$$

with boundary and interface conditions analogous to those in (4.2) and (4.3) and

$$\tilde{\sigma}^e_{jk} = \hat{\sigma}^e_{(j)(k)} \phi_j \phi_k ,$$

where ϕ_j , $j = 1,\dots,N$, $k = 1,\dots,N$, are positive functions constant within G_m , are related as follows

$$||L^{-1}B - \tilde{L}^{-1}\tilde{B}||_W < \tau ,$$

then the solution to the problem $\tilde{L}u = \tilde{B}u + \tilde{q}$ is obtained as a limit of the iterants of the aggregated system described in Theorem 3.1.

Moreover, for the exact solution u^* to $Lu = Bu + q$ we have that

$$||u^* - \tilde{u}|| \leq \omega(\tau) ,$$

where $\omega(\tau)$ tends to zero as soon as τ does too; in other words, $||u^* - u^{(k)}|| \leq 2\omega(\tau)$ for sufficiently large k's .

At conclusion we should remark how to choose the so far in a sense arbitrary element w in the modified process. It is recommendable to choose w close to the principal eigenelement u_0^+ of the adjoint problem

$$L^*v = B^*v + \frac{1}{\lambda} C^*v .$$

There are at least two reasons why to make this choice. First, if w is close to u_0^+ then the closeness of the aggregated operators to the original ones is guaranteed as well as the contractivity of the aggregated equation. Second, it is well known that u_0^+ depends upon the data in a rather weak manner and this fact is reflected in numerical stability of the process considered.

R e f e r e n c e s

[1] Habetler, G. J., and M.A. Martino: Existence theorems and spectral theory for the multigroup diffusion models. Proc. Symp. in Appl. Math. Vol. XI. Amer. Math. Soc., Providence, R. I. 1961.

[2] Habetler, G. J., and M.A. Martino: The multigroup diffusion equations of reactor physics. Report KAPL-1886, July 1958, Knolls Atomic Power Lab.

[3] Khromov, V. V., I. S. Slezarev, and A. M. Kuzmin: Variational method of synthesis of a multigroup neutron field in a multizone reactor. In; Problems of Engineering and Physics of Nuclear Reactors, Ed. L. N. Yurovay , Atomizdat, Moscow 1966. (In Russian.)

[4] Krasnoselski, M. A., E. A. Lifschitz, and A. V. Sobolev: Positive Linear Systems. Nauka, Moscow 1985. (In Russian.)

[5] Krein, M. G., and M. A. Rutman: Linear operators leaving invariant a cone in a Banach space. Usp. mat. nauk III (1948), Nr. 1, 3-95. (In Russian.) English translation in Amer. Math. Soc. Transl., Nr. 26 (1958), 128 pp.

[6] Mandel, J., and B. Sekerka: A local convergence proof for the iterative aggregation method. Linear Algebra and its Applications 51 (1983), 163-172.

[7] Marchouk, G. I.: Numerical Methods of Computation of Nuclear Reactors. Atomizdat, Moscow 1958. (In Russian.)

[8] Marek, I.; Frobenius theory of positive operators: Comparison theorems and applications. SIAM J. Appl. Math. 19 (1970), 607-628.

[9] Marek, I.: On some spectral properties of Radon-Nikolski operators and their generalizations. Comment. Math. Univ. Carol 3 (1962), Nr. 1, 20-30.

[10] Marek, I.: Some mathematical problems of nuclear reactors on fast neutrons. Aplikace matematiky 8 (1963), 442-470. (In Russian.)

[11] Marek, I.: u_o-positive operators and some of their applications. SIAM J. Appl. Math. 15 (1967), 484-494.

[12] Marek, I., and K. Žitný: Matrix Analysis for Applied Sciences. Teubner Texte zur Mathematik, Leipzig 1983.

[13] Multigrid Methods. Springer Lecture Notes in Mathematics 960. Editors W. Hackbusch and U. Trottenberg. Berlin - Heidelberg - New York 1982.

[14] Shikhov, S. B.: Lectures in Mathematical Theory of Nuclear Reactors. I. Linear Theory. Atomizdat, Moscow, 1973. (In Russian.)

[15] Vladimirov, V. S.: Integro-differential equation for the neutron transport. Izvestija Akad. Nauk USSR - math. series, 21 (1957), 681-710. (In Russian.)

[16] Vladimirov, V. S.: On 2n integro-differential equation. Izvestija Akad. Nauk USSR - math. series, 21 (1957), 3-52. (In Russian.)

THE ROTHE METHOD AND MONOTONE DISCRETIZATION FOR PARABOLIC EQUATIONS

H.-G.Roos [1]

1. Introduction

Concerning the numerical solution of boundary value problems it is not only interesting to obtain a numerical approximation of the solution with a certain accuracy but it is also of practical importance to construct upper and lower bounds of the solution itself. Such inclusion generating discretizations for parabolic boundary value problems have been proposed in [1],[2],[6],[8]. The aim of the present paper consists of deriving a new principle to generate uniformly enclosing discretization techniques combining the Rothe method and the monotone discretization technique proposed in [4],[5].

The organization of the paper is as follows: In section 2 we describe the modification of the Rothe method and place at disposal some basic a-priori estimates. The convergence of the modified Rothe method is proved in section 3.

In section 4 the two-point boundary value problems arising in every time step are solved by a monotone discretization technique and the convergence of the total process is proved.

2. A modification of the Rothe method

In this paper we deal with the numerical solution of the linear parabolic boundary value problem

$$\frac{\partial u}{\partial t} - \frac{\partial^2 u}{\partial x^2} = f(x) \qquad \text{in } Q = \Omega \times (0,T) \text{ with } \Omega = (0,1) \tag{2.1}$$

$$u|_{t=0} = 0$$

$$u|_{x=0} = u|_{x=1} = 0 .$$

Let us start from a weak formulation of this problem: Find some $u \in L^2((0,T),H^1(\Omega))$ with $u' \in L^2((0,T),L^2(\Omega))$ such that

$$(u'(t),v)+a(u,v) = (f,v) \quad \forall v \in H_0^1(\Omega) \tag{2.2}$$
$$u(0) = 0 \qquad \text{in } C((0,T),L^2(\Omega)),$$

here $(u,v) = \int_\Omega uv d\Omega$ denotes the inner product in $L^2(\Omega)$ and

---

[1] Technische Universität Dresden, Sektion Mathematik, Wissenschaftsbereich Numerische Mathematik, DDR-8027 Dresden, Mommsenstr. 13

$a(u,v) = (u',v')$ a bilinearform on $H^1(\Omega) \times H^1(\Omega)$. The results derived for problem (2.1) could be extended to such problems where the bilinearform $a(.,.)$ and the function f fulfill the following conditions:

(A1) (i) $a(v,v) \geq \alpha\|v\|_1^2$ for all $v \in H_0^1(\Omega)$

(ii) $a(v,v) \geq 0$ for all $v \in H^1(\Omega)$

(A2) $|a(u,v)| \leq \beta\|u\|_1 \cdot \|v\|_1$ for all $u,v \in H^1(\Omega)$

(A3) (i) $f \in H_0^1(\Omega) \cap H^2(\Omega)$

(ii) $|a(f,v)| \leq M\|f\|_2\|v\|_0$ for all $v \in H_0^1(\Omega)$.

Here we denote by $H^k(\Omega)$ the Sobolew space of functions which k-th derivatives are square integrable, equipped with the norm $\|\cdot\|_k$. $\|\cdot\|_0$ denotes the norm in $L^2(\Omega)$.

It is obvious that under the assumption (A3),(i) all the conditions for problem (2.2) are fulfilled.

The conditions (A1),(A2),(A3) guarantee that problem (2.2) admits a unique solution with

$$u \in C((0,T),H_0^1(\Omega))$$

$$u' \in L^2((0,T),H_0^1(\Omega)) \cap C((0,T),L^2(\Omega))$$

(compare [7], Theorem 11.1,12.1). In the one-dimensional case the function $u(x,t)$ is continuous.

Now we choose $p_1 \in N, p_1 \geq 3$, $p_k = 2^{k-1}p_1$ $(k=2,3,\dots)$ and divide the t-interval into $2^{k-1}p_1$ subintervals $[t_{i-1}^k,t_i^k]$ by the definition $t_\mu^k = \tau_k\mu$ with $\tau_k = \frac{T}{p_k}$. Let us set

$$\varphi_i^k(t) = \begin{cases} \dfrac{t-t_{i-1}^k}{\tau_k} & \text{for } t \in [t_{i-1}^k,t_i^k] \\ \dfrac{t_{i+1}^k-t}{\tau_k} & \text{for } t \in [t_i^k,t_{i+1}^k] \\ 0 & \text{otherwise} \end{cases} .$$

The discretization in time is realized by means of the representation

$$u^{\tau,k}(x,t) = \sum_{i=1}^{p_k} z_i^k(x)\varphi_i^k(t)$$

for an approximate solution $u^{\tau,k}(x,t)$ of problem (2.2).

It holds on the interval (t_{i-1}^k,t_i^k)

$$((u^{\tau,k})',v)+a(u^{\tau,k},v)=(\frac{z_i^k-z_{i-1}^k}{\tau_k},v)+\frac{t-t_{i-1}^k}{\tau_k}a(z_i^k,v)+\frac{t_i^k-t}{\tau_k}a(z_{i-1}^k,v)$$

$$=\frac{t-t_{i-1}^k}{\tau_k}[(\frac{z_i^k-z_{i-1}^k}{\tau_k},v)+a(z_i^k,v)]+$$

$$+\frac{t_i^k-t}{\tau_k}[(\frac{z_{i-1}^k-z_{i-2}^k}{\tau_k},v)+a(z_{i-1}^k,v)]+$$

$$+\frac{t_i^k-t}{\tau_k}(\frac{z_i^k-2z_{i-1}^k+z_{i-2}^k}{\tau_k},v).$$

The Rothe method consists in determining the functions $z_i^k(x)$ from

$$(\frac{z_i^k-z_{i-1}^k}{\tau_k},v)+a(z_i^k,v)=(f,v)\text{ for all }v\in H_0^1(\Omega).$$

We choose some suitable constant C^* and determine the functions $z_i^k(x)$ from

$$(\frac{z_i^k-z_{i-1}^k}{\tau_i},v)+a(z_i^k,v)=(f+\tau_kC^*,v)\text{ for all }v\in H_0^1(\Omega), \tag{2.3}$$

let $z_o\equiv 0$.

Thus, it follows on (t_{i-1}^k,t_i^k)

$$((u^{\tau,k})',v)+a(u^{\tau,k},v)=(f,v)+\tau_k(C^*+\frac{z_i^k-2z_{i-1}^k+z_{i-2}^k}{\tau_k^2}\frac{t_i^k-t}{\tau_k},v). \tag{2.4}$$

In the sequel we will show that it is possible to choose C^* in such a way that

$$((u^{\tau,k})',v)+a(u^{\tau,k},v)\geq(f,v)\text{ for all }v\geq 0,\ v\in H_0^1(\Omega). \tag{2.5}$$

Consequently, the weak maximum principle for parabolic equations yields the desired bounding property $u^{\tau,k}(x,t)\geq u(x,t)$. In the following for $k=1$ we omit the superscript k.

Now, additionally we assume

(A3) (iii) $f\in C^4(\bar{\Omega})$.

<u>Lemma 1</u>: There exists some constant $\mu(f)$ such that for $\tau\leq\tau_o$ the estimate

$$\frac{z_{i+1}-2z_i+z_{i-1}}{\tau^2}\geq\mu(f)-\frac{1}{e}C^*$$

holds.

Proof: From (2.3) it follows

$$-\tau z_i''(x) + z_i(x) = z_{i-1}(x) + \tau f(x) + \tau^2 C^* \qquad (i \geq 1) \tag{2.6}$$
$$z_i(0) = z_i(1) = 0 ,$$

we have $z_0 \equiv 0$. We interpret τ as a perturbation parameter and construct an asymptotic expansion of z_1, z_2. A standard technique yields

$$z_1 = f\tau + (f'' + C^*)\tau^2 - z_L^{(1)}\tau^2 - z_R^{(1)}\tau^2 + R_1\tau^2 . \tag{2.7}$$

$z_L^{(1)}$ and $z_R^{(1)}$ are boundary layer corrections in the neighbourhood of $x=0$, $x=1$, respectively. In the following we restrict ourselver on the investigation of these corrections near $x=0$ in detail. $z_L^{(1)}$ satisfies

$$-\frac{d^2 z_L^{(1)}}{d\zeta^2} + z_L^{(1)} = 0, \quad z_L^{(1)}\Big|_{\zeta=0} = f''(0) + C^* \quad \text{with } \zeta = x/\sqrt{\tau}.$$

One obtains explicitly

$$z_L^{(1)} = (C^* + f''(0))\exp(-\zeta).$$

The function R_1 of the remainder term satisfies

$$-\tau R_1'' + R_1 = \tau f^{(4)}$$
$$R_1(0) = z_R^{(1)}(0), \quad R_1(1) = z_L^{(1)}(1).$$

Definitely, $z_R^{(1)}(0)$ and $z_L^{(1)}(1)$ are exponentially small. Therefore the maximum principle yields the estimate (with $M^{(4)} > |||f^{(4)}|||$)

$$|||R_1||| \leq M^{(4)}\tau \quad \text{for } \tau \leq \tau_0,$$

here $|||\cdot|||$ denotes the maximum norm. In a similar way we construct an asymptotic expansion of z_2. One gets

$$z_2 = 2f\tau + (3f'' + 2C^*)\tau^2 - z_L^{(2)}\tau^2 - z_R^{(2)}\tau^2 + R_2\tau^2 \tag{2.8}$$

with
$$z_L^{(2)} = (3f''(0) + 2C^*)e^{-\zeta} + \frac{1}{2}(f''(0) + C^*)\zeta e^{-\zeta}$$

and $|||R_2||| \leq 4M^{(4)}\tau$.

Combining (2.7), (2.8), we obtain

$$\frac{z_2 - 2z_1 + z_0}{\tau^2} = f'' - f''(0)e^{-\zeta} - f''(1)e^{-\zeta^*} - \frac{1}{2}f''(0)\zeta e^{-\zeta} - \frac{1}{2}f''(1)\zeta^* e^{-\zeta^*} -$$
$$-\left(\frac{1}{2}\zeta e^{-\zeta} + \frac{1}{2}\zeta^* e^{-\zeta^*}\right)C^* + (R_2 - 2R_1),$$

ζ^* denotes the boundary layer variable $\zeta^* = (1-x)/\sqrt{\tau}$. We have $0 \leq \zeta e^{-\zeta} \leq \frac{1}{e}$ for $\zeta \in [0,\infty)$, therefore for $i=1$ the assertion of Lemma 1 immediately follows.

Now we proceed by induction. Setting $\tilde{z}_i = \frac{-z_{i+1}+2z_i-z_{i-1}}{\tau^2}$ $(i=1,2,\ldots)$, the functions $\tilde{z}_i$ satisfy

$$-\tau(\tilde{z}_i)'' + \tilde{z}_i = \tilde{z}_{i-1} \qquad (i=2,3,\ldots) \tag{2.9}$$
$$\tilde{z}_i(0) = \tilde{z}_i(1) = 0 \ .$$

Until now we proved $\tilde{z}_1 \le \frac{1}{e} C^* - \mu(f)$. Choosing C^* in such a way that the constant $\frac{1}{e} C^* - \mu(f)$ is positive this constant is a bounding function for $\tilde{z}_i$, too. Thus, proceeding step by step, the assertion of Lemma 1 follows.

Lemma 2: One can choose C^* in such a way that the bounding property $u^{\tau,k}(x,t) \ge u(x,t)$ holds.

Lemma 2 follows from Lemma 1 and (2.4) because one has to guarantee

$$\rho_i C^* + \mu(f) \frac{t_i - t}{\tau} \ge 0 \quad \text{on } (t_{i-1}, t_i) \tag{2.10}$$

with $\rho_i = 1 - \frac{1}{e} \frac{t_i - t}{\tau}$. Obviously, it is possible to choose C^* in an appropriate way.

3. The convergence of the modified Rothe method

In this chapter we study the convergence properties of the modified Rothe method

$$\left(\frac{z_i^k - z_{i-1}^k}{\tau_k}, v\right) + a(z_i^k, v) = (f + \tau_k C^*, v) \text{ for all } v \in H_0^1(\Omega) \tag{3.1}$$

under the assumptions (A1),(A2),(A3),(i),(ii) in the n-dimensional case, C^* being an arbitrary constant.

It is necessary to place at our disposal some a-priori estimates for the solution of (3.1). Let us denote

$$Z_j^k = \frac{z_j^k - z_{j-1}^k}{\tau_k} \qquad j=1,2,\ldots,p$$
$$s_j^k = \frac{Z_j^k - Z_{j-1}^k}{\tau_k} \qquad j=2,3,\ldots,p \ .$$

Again we omit the superscript k for k=1. Let us set $\gamma = \max_{0<\tau\le\tau_0} \|f + \tau C^*\|_0$.

Lemma 3: For the solution of (3.1) the following a-priori estimates are valid:

(i) $\|z_j\|_0 \le T\gamma$, $\|z_j\|_1 \le \sqrt{\frac{2T}{\alpha}}\gamma$, (ii) $\|Z_j\|_0 \le \gamma$,

(iii) $\|s_j\|_0 \le 2C^* + \|f\|_2$.

<u>Proof:</u>

Because of $z_o = 0$ we have

$$a(z_1,v) + \frac{1}{\tau}(z_1,v) = (f+\tau C^*,v).$$

Setting $v = z_1$ it follows immediately

$$\|z_1\|_o \le \tau\gamma.$$

Subtracting from (3.1) the same identity written for i-1 we obtain

$$a(z_j-z_{j-1},v) + \frac{1}{\tau}(z_j-z_{j-1},v) = \frac{1}{\tau}(z_{j-1}-z_{j-2},v).$$

At once we get

$$\|z_j-z_{j-1}\|_o \le \|z_{j-1}-z_{j-2}\|_o$$

and therefore

$$\|z_j-z_{j-1}\|_o \le \tau\gamma.$$

Thus, we proved (ii). Further it follows immediately

$$\|z_j\|_o \le j\tau\gamma, \text{ thus } \|z_j\|_o \le T\gamma.$$

(3.1) is equivalent to the identity

$$a(z_j,v) + (Z_j,v) = (f+\tau C^*,v) \quad \forall\, v \in H_o^1(\Omega).$$

Setting $v = z_j$ we get

$$\alpha\|z_j\|_1^2 \le \|Z_j\|_o\|z_j\|_o + \gamma\|z_j\|_o,$$

implying finally

$$\|z_j\|_1 \le \sqrt{\frac{2T}{\alpha}}\,\gamma.$$

In order to verify (iii) we start from

$$a(Z_1-f,v) + \frac{1}{\tau}(Z_1-f,v) = \frac{1}{\tau}(C^*\tau,v)-a(f,v).$$

Setting $v = Z_1-f$ one obtains, taking into account (A3),(ii),

$$\|Z_1-f\|_o \le (C^*+\|f\|_2)\tau.$$

The identity

$$a(Z_2-Z_1,v) + \frac{1}{\tau}(Z_2-Z_1,v) = \frac{1}{\tau}(Z_1-f,v)-(C^*,v)$$

yields

$$\|Z_2-Z_1\|_o \le \|Z_1-f\|_o + \tau C^* \le (2C^* + \|f\|_2)\tau.$$

Analogously as for the z_j we have

$$\|Z_j-Z_{j-1}\| \le \|Z_{j-1}-Z_{j-2}\|.$$

Thus the assertion (iii) of Lemma 3 follows at once.

It is essential to note that for $\tau \le \tau_0$ all the bounds of Lemma 3 are independent of τ and only depend on τ_0. Finally we remark: what has been derived for the case of the division p_1 can be derived, in a quite similar way, for the case of an arbitrary devision p_k.
In the first step of our convergence analysis it is no problem to transfer some ideas of the proof of theorem 11.1 in [7] to our problem (3.1). In such a way it follows that $\{u^{\tau,k}(t)\}$ for $k \to \infty$ tends to the solution of the original problem (2.2) in $L^2((0,T), H_0^1(\Omega))$ and in $C((0,T), L^2(\Omega))$. It is only necessary to use the a-priori estimates of Lemma 3, the term $(C^*\tau_k, v)$ causes no difficulties with respect to the limit $k \to \infty$. Taking its pattern from [7] we also prove

<u>Theorem 1</u>: The following error estimations are valid:

$$\text{(i)} \quad \|u(t_k)-u^\tau(t_k)\|_0 \le k(\tfrac{1}{2}\|f\|_2 + 2C^*)\tau^2 \tag{3.2}$$

$$\text{(ii)} \quad \|u(t)-u^\tau(t)\|_0 \le M_1\tau \text{ with} \tag{3.3}$$

$$M_1 = (\tfrac{1}{2}\|f\|_2 + 2C^*)T + 2C^* + \|f\|_2.$$

<u>Proof</u>: We start from

$$a(z_j^1, v) + \tfrac{1}{\tau}(z_j^1 - z_{j-1}^1, v) = (f + C^*\tau, v) \qquad j=1,\dots,p$$

$$a(z_{2j}^2, v) + \tfrac{2}{\tau}(z_{2j}^2 - z_{2j-1}^2, v) = (f + C^*\tfrac{\tau}{2}, v) \quad j=1,\dots,2p.$$

Subtracting these relations it follows

$$a(z_{2j}^2 - z_j^1, v) + \tfrac{1}{\tau}(z_{2j}^2 - z_j^1 - z_{2j-2}^2 - z_{j-1}^1, v) =$$

$$= -\tfrac{1}{\tau}(z_{2j}^2 - 2z_{2j-1}^2 + z_{2j-2}^2, v) - \tfrac{1}{2}C^*(\tau, v).$$

Setting $q_j^k = z_{2j}^{k+1} - z_j^k$ one obtains

$$a(q_j^1, v) + \tfrac{1}{\tau}(q_j^1 - q_{j-1}^1) = -\tfrac{\tau}{4}(s_{2j}^2, v) - \tfrac{1}{2}C^*(\tau, v) \tag{3.4}$$

$q_0^1 = 0$ yields

$$\|q_1^1\|_0 \le \tfrac{\tau^2}{4}\|s_{2j}^2\|_0 + \tfrac{1}{2}C^*\tau^2.$$

Proceeding step by step (3.4) implies

$$\|q_j^1\|_0 \le j\,\tfrac{\tau^2}{4}\|s_{2j}^2\|_0 + \tfrac{1}{2}\,j\,C^*\tau^2.$$

According to Lemma 3 it follows

$$\|q_j^1\|_0 \le j(\tfrac{1}{4}\|f\|_2 + C^*)\tau^2.$$

Analogously we get

$$\|q_j^2\|_0 \le j(\tfrac{1}{4}\|f\|_2 + C^*)(\tfrac{\tau}{2})^2$$

and, in general,

$$\|q_j^n\|_0 \le j(\tfrac{1}{4}\|f\|_2 + C^*)(\tfrac{\tau}{2^{n-1}})^2 \qquad j=0,1,\dots,2^{n-1}p. \tag{3.5}$$

We have $t_k = t^n_{2^{n-1}k}$. Thus one obtains

$$\|u^{\tau,n}(t_k)-u^\tau(t_k)\|_0 \le \|u^{\tau,n}(t_k)-u^{\tau,n-1}(t_k)\|_0 + \dots + \|u^{\tau,2}(t_k)-u^\tau(t_k)\|$$

$$\le \sum_{j=1}^{n-1} \|q^j_{2^{j-1}k}\|_0 \le \sum_{j=1}^{n-1} 2^{j-1}k(\tfrac{1}{4}\|f\|_2 + C^*)(\tfrac{\tau}{2^{j-1}})^2$$

$$\|u^{\tau,n}(t_k)-u^\tau(t_k)\|_0 \le k(\tfrac{1}{2}\|f\|_2 + 2C^*)\tau^2 . \tag{3.6}$$

Now, the sequence $\{u^{\tau,n}(t)\}$ converges to $u(t)$ in $C((0,T),L^2(\Omega))$. Thus, the first part of Theorem 1 follows at once. Let us fix $t \in [t_k,t_{k+1}]$ and let

$$\omega^n(t) = Z^n_j \text{ for } t \in (t^n_{j-1},t^n_j).$$

Then it holds

$$\|u^{\tau,n}(t)-u^\tau(t)\|_0 = \|u^{\tau,n}(t_k) + \int_{t_k}^{t} \omega^n(\tau)d\tau - u^\tau(t_k) - \int_{t_k}^{t} \omega(\tau)d\tau\|_0$$

$$\le \|u^{\tau,n}(t_k)-u^\tau(t_k)\|_0 + \|\int_{t_k}^{t}(\omega^n(\tau)-\omega(\tau))d\tau\|_0 .$$

(3.6) and Lemma 3,(iii) yield

$$\|u^{\tau,n}(t)-u^\tau(t)\|_0 \le k(\tfrac{1}{2}\|f\|_2 + 2C^*)\tau^2 + (2C^* + \|f\|_2)\tau .$$

Taking the above mentioned convergence of the sequence $\{u^{\tau,n}(t)\}$ into consideration theorem 1 is proved.

Now, we derive an error estimation in the H^1-norm.

<u>Theorem 2</u>: The following error estimate is valid:

$$\|u(t)-u^\tau(t)\|_1 \le \sqrt{\tfrac{2\gamma}{\alpha}}\, M_1\, \tau^{\frac{1}{2}} .$$

<u>Proof</u>: Analogously to

$$u^{\tau,n}(t) = z^n_{j-1} + \frac{z^n_j - z^n_{j-1}}{\tau_n}(t-t^n_{j-1}) \text{ for } t \in [t^n_{j-1},t^n_j]$$

let us define

$$U^{\tau,n}(t) = Z^n_{j-1} + \frac{Z^n_j - Z^{n-1}_j}{\tau_n}(t-t^n_{j-1}) \text{ for } t \in [t^n_{j-1},t^n_j].$$

It holds

$$(\dot{U}^{\tau,n}(t),v) + a(u^{\tau,n}(t),v) = (f+C^*\tau,v). \tag{3.7}$$

Subtracting this identity from that for the m-th division and choosing an arbitrary but fixed t we get

$$a(u^{\tau,n}(t)-u^{\tau,m}(t),v)+(U^{\tau,n}(t)-U^{\tau,m}(t),v) = 0 .$$

Setting $v = u^{\tau,n}(t)-u^{\tau,m}(t)$ we obtain

$$\|u^{\tau,n}(t)-u^{\tau,m}(t)\|_1^2 \le \frac{1}{\alpha} \|U^{\tau,n}(t)-U^{\tau,m}(t)\|_o\|u^{\tau,n}(t)-u^{\tau,m}(t)\|_o .$$

According to Lemma 3,(ii) we have

$$\|U^{\tau,n}(t)\|_o \le (1-\frac{t-t^n_{j-1}}{\tau_n})\|z^n_{j-1}\|_o + (\frac{t-t^n_{j-1}}{\tau_n})\|z^n_j\|_o \le \gamma .$$

Thus,

$$\|u^{\tau,n}(t)-u^{\tau,m}(t)\|_1^2 \le \frac{2\gamma}{\alpha} \|u^{\tau,n}(t)-u^{\tau,m}(t)\|_o . \tag{3.8}$$

It follows that $\{u^{\tau,n}(t)\}$ is Cauchy sequence in $H^1_o(\Omega)$ uniformly on (0,T) and that $u^{\tau,n}(t)$ tends to u(t) in $C((0,T),H^1_o(\Omega))$, too. The limit $n \to \infty$ in (3.8) yields the assertion of theorem 2.

Now let us return to the one-dimensional case under the assumption (A3),(iii). Then, $|\frac{z_i-2z_{i-1}+z_{i-2}}{\tau^2}|$ is bounded and the difference $u-u^\tau$ satisfies (compare (2.4))

$$((u-u^\tau)',v) + a(u-u^\tau,v) = \tau(\phi,v) ,$$

here $|\phi|$ is bounded. Thus, the maximum principle for parabolic equations leads to

<u>Lemma 4</u>: In the one-dimensional case there exists a constant M_2 such that

$$|||u(t)-u^\tau(t)||| \le M_2\tau .$$

4. <u>Uniformly enclosing discretization of the boundary value problems generated by the modified Rothe method</u>

The solution by the modified Rothe method leads in the one-dimensional case to the problem of finding, successively for j=1,...,p, functions $z_j(x)$ with

$$-\tau z_j'' + z_j = z_{j-1}+\tau f + \tau^2 C \tag{4.1}$$

$$z_j(0) = z_j(1) = 0 ,$$

we have $z_o \equiv 0$.

In general, approximate methods are to be applied to solve the problems (4.1). We use uniformly enclosing discretization methods proposed in [4],[5] to approximate the solutions of (4.1).

The original problem in [4],[5] is an operator equation of the following form: Find $u \in H_0^1(\Omega)$ such that

$$Lu + Gu = 0 \,, \tag{4.2}$$

where $L,G : H_0^1(\Omega) \to (H_0^1(\Omega))^*$ are defined by

$$\langle Lu,v\rangle := b(u,v)$$

$$\langle Gu,v\rangle := \int_\Omega g(x,u(x))v(x)dx \,,$$

$\langle .,.\rangle$ denoting the dual pairing. $b(u,v)$ is a given continous and coercive bilinear form (let the ellipticity constant be β), g a given continuous, monotone function.
There are at least two possibilities to interprete (4.1) as a problem of the form (4.2) by setting

$$b_1(u,v) = \tau(u',v'); \; g_{1,i}(x,u) = u-(z_{i-1}(x)+\tau f(x)+\tau^2C^*) \tag{4.3}$$

or

$$b_2(u,v) = \tau(u',v')+(u,v); \; g_{2,i}(x,u) = -(z_{i-1}(x)+\tau f(x)+\tau^2C^*). \tag{4.4}$$

According to Lemma 3,(i) and the imbedding $H_0^1(\Omega) \to C^{1/2}(\bar\Omega)$ the central assumption of [5]

$$|g(x,s)-g(y,t)| \le \ell(r)(|x-y|^{\frac12}+|s-t|) \quad \forall\, x,y\in\bar\Omega,\; |s|,|t| \le r$$

is fulfilled for our functions $g_{1,i}, g_{2,i}$ with a constant ℓ independent of i.

In the case (4.3) let us set

$$z_{i,h}(x) = \sum_{\ell=1}^{N-1} w_\ell^{(1)}\varphi_\ell(x) + \sum_{i=1}^{N} w_\ell^{(2)}\psi_\ell(x) \tag{4.5}$$

with

$$\varphi_\ell(x) = \begin{cases} (x-x_{\ell-1})/h_\ell & x \in \bar\Omega_\ell \\ (x_{\ell+1}-x)/h_{\ell+1} & x \in \bar\Omega_{\ell+1} \\ 0 & \text{otherwise} \end{cases} \qquad \ell=1(1)N-1$$

and

$$\psi_\ell(x) = \begin{cases} (x-x_{\ell-1})(x_\ell-x)/2 & x \in \bar\Omega_\ell \quad \ell=1(1)N \\ 0 & \text{otherwise} \,. \end{cases}$$

$z_{i,h}(x)$ is an upper solution of (4.1) if the constants $w_\ell^{(1)}, w_\ell^{(2)}$ satisfy the nonlinear system

$$\tau w^{(2)} + \min_{y\in\bar\Omega_\ell} g_{1,i}(y,z_{i,h}(y)) = 0 \,, \qquad \ell=1(1)N$$

$$\frac{w_\ell^{(1)}-w_{\ell-1}^{(1)}}{h_\ell} - \frac{w_{\ell+1}^{(1)}-w_\ell^{(1)}}{h_{\ell+1}} - (w_\ell^{(1)}w_\ell^{(2)}+w_{\ell+1}^{(1)}w_{\ell+1}^{(2)})/2 = 0, \; \ell=1(1)N-1 \tag{4.6}$$

$$w_0^{(1)} = w_N^{(1)} = 0$$

In the case (4.4) let us set

$$z^*_{i,h}(x) = \sum_{\ell=1}^{N-1} w^{(1)}_\ell \varphi^*_\ell(x) + \sum_{i=1}^{N} w^{(2)}_\ell \phi^*_\ell(x) \tag{4.7}$$

where the functions $\varphi^*_\ell(x), \phi^*_\ell(x)$ satisfy ($Lu \equiv -\tau u'' + u$)

$$L\varphi^*_\ell = 0 \text{ in } \Omega_k \quad ; \quad L\phi^*_\ell = \delta_{\ell k} \text{ in } \Omega_k$$

$$\varphi^*_\ell(x_k) = \delta_{\ell k} \qquad \phi^*_\ell(x_k) = 0 \quad .$$

Now, $z^*_{i,h}(x)$ is an upper solution of (4.1) if the constants $w^{(1)}_\ell, w^{(2)}_\ell$ are the solutions of the linear tridiagonal system

$$w^{(2)}_\ell = \max_{y \in \bar\Omega_\ell} g_{2,i}(y)$$

$$-\frac{\mu_1}{\mu_2} w^{(1)}_{\ell-1} + \left(1 + \frac{2}{\mu_2}\right) w^{(1)}_\ell - \frac{\mu_1}{\mu_2} w^{(1)}_{\ell+1} = \frac{\mu_1 - 1}{\mu_1 + 1}\left(w^{(2)}_\ell + w^{(2)}_{\ell+1}\right)/2 \tag{4.8}$$

with $\mu_1 = \exp(h/\sqrt{\tau})$, $\mu_2 = \exp(2h/\sqrt{\tau} - 1)$. In both cases (4.3), (4.4) we obtain

<u>Lemma 5</u>: There exists some constant M_3 such that the approximate problems (4.6), (4.8) admit a solution for all $h \in (0, M_3\tau^2)$. Further, the error estimate

$$|||z_i - z_{i,h}||| \le M_4 h^{1/2}/\tau \tag{4.9}$$

holds with a constant M_4 independent of h, τ.

The proof of Lemma 5 immediately follows by analysing the general proof of the existence of the discrete solutions for uniformly enclosing discretization methods (theorem 1.1 in [5]) taking into account the validity of

$$\|z_i - z_{i,h}\| \le \frac{1}{\beta} \|G_h z_i - G z_i\|_* \ .$$

Here $\|\cdot\|_*$ denotes the norm in $(H^1_0(\Omega))^*$, the ellipticity constant β is equal to τ in our case and G_h is the simplest bounding operator defined by

$$[G_h v](x) := \min_{y \in \bar\Omega_i} g(y, v(y)) \text{ for } x \in \Omega_i .$$

Thus, it is possible to estimate $\|G_h z_i - G z_i\|_*$ taking into consideration the uniform H^1-boundness of z_i and the embedding $H^1 \to C^{1/2}(\bar\Omega)$.

In order to combine the modified Rothe method and the enclosing discretization method let us set

$$\bar u(x,t) = \sum_{i=1}^{P} z_{i,h}(x) \varphi_i(t). \tag{4.10}$$

From lemma 4 and lemma 5 one obtains using the triangle inequality

$$|||u-\bar{u}||| \leq M_2\tau + M_4 h^{1/2}/\tau .$$

Altogether we summarize

Theorem 3: Let us choose C^* in such a way, that (2.10) is fulfilled and let us control the step size h by $h \leq M_5\tau^{2+\varepsilon}$ ($\varepsilon > 0$, arbitrary). Then, our modified Rothe function (4.10) is well defined, we have

$$\bar{u}(x,t) \geq u(x,t) \quad \text{for all } x,t \in \bar{Q}$$

and $\bar{u}(x,t)$ tends for $\tau \to 0$ to the exact solution of the original problem.

References

[1] Adams,E., Spreuer,H.: Konvergente numerische Schrankenkonstruktionen mit Spline-Funktionen für nichtlineare gewöhnliche bzw. lineare parabolische Randwertaufgaben. "Intern.Math." (ed.K.Nickel) Springer Verlag, Berlin 1975.

[2] Faaß,E.: Beliebig genaue numerische Schranken für die Lösung parabolischer Randwertaufgaben. Diss., Karlsruhe 1975.

[3] Goering,H.u.a.: Singularly perturbed differential equations. Akademie-Verlag, Berlin 1983.

[4] Großmann,Ch., Krätzschmar,M., Roos,H.-G.: Gleichmäßig einschließende Diskretisierungsverfahren für schwach nichtlineare Randwertaufgaben (to appear in Num.Mathematik).

[5] Großmann,Ch., Krätzschmar,M., Roos,H.-G.: Uniformly enclosing discretization methods of high order for boundary value problems (eingereicht bei Math. of Computation)

[6] Meyn,K.H.: Monotonieaussagen für elliptische und parabolische Randwertaufgaben und Anwendung auf Finite-Elemente-Funktionen. Diss., Hamburg 1979.

[7] Rektorys,K.: The method of discretization in time and partial differential equations. Dordrecht/Boston 1982.

[8] Scheu,G.: Schrankenkonstruktionen für die Randwertaufgabe der linearen Wärmeleitungsgleichung. Diss., Karlsruhe 1973.

ON THE BEHAVIOUR OF m-STAGE ROW METHODS WITH GLOBAL ERROR ESTIMATION IN THE SMOOTH PHASE OF STIFF PROBLEMS

Siegfried Scholz [1)]

1. Introduction

In spite of the development which the construction of error b o u n d s for numerical approximations has taken in the last years the search for easily available error e s t i m a t i o n s remains important. In the numerical solution of ordinary differential equations such error estimations for discretization methods have been applied already for a long time. They are based on the assumed existence of an asymptotic expansion of the global error in powers of the discretization parameter h. For $h \to 0$ it is possible to take the first term of this expansion as a representative of the whole global error. Now, the error estimation consists in estimating this leading term. On this basis Stetter [11] developed an error estimation which for the purpose of further investigations is described briefly in section 2.

This approach may be problematical in the treatment of stiff differential equations, where the boundedness of the coefficients of the classical error expansion is no longer guaranteed. In the smooth phase of the problem, where the transient components of the solution have disappeared, a suitably chosen discretization method, e.g. the ROW methods presented in section 3, is executed with a relatively large stepsize ($h >> 0$). Therefore the assumed dominance of the leading term of the classical asymptotic error expansion is usually not given. On the contrary, because of the stiffness of the problem, the terms of the expansion may increase with increasing powers of h. Hence the basis of the described error estimation is no longer valid here.

The main purpose of this paper consists in the investigation of the behaviour of ROW methods with Stetter's global error estimation as applied to the stiff model problem (3.7). The aim is to find conditions which ensure the error estimation to be valid also in the smooth phase of the problem. For this end in the fourth section, first of all, an expansion of the global error of m-stage ROW methods in the smooth phase of the model (3.7) is given, the coefficients of which are bounded also for arbitrary stiffness and $h >> 0$. The leading term of this expansion for stepsizes h, chosen in correspondence to the smooth solution, may again serve as a representation of the global error.

---

1) Technische Universität Dresden, Sektion Mathematik, Wissenschaftsbereich Numerische Mathematik, DDR-8027 Dresden, Mommsenstraße 13

By means of this adapted expansion, in section 5 the effective error order in the smooth phase is studied. It is proven that m-stage ROW methods with order of consistency $p = m+1$ and methods with $p = m$, which satisfy the conditions of Stetters error estimation, have an effective error order not greater than 2. For the case $m = p = 2$, it was shown in [9] that Stetter's error estimation is valid also in the smooth phase of the model (3.7) if the coefficients of the method satisfy a certain "fitting condition". To get this property for $p > 2$ also, one has to choose $m > p$ and to satisfy a large number of fitting conditions for the coefficients of the method. Nevertheless for $m = 4$, $p = 3$, there exist ROW methods with Stetter's global error estimation which are valid in the transient as well as in the smooth phase.

2. Stetter's global error estimation for one-step methods

Let the given initial value problem

$$\frac{dy}{dx} = f(y),\ y(x_o) = y_o \in R^d,\ x \in [x_o, X] \tag{2.1}$$

possess the unique solution $y(x)$.
By means of a one-step method

$$y_{n+1} = y_n + h\,\emptyset\,(y_n, h),\ y_o = y(x_o),\ n=0(1)N-1, \tag{2.2}$$

one obtains approximations $y_{n+1} \approx y(x_{n+1})$ at the points $x_{n+1} = x_o+(n+1)h$, $n=0(1)N-1$, $x_N = X$, forming the grid G_N.
If f is sufficiently often differentiable and (2.2) has the order of consistency $p \geq 1$, then the local truncation error

$$LTE(x) := h\,\emptyset\,(y(x),h)-y(x+h)+y(x)$$

of the method has an asymptotic expansion

$$LTE(x) = \sum_{r=p}^{L'} h^{r+1}\varphi_r(y(x)) + O(h^{L'+2}) \tag{2.3}$$

for $h \to 0$, $x \in G_N$.

Under certain conditions on $\emptyset$ (see e.g.[12]), (2.3) implies for the global error

$$\varepsilon(x_n) := y_n - y(x_n) \tag{2.4}$$

the existence of an asymptotic expansion of the form

$$\varepsilon(x) = \sum_{r=p}^{L} h^r e_r(x) + O(h^{L+1}) \text{ for } h \to 0,\ L \leq L',\ x \in G_N \tag{2.5}$$

Here the functions e_r are independent of h and satisfy certain linear differential equations. For the function e_p, especially, there holds

$$\frac{d}{dx} e_p(x) - f_y(y(x))e_p(x) = \varphi_p(y(x)). \tag{2.6}$$

In [11] Stetter proved: If the coefficients of the method (2.2) are chosen in such a way that φ_p is built up by elementary differentials according to

$$\varphi_p = \sum_{j=1}^{n_p} \mu_j\left(\frac{d}{dy} F_j^{(p)}(y)f(y) - \frac{df}{dy}(y)F_j^{(p)}(y)\right), \tag{2.7}$$

where $F_j^{(p)}$ is the j-th elementary differential of the order p (see [1] and table 1),

then the differential equation (2.6) is exact and has the solution

$$e_p(x) = \sum_{j=1}^{n_p} \mu_j F_j^{(p)}(y(x)), \tag{2.8}$$

which, of course, has to satisfy the initial condition

$$e_p(x_o) = \sum_{j=1}^{n_p} \mu_j F_j^{(p)}(y(x_o)). \tag{2.9}$$

Now, it is not neccessary to obtain e_p by use of the elementary differentials, rather one calculates another approximation $\tilde{y}_{n+1}$ via the procedure

$$\tilde{y}_{n+1} = y_n + h\,\phi_e(y_n,h), \tag{2.10}$$

where ϕ_e is to be chosen in such a way that there holds

$$y_{n+1}-\tilde{y}_{n+1} = h^p \sum_{j=1}^{n_p} \mu_j F_j^{(p)}(y_{n+1}) + O(h^{p+1}) \quad . \tag{2.11}$$

To satisfy the initial condition (2.9), one has to start with a suitable procedure replacing (2.2),

$$y_1 = y_o + h\,\phi_s(y_o,h)$$

for which (2.9) holds.

In this way the quantity

$$y_{n+1}-\tilde{y}_{n+1} = h^p e_p(x_{n+1}) + O(h^{p+1})$$

is an adequate global error estimation for $h \to 0$.

3. ROW methods

If the problem (2.1) is stiff - this means that the solution y(x) consists of "very transient" and "very smooth" components - adapted methods have to be used for its effective treatment. Such methods are e.g. the ROW methods.

An m-stage ROW method has the form

$$y_{n+1} = y_n + h \sum_{i=1}^{m} w_i k_i(y_n), \tag{3.1a}$$

where

$$L_n k_i(y_n) = f(y_n + h\sum_{j=1}^{i-1} \alpha_{ij} k_j(y_n)) + h\, f_y(y_n) \sum_{j=1}^{i-1} \gamma_{ij} k_j(y_n), \quad i=1(1)m, \tag{3.2}$$

$$L_n = I - h\beta f_y(y_n)\ , \quad \beta \neq 0\ , \tag{3.3}$$

$f_y = \frac{df}{dy} = f'$ is the Jacobian of f, and the w_i, α_{ij}, γ_{ij}, β are parameters, characterizing the method.

The method (3.1a) has the order of consistency p if $c_j^{(r)} = 0$ for $r = 1(1)p$, where

$$\varphi_r = \sum_{j=1}^{n_r} c_j^{(r)} F_j^{(r)} \tag{3.4}$$

are the coefficients of h^{r+1} in the asymptotic expansion (2.3) of LTE(x). For r=1(1)4 the elementary differentials $F_j^{(r)}$ and the weights $c_j^{(r)}$ are given in table 1 (cf.[3],e.g.).

Table 1

order p	elementary differential	related weight
1	$F_1^{(1)} = f$	$c_1^{(1)} = \Sigma\ w_i - 1$
2	$F_1^{(2)} = f'f$	$c_1^{(2)} = \Sigma\ w_i\beta_i - (\frac{1}{2} - \beta)$
3	$F_1^{(3)} = f''(f,f)$	$c_1^{(3)} = \frac{1}{2}\Sigma\ w_i\alpha_i^2 - \frac{1}{6}$
	$F_2^{(3)} = f'f'f$	$c_2^{(3)} = \Sigma\ w_i\beta_{ij}\beta_j - (\beta^2 - \beta + \frac{1}{6})$
4	$F_1^{(4)} = f'''(f,f,f)$	$c_1^{(4)} = \frac{1}{6}\Sigma\ w_i\alpha_i^3 - \frac{1}{24}$
	$F_2^{(4)} = f''(f'f,f)$	$c_2^{(4)} = \Sigma\ w_i\alpha_i\alpha_{ij}\beta_j - (\frac{1}{8} - \frac{\beta}{3})$
	$F_3^{(4)} = f'f''(f,f)$	$c_3^{(4)} = \frac{1}{2}\Sigma\ w_i\beta_{ij}\alpha_j^2 - (\frac{1}{24} - \frac{\beta}{6})$
	$F_4^{(4)} = f'f'f'f$	$c_4^{(4)} = \Sigma\ w_i\beta_{ij}\beta_{jk}\beta_k -$
		$-(-\beta^3 + \frac{3}{2}\beta^2 - \frac{1}{2}\beta + \frac{1}{24})$

Here the sums run from 1 to m (in the sense of sum convention), and the following abbreviations have been used:

$$\beta_{ij} = \alpha_{ij} + \gamma_{ij}\ , \quad \alpha_i = \sum_{j=1}^{i-1} \alpha_{ij}, \quad \beta_i = \sum_{j=1}^{i-1} \beta_{ij}\ . \tag{3.5}$$

(Recall that $\alpha_{ij} = \gamma_{ij} = 0$ for $i \leq j$.)

For the purpose of Stetters global error estimation, in addition to y_{n+1} we calculate

$$\tilde{y}_{n+1} = y_n + h \sum_{i=1}^{m} W_i k_i(y_n),\ k_i \text{ from } (3.2). \tag{3.1b}$$

Herein the weights W_i are chosen in such a way that (2.11) holds.

As a starting procedure we use

$$y_1 = y_0 + h \sum_{i=1}^{m} (w_i + \hat{w}_i)\ k_i(y_0) \tag{3.1c}$$

instead of (3.1a), where the weights $\hat{w}_i$ are chosen in such a way that

$$h \sum_{i=1}^{m} \hat{w}_i k_i(y_0) = h^p \sum_{j=1}^{n_p} \mu_j F_j^{(p)}(y(x_0)) + O(h^{p+1}) \tag{3.6}$$

is valid.

ROW methods with built-in global error estimation of order $p = 2$ and $p = 3$ are given in [8].

In order to investigate whether this error estimation is also valid in the case of arbitrary stiffness (where the order of convergence of ε may decrease, see e.g. [7], [2]) and in the smooth phase of the stiff problem (where stepsizes $h \gg 0$ are common), we shall apply the methods (3.1) to the stiff model problem

$$y' = A(y-g(x)) + g'(x),\ y(x_0) = \eta \in R^d,\ A \in L(R^d),\ d \geq 1. \tag{3.7}$$

Here $g|R \to R^d$ is a smooth function, and (3.7) has the solution

$$y(x) = e^{xA}C + g(x),\ C = \eta - g(x_0). \tag{3.8}$$

By use of a simple technique, in the next section we shall give an error expansion the coefficients of which are bounded even in the case of arbitrary stiffness and $h \gg 0$.

4. An adapted representation of the global error and the error estimation of (3.1),(3.2) as applied to the model (3.7)

Applying (3.1),(3.2) to a non-autonomous problem we obtain the k_i via

$$L_n k_i(x_n,y_n) = f(x_n+\alpha_i h, y_n+h \sum_{j=1}^{i-1} \alpha_{ij} k_j(x_n,y_n)) + h\ \bar{b}_i f_x(x_n,y_n) + h\ f_y(x_n,y_n) \sum_{j=1}^{i-1} \gamma_{ij} k_j(x_n,y_n) \tag{4.1}$$

where $\bar{b}_i = \beta + \beta_i - \alpha_i$, α_i, β_i from (3.5). (4.2)

Proposition 1:

In the case of the model problem (3.7), hk_i has the following representation:

$$hk_i(x,y) = u_{i0}(S)(y-g(x)) + u_{i1}(S)hg'(x) + u_{i2}(S)h^2g''(x) + \sum_{k=2}^{i} v_{i,k,1}(S)(g(x)-g_k(x)) + \sum_{k=2}^{i} v_{i,k,2}(S)hg_k'(x) \ , \ i=1(1)m \tag{4.3}$$

where

$$S := (I - h\beta A)^{-1} \ , \ A \text{ from } (3.7), \tag{4.4}$$

$g_k(x) := g(x+\alpha_k h)$,

$u_{10}(S) = hAS$, $u_{11}(S) = I$, $u_{12}(S) = \beta S$, $v_{1,1,l}(S) = 0$, $l=1,2$,

and for $i=2(1)m$ (4.5)

$u_{i0}(S) = hAS\ (I + \sum_{j=1}^{i-1} \beta_{ij}u_{j0}(S))$, $u_{i1}(S) = hAS\ (\sum_{j=1}^{i-1} \beta_{ij}u_{j1}(S)-\bar{b}_iI)$,

$u_{i2}(S) = hAS \sum_{j=1}^{i-1} \beta_{ij}u_{j2}(S) + \bar{b}_iS$, $v_{i,i,1}(S) = hAS$, $v_{i,i,2}(S) = S$,

$v_{i,k,l}(S) = hAS \sum_{j=k}^{i-1} \beta_{ij}v_{j,k,l}$, $k=2(1)i-1$.

Proof: By induction on i, using (3.2) with $L_n = S^{-1}$.

Proposition 2:

The global error $\varepsilon(x_{n+1}) =: \varepsilon_{n+1}$ of the method (3.1a), applied to (3.7) with constant stepsize h, satisfies the recursion

$$\varepsilon_{n+1} = R(S)\varepsilon_n + (R(S) - e^{hA})e^{nhA}C + \ell_n \ , \tag{4.6}$$

where

$$R(S) := I + \sum_{i=1}^{m} w_iu_{i0}(S) = \hat{R}(hA) \tag{4.7}$$

denotes the stability function of the method (3.1a). Further ℓ_n is defined by

$$\ell_n := \sum_{i=1}^{m} w_i[u_{i1}(S)hg'(x_n) + u_{i2}(S)h^2g''(x_n) + \sum_{k=2}^{i} v_{i,k,1}(S)(g(x_n)-g_k(x_n)) + \sum_{k=2}^{i} v_{i,k,2}g_k'(x_n)] + g(x_n) - g(x_n+h) \tag{4.8}$$

$$= \sum_{\sigma=1}^{\infty} (\sum_{i=1}^{m} w_iK_i^{(\sigma)}(S) - \frac{1}{\sigma!} I)\ h^{\sigma}g^{(\sigma)}(x_n) \ , \tag{4.9}$$

where

$$K_i^{(\sigma)}(S) = u_{i\sigma}(S) + \sum_{k=2}^{i} (v_{i,k,2}(S) - v_{i,k,1}(S) \frac{\alpha_k}{\sigma}) \frac{\alpha_k^{\sigma-1}}{(\sigma-1)!} , \sigma \geq 1, \quad (4.10)$$

$$u_{i\sigma} = 0 \text{ for } \sigma > 2 .$$

<u>Proof:</u> With (3.8),(3.1a),(4.3) we have

$$\varepsilon_{n+1} = y_{n+1} - g(x_{n+1}) - e^{(n+1)hA}C = y_n + h \sum_{i=1}^{m} w_i k_i(x_n, y_n) - g(x_{n+1}) - e^{(n+1)hA}C$$

$$= y_n - g(x_n) - e^{nhA}C + (I - e^{hA} + \sum_{i=1}^{m} w_i u_{io}(S)) e^{nhA}C$$

$$+ \sum_{i=1}^{m} w_i u_{io}(S)(y_n - g(x_n) - e^{nhA}C) + \ell_n$$

$$= R(S)\varepsilon_n + (R(S) - e^{hA}) e^{nhA}C + \ell_n, \ \ell_n \text{ from (4.8).}$$

(4.9),(4.10) follow from (4.8) by Taylor expansion of $g(x_{n+1}), g_k(x_n), g_k'(x_n)$.

In a similar manner one finds a representation of $\tilde{y}_{n+1} - g(x_{n+1}) - e^{x_{n+1}A}C$. Upon subtraction from (4.6), one gets:

<u>Proposition 3:</u>

The error estimation according (3.1a,b) satisfies the relation

$$y_{n+1} - \tilde{y}_{n+1} = (R(S) - \tilde{R}(S))\varepsilon_n + (R(S) - \tilde{R}(S)) e^{nhA}C + \ell_n - \tilde{\ell}_n , \quad (4.11)$$

where $\tilde{R}(S)$ and $\tilde{\ell}_n$ result from $R(S)$ and ℓ_n, respectively, replacing w_i by W_i, $i=1(1)m$.

<u>Remark:</u> Because of the boundedness of the norm of S (for the eigenvalues $\varkappa_i$ of S, there holds $0 < |\varkappa_i| < 1$ for arbitrary stiffness) and of

$$hAS = \frac{1}{\beta} (S - I) , \quad (4.12)$$

it is obvious that the coefficients of $h^j g^{(j)}(x_n)$, $j=1,2,\dots$, in the expansion of ℓ_n and $\tilde{\ell}_n$ are bounded.

Assume that in the smooth phase of (3.7)

- the solution (3.8) is represented essentially by $g(x)$, i.e. $e^{nhA}C \to 0$ for $h >> 0$ and sufficiently large n;
- $\ell_n, \tilde{\ell}_n$ vary only slowly with x, such that $\ell_n \approx \ell$, $\tilde{\ell}_n \approx \tilde{\ell}$ for all sufficiently large n.

Then the following proposition holds:

Proposition 4:

If the stability function $\hat{R}(hA)$ of (3.1a) is strongly A-acceptable, then for $h \gg 0$ and large n, there hold

$$\varepsilon_{n+1} \approx -(R(S)-I)^{-1}\ell_n \tag{4.13}$$

and

$$y_{n+1}-\tilde{y}_{n+1} \approx -(R(S)-\tilde{R}(S))(R(S)-I)^{-1}\,\ell_n + \ell_n - \tilde{\ell}_n. \tag{4.14}$$

Proof: Due to the above assumption on ℓ_n, the solution of the difference equation (4.6) can be expressed by

$$\varepsilon_{n+1} \approx R^{n+1}(S)\varepsilon_o + (R^{n+1}(S)-e^{(n+1)hA})C + (R(S)-I)^{-1}(R^{n+1}(S)-I)\ell_n.$$

Because of the strong A-acceptability of $\hat{R}(hA) = R(S)$ and the stiffness of the problem, $R^n(S)$ and $e^{nhA}C$ tend to zero for large n and $h \gg 0$. Hence (4.13) follows.

Substituting ε_n from (4.13) into (4.11), just the same arguments yield (4.14).

5. The effective error order and the validity of the error estimation of (3.1a,b),(3.2) in the smooth phase of (3.7)

Now, with (2.5),(2.8) on the one hand and (4.13) on the other hand we possess two different representations of the global error. The first is suitable for the transient phase, the second for the smooth phase of the model (3.7). In order to examine if the error estimation (2.11) due to Stetter, which rests on the asymptotic expansion (2.5), is also suitable for the estimation of the error of the form (4.13), first we must get insight into the effective h-order of ε_{n+1} from (4.13). For this purpose we now study in detail the coefficients $C_\sigma(S)$ of the expansion (4.9) of

$$\ell_n = \sum_{\sigma=1}^{\infty} C_\sigma(S)h^\sigma g^{(\sigma)}(x_n), \tag{5.1}$$

where

$$C_\sigma(S) = \sum_{i=1}^{m} w_i K_i^{(\sigma)}(S) - \frac{1}{\sigma!} I, \tag{5.2}$$

$K_i^{(\sigma)}(S)$ from (4.10).

Lemma 1: For all consistent ROW methods, $C_1(S) \equiv 0$ is valid.

Proof: Due to (4.10),

$$K_i^{(1)} = u_{i1}(S) + \sum_{k=2}^{i} (v_{i,k,2}(S) - v_{i,k,1}(S)\alpha_k)$$

$$= hAS(\sum_{j=1}^{i-1} \beta_{ij} u_{j1}(S) - \bar{b}_i I) + hAS \sum_{k=2}^{i-1} \sum_{j=k}^{i-1} \beta_{ij}(v_{j,k,2}(S) - v_{j,k,1}(S)\alpha_k)$$

$$- hAS\alpha_i + S \quad , \quad \text{using (4.5)} .$$

Because of

$$\sum_{k=2}^{i-1} \sum_{j=k}^{i-1} \ldots = \sum_{j=2}^{i-1} \sum_{k=2}^{j} \ldots \quad , \tag{5.3}$$

it follows that

$$K_i^{(1)} = hAS \sum_{j=2}^{i-1} \beta_{ij}(u_{j1}(S) + \sum_{k=2}^{j} (v_{j,k,2}(S) - v_{j,k,1}(S)\alpha_k))$$

$$+ hAS(\beta_{i1}u_{11} - (\bar{b}_i + \alpha_i)I) + S$$

$$= hAS \sum_{j=2}^{i-1} \beta_{ij} K_j^{(1)} + hAS(\beta_{i1} K_1^{(1)} - (\beta + \beta_i)I) + S.$$

Using (4.12) and (3.5) we get

$$K_i^{(1)} = hAS \sum_{j=1}^{i-1} \beta_{ij}(K_j^{(1)} - I) + I .$$

From $K_1^{(1)} = I$, there follows immediately by induction that $K_i^{(1)} = I$, $i=1(1)m$. Because of $\sum_{i=1}^{m} w_i = 1$ for consistent methods (see table 1), the statement of the lemma follows from (5.2) with $\sigma=1$.

<u>Lemma 2:</u>

$$C_2(S) = \sum_{\nu=0}^{m-1} [-N^{(2)}(\nu-1) + (\beta M(\nu-1) + M(\nu))S]H^{\nu} \tag{5.4}$$

$$C_\sigma(S) = \sum_{\nu=0}^{m-1} [-N^{(\sigma)}(\nu-1) + N^{(\sigma-1)}(\nu)S]H^{\nu}, \ \sigma>2 , \tag{5.5}$$

where

$$H = hAS \quad , \tag{5.6}$$

$$M(\nu) = \sum_{i=1}^{m} w_i M_i(\nu) \quad , \quad N^{(\sigma)}(\nu) = \sum_{i=1}^{m} w_i N_i^{(\sigma)}(\nu) \quad , \quad \sigma \geq 2 , \tag{5.7}$$

and

$$M_i(\nu) = \begin{cases} 1 & \text{if } \nu \text{ is a negative integer,} \\ \beta_i & \text{if } \nu = 0, \\ \Sigma\, \beta_{ij_1}\beta_{j_1j_2} \cdots \beta_{j_{\nu-1}j_\nu}\beta_{j_\nu} & \text{if } \nu=1(1)i-2\,, \\ 0 & \text{if } \nu \geq i-1\,, \end{cases} \tag{5.8}$$

$$N_i^{(\sigma)}(\nu) = \begin{cases} \frac{1}{\sigma!} & \text{if } \nu \text{ is a negative integer,} \\ \frac{1}{\sigma!}\alpha_i^\sigma & \text{if } \nu=0, \\ \frac{1}{\sigma!}\Sigma\beta_{ij_1}\beta_{j_1j_2} \cdots \beta_{j_{\nu-1}j_\nu}\,\alpha_{j_\nu}^\sigma & \text{if } \nu=1(1)i-2\,, \\ 0 & \text{if } \nu \geq i-1\,. \end{cases} \tag{5.9}$$

<u>Proof:</u> Due to (4.10), (4.5) we have

$$K_1^{(2)}(S) = \beta S\,,$$

$$K_i^{(2)}(S) = H\sum_{j=1}^{i-1} \beta_{ij}u_{j2}(S) + \overline{b}_i S + H\sum_{k=2}^{i-1}\sum_{j=k}^{i-1} \beta_{ij}(v_{j,k,2}(S) - v_{j,k,1}(S)\frac{\alpha_k}{2})\alpha_k$$

$$+ \alpha_i S - \frac{1}{2}\alpha_i^2 H\,.$$

Using (5.3) one gets

$$K_i^{(2)}(S) = H\beta_{i1}u_{12}(S) + H\sum_{j=2}^{i-1} \beta_{ij}(u_{j2}(S) + \sum_{k=2}^{j}(v_{j,k,2}(S) - v_{j,k,1}(S)\frac{\alpha_k}{2}))\alpha_k$$

$$+ (\overline{b}_i + \alpha_i)S - \frac{1}{2}\alpha_i^2 H\,.$$

Observing (4.10), (4.2) and the identity $u_{12}(S) = \beta S = K_1^{(2)}(S)$, this yields

$$K_i^{(2)}(S) = H(\sum_{j=1}^{i-1} \beta_{ij}K_j^{(2)}(S) - \frac{1}{2}\alpha_i^2 I) + (\beta + \beta_i)S\,. \tag{5.10}$$

From (5.10) the following independent formula for $K_i^{(2)}(S)$ is available:

$$K_i^{(2)}(S) = (\beta+\beta_i)S + \sum_{\nu=1}^{i-1}[-N_i^{(2)}(\nu-1)I + (\beta M_i(\nu-1) + M_i(\nu))S]H^\nu\,, \tag{5.11}$$

which we proof by induction on i:

For i=1, (5.11) holds because of $K_1^{(2)}(S) = \beta S$ and $\beta_1=0$. Suppose, (5.11) is true for i=2(1)n. Then due to (5.10)

$$K_{n+1}^{(2)}(S) = H(\sum_{j=1}^{n} \beta_{n+1,j}K_j^{(2)}(S) - \frac{1}{2}\alpha_{n+1}^2 I) + (\beta + \beta_{n+1})S\,,$$

and with the assumption of induction, there follows

$$K_{n+1}^{(2)}(S) = H\Big(\sum_{j=1}^{n} \beta_{n+1,j}[(\beta+\beta_j)S + \sum_{\nu=1}^{j-1} \{-N_j^{(2)}(\nu-1)I+(\beta M_j(\nu-1)+M_j(\nu)S\}H^\nu$$

$$- \frac{1}{2}\alpha_{n+1}^2 I]\Big) + (\beta+\beta_{n+1})S \, .$$

Since $\frac{1}{2}\alpha_{n+1}^2 = N_{n+1}^{(2)}(0)$ and

$$\Sigma\beta_{n+1,j} = \beta_{n+1} = M_{n+1}(0) \, , \quad \Sigma\beta_{n+1,j}\beta_j = M_{n+1}(1), \tag{5.12}$$

$$\Sigma\beta_{n+1,j}N_j^{(2)}(\nu-1) = N_{n+1}^{(2)}(\nu) \, , \quad \Sigma\beta_{n+1,j}M_j(\nu-1) = M_{n+1}(\nu) \, , \tag{5.13}$$

one gets

$$K_{n+1}^{(2)}(S) = H([\beta M_{n+1}(0) + M_{n+1}(1)]S - N_{n+1}^{(2)}(0)I)$$

$$+ H\sum_{\nu=1}^{n-1}\{-N_{n+1}^{(2)}(\nu)I + (\beta M_{n+1}(\nu)+M_{n+1}(\nu+1))S\}H^\nu + (\beta+\beta_{n+1})S$$

$$= (\beta+\beta_{n+1})S + \sum_{\nu=1}^{n}\{-N_{n+1}^{(2)}(\nu-1)I + (\beta M_{n+1}(\nu-1) + M_{n+1}(\nu))S\}H^\nu \, .$$

From (5.11) using (5.7), (5.4) follows immediately.
For the proof of (5.5) one proceeds analogously. First for $K_i^{(\sigma)}$, $\sigma>2$, one gets the following representation from (4.10) via (4.5),(5.3):

$$K_i^{(\sigma)} = H\Big(\sum_{j=2}^{i-1} \beta_{ij}K_j^{(\sigma)} - \frac{\alpha_i^\sigma}{\sigma!} I\Big) + \frac{\alpha_i^{\sigma-1}}{(\sigma-1)!} S \, . \tag{5.14}$$

Using this, one proves by induction

$$K_i^{(\sigma)} = \sum_{\nu=1}^{i-1} \{-N_i^{(\sigma)}(\nu-1) + N_i^{(\sigma-1)}(\nu)S\}H^\nu + N_i^{(\sigma-1)}(0)S \, , \tag{5.15}$$

and finally with (5.2),(5.15) the statement (5.5) follows.

<u>Lemma 3</u>: For the stability function R(S) of every consistent ROW method, there holds

$$R(S) = \sum_{\nu=0}^{m} M(\nu-2)H^\nu, \tag{5.16}$$

where H and M are defined in (5.6) and (5.7), respectively.

Proof: First we prove by induction that

$$u_{io}(S) = H\{I + \sum_{\nu=o}^{i-2} M_i(\nu)H^{\nu+1}\} = \sum_{\nu=1}^{i} M_i(\nu-2)H^{\nu}, \qquad (5.17)$$

with M_i from (5.8):

For i=1,(5.17) is true because of (4.5).

Assume, (5.17) holds for i=2(1)n. Due to (4.5)

$$u_{n+1,o}(S) = H + h \sum_{j=1}^{n} \beta_{n+1,j} u_{jo}(S).$$

From this we get via the assumption of the induction

$$u_{n+1,o}(S) = H + H \sum_{j=1}^{n} \beta_{n+1,j}[H + \sum_{\nu=2}^{j} M_j(\nu-2)H^{\nu}]$$

$$= H + \sum_{j=1}^{n} \beta_{n+1,j}H^2 + \sum_{\nu=2}^{n} \sum_{j=\nu}^{n} \beta_{n+1,j}M_j(\nu-2)H^{\nu+1},$$

and with (5.8),(5.12),(5.13) we get

$$u_{n+1,o}(S) = \sum_{\nu=1}^{n+1} M_{n+1}(\nu-2)H^{\nu}.$$

The consistency of the method and (4.7),(5.17) yield immediately the validity of (5.16).

Theorem 1: Every m-stage ROW method with order of consistency p = m+1 has a non-vanishing $C_2(S)$.

Proof: Because of $S = \beta H+I$ (cf.(4.4),(5.6)), $C_2(S)$ from (5.4) can be written as

$$C_2(S) = \hat{C}(H) = \sum_{\nu=o}^{m} A_{\nu}H^{\nu},$$

where

$$\begin{aligned} A_o &= -N^{(2)}(-1) + \beta M(-1) + M(0) \\ A_{\nu} &= -N^{(2)}(\nu-1)+\beta M(\nu-1)+M(\nu)+\beta(\beta M(\nu-2)+M(\nu-1)), 1 \leq \nu \leq m-1 \\ A_m &= \beta(\beta M(m-2)+M(m-1)). \end{aligned} \qquad (5.18)$$

For $C_2(S)$ to vanish, necessarily $A_{\nu} = 0$ for ν=0(1)m.

In the case of $A_m = 0$, this implies

$$\beta M(m-2) + M(m-1) = 0 . \qquad (5.19)$$

Because of $\alpha_2 = 0$, $\beta_1 = 0$ and $\beta_{ij} = 0$ if j > i (cf.(3.2),(3.5)) we obtain due to (5.8),(5.9)

$$\left.\begin{aligned} M(\nu) &= 0 \\ N^{(\sigma)}(\nu) &= 0 \end{aligned}\right\} \text{ if } \nu \geq m-1,\ m \geq 1. \qquad (5.20)$$

Hence from (5.19), there follows

$$M(m-2) = 0 . \qquad (5.21)$$

From [4,5] one concludes that

$$M(\nu) = \rho_{\nu+2}^{(1)}(\beta) \text{ for } \nu = -1(1)p-2, \qquad (5.22)$$

$$N^{(\sigma)}(\nu) = \rho_{\nu+\sigma+1}^{(\sigma+1)}(\beta) \text{ for } \nu = 0(1)p-\sigma-1. \qquad (5.23)$$

Here $\rho_n^{(m)}$ denotes the modified Laguerre polynomial of order m and degree n-m which is given by

$$\rho_n^{(m)}(\beta) = (-1)^n \frac{(n-m)!}{n!} \beta^{n-m} \frac{d^m}{dx^m} L_n(x)\Big|_{x=1/\beta}, \qquad (5.24)$$

where $L_n(x)$ is the Laguerre polynomial of degree n.
In this notation, (5.21) is equivalent to

$$\rho_m^{(1)}(\beta) = 0 . \qquad (5.25)$$

Next we consider A_{m-1}. Observing (5.20), (5.21), (5.22), we get

$$A_{m-1} = -\rho_{m+1}^{(3)}(\beta)+\beta(\beta\rho_{m-1}^{(1)}(\beta)+\rho_m^{(1)}(\beta)) = -\rho_{m+1}^{(3)}(\beta)+\beta^2\rho_{m-1}^{(1)}(\beta).$$

In [6] the following relation is proven:

$$\rho_n^{(m+1)}(\beta) = \beta\,\rho_{n-1}^{(m)}(\beta) + \rho_n^{(m)}(\beta). \qquad (5.26)$$

Accordingly we have

$$\begin{aligned} A_{m-1} &= -(\beta\rho_m^{(2)}(\beta)+\rho_{m+1}^{(2)}(\beta)) + \beta^2\rho_{m-1}^{(1)}(\beta) \\ &= -(\beta^2\rho_{m-1}^{(1)}(\beta)+2\beta\rho_m^{(1)}(\beta)+\rho_{m+1}^{(1)}(\beta)) + \beta^2\rho_{m-1}^{(1)}(\beta) \\ &= -\rho_{m+1}^{(1)}(\beta). \end{aligned}$$

Because $\rho_m^{(1)}(\beta)$ and $\rho_{m+1}^{(1)}(\beta)$ have no common roots, A_m and A_{m-1} cannot vanish simultaneously. From this, the theorem follows.

__Remark:__ The supposition p=m+1 in the theorem is essential. E.g. for the strongly A-stable 3-stage ROW method with p=3 holds $C_2 \equiv 0$ if its coefficients satisfy

$$w_1 + w_2 + w_3 = 1 , \quad \beta^2-\beta+\tfrac{1}{6} = 0 , \quad \beta_2 = 0 , \quad w_3\beta_3 = \tfrac{1}{2} - \beta ,$$

$$\tfrac{1}{2}(w_2\alpha_2^2 + w_3\alpha_3^2) = \tfrac{1}{6} , \quad w_3\beta_{32}\alpha_2^2 = \tfrac{1}{3} - \beta .$$

Theorem 2: Every m-stage ROW method of order $p = m$, which satisfies Stetter's condition (2.7), has a non-vanishing $C_2(S)$.

Proof: Let $F_{n_p}^{(p)}$ denote the elementary differential of order p with the "tall tree", namely $F_{n_p}^{(p)} = \underbrace{f'f'\ldots f'}_{p-1}f$.

First we prove that the elementary differential $F_{n_{p+1}}^{(p+1)}$ does not appear in the representation (2.7):

Among the ones arising from $\frac{d}{dy} F_j^{(p)}$, $j=1(1)n_p$, the elementary differential $F_{n_{p+1}}^{(p+1)}$ occurs once and exactly in the expression for $\frac{d}{dy} F_{n_p}^{(p)}$. Therefore $F_{n_{p+1}}^{(p+1)}$ may only occur in the term $\mu_{n_p}(\frac{d}{dy} F_{n_p}^{(p)} - \frac{df}{dy} F_{n_p}^{(p)})$. Because of $\frac{df}{dy} F_{n_p}^{(p)} = F_{n_{p+1}}^{(p+1)}$, however, this elementary differential drops out. This means that in the representation (3.4) of φ_p, the coefficient $c_{n_{p+1}}^{(p+1)}$ of $F_{n_{p+1}}^{(p+1)}$ is zero. As is well known (cf. table 1, (5.8),(5.22)),

$$c_{n_{p+1}}^{(p+1)} = M(p-1) - \rho_{p+1}^{(1)}(\beta) .$$

Due to (5.20), $M(p-1) = M(m-1) = 0$ and consequently, $\rho_{p+1}^{(1)}(\beta) = 0$. On the other hand, for the vanishing of $C_2(S)$ in a p-stage method, it is necessary that $\rho_p^{(1)}(\beta) = 0$ (see (5.25)). Since $\rho_p^{(1)}$ and $\rho_{p+1}^{(1)}$ have no common roots, the theorem follows.

From (4.13), theorem 2 and lemma 3 we get the

Conclusion: In the smooth phase of the model problem (3.7) the effective error order of an m-stage ROW method with built-in global error estimation and order of consistency $p = m$ cannot exceed 2.

In [9], it was schown that a 2-stage ROW method with $p = 2$ yields valid error estimations in the transient as well in the smooth phase of model (3.7) if its coefficients, additionally to Stetter's condition (2.7) and the order conditions for $p = 2$ (due to table 1), satisfy the "fitting condition"

$$\frac{1}{2} w_2\alpha_{21}^2 - \frac{1}{6} = -\beta^2.$$

However, if one wants to construct ROW methods with $p > 2$ and with Stetter's built-in global error estimation, whose error order is to be preserved and correctly reproduced by the error estimation in the smooth phase of model (2.7), too, one has necessarily to choose the stage-number m larger than p. Further, besides of the usual requirements, i.e.,

(i) conditions for the order of consistency p due to table 1,

(ii) Stetter's condition (2.7),

(iii) condition (2.11) for the validity of the error estimation $y_{n+1} - \tilde{y}_{n+1}$ in the transient phase,

one has to require additionally

(iv) conditions for the vanishing of $C_2, \ldots, C_{p-1}$,

(v) "fitting conditions" for the validity of the error estimation $y_{n+1} - \tilde{y}_{n+1}$ in the smooth phase.

In the case $p = 3$, $m = 4$, the following theorem holds:

<u>Theorem 3:</u> For every $\mu \neq 0$, there exists a family of strongly A-stable 4-stage ROW methods (3.1) of order 3 with $\mu_1 = \mu_2 = \mu$ satisfying the conditions (ii) and (iii) as well as (iv) and (v). The coefficients of these methods are the following:

$\beta = 1.0685\ 7902$ (the largest root of $\rho_4^{(1)} = 0$)

$$\alpha_{21} = \alpha_2 = 3(\beta + \frac{\mu}{\beta^2}), \quad \gamma_{21} = -\alpha_{21}, \quad \beta_2 = 0,$$

$$\alpha_3 = \alpha_{31} + \alpha_{32} \quad \text{arbitrary} \neq 0, \neq \alpha_2,$$

$$\alpha_4 = \alpha_{41} + \alpha_{42} + \alpha_{43} \quad \text{arbitrary} \neq 0,$$

$$\beta_{32} = \frac{\beta^4 \alpha_3^2 r_3(\alpha_2 - \alpha_3)}{3\alpha_2^2 [\mu(r_4 - \beta^2 r_2) - \beta^4 r_3]}, \quad \text{where } r_2 := \rho_2^{(1)}, \ r_3 := \rho_3^{(1)}, \ r_4 := \rho_4^{(3)},$$

$$\beta_3 = \frac{\beta^2 \alpha_3^2 r_3(\alpha_2 - \alpha_3)}{6[\mu(r_4 - \beta^2 r_2) - \beta^4 r_3]}, \quad \beta_{31} = \beta_3 - \beta_{32}, \quad \beta_{43} \ \text{arbitrary} \neq 0,$$

$$\beta_{42} = [(\frac{2r_4}{\alpha_2^2 r_3} + \frac{(\frac{1}{3}\alpha_2 - \frac{1}{4} - 6\mu)\beta_{32}}{\alpha_3^2 r_3(\alpha_3 - \alpha_2)})\beta_3 - \frac{\alpha_3^2}{\alpha_2^2}]\beta_{43} + \frac{\alpha_4^2(\alpha_4 - \alpha_2)\beta_{32}}{\alpha_3^2(\alpha_3 - \alpha_2)}$$

$$\beta_4 = [\frac{r_2}{r_3} + \frac{(\frac{1}{3}\alpha_2 - \frac{1}{4} - 6\mu)\beta_3}{\alpha_3^2 r_3(\alpha_3 - \alpha_2)}]\beta_3\beta_{43} + \frac{\alpha_4^2(\alpha_4 - \alpha_2)\beta_3}{\alpha_3^2(\alpha_3 - \alpha_2)},$$

$$\beta_{41} = \beta_4 - \beta_{43} - \beta_{42}, \quad \gamma_{3i} = \beta_{3i} - \alpha_{3i}, \quad \gamma_{4i} = \beta_{4i} - \alpha_{4i}, \ i=1,2;$$

$$w_2 = \frac{1}{\alpha_2^2}[\frac{1}{3} - \frac{r_2\alpha_3^2}{\beta_3} - \frac{r_3}{\beta_{43}\beta_3^2}(\alpha_4^2\beta_3 - \alpha_3^2\beta_4)],$$

$$w_3 = \frac{1}{\beta_{43}\beta_3^2} (r_2\beta_{43}\beta_3 - r_3\beta_4) \ , \quad w_4 = \frac{r_3}{\beta_{43}\beta_3} \ , \quad w_1 = 1 - w_2 - w_3 - w_4 ,$$

$$W_2 = \frac{1}{\alpha_2^2} [\frac{1}{3} - 2\mu - \frac{r_2\,\alpha_3^2}{\beta_3} - \frac{(r_3 - \mu)}{\beta_{43}\beta_3^2} (\alpha_4^2\beta_3 - \alpha_3^2\beta_4)] ,$$

$$W_3 = \frac{1}{\beta_{43}\beta_3^2} [r_2\beta_{43}\beta_3 - (r_3 - \mu)\beta_4] , \quad W_4 = \frac{r_3 - \mu}{\beta_{43}\beta_3} , \quad W_1 = 1 - W_2 - W_3 - W_4 ,$$

$$\bar{w}_2 = \frac{1}{\alpha_2^2} [\frac{1}{3} + 2\mu - \frac{r_2\alpha_3^2}{\beta_3} - \frac{(r_3 + \mu)}{\beta_{43}\beta_3^2} (\alpha_4^2\beta_3 - \alpha_3^2\beta_4)] ,$$

$$\bar{w}_3 = \frac{1}{\beta_{43}\beta_3^2} [r_2\beta_{43}\beta_3 - (r_3 + \mu)\beta_4] , \quad \bar{w}_4 = \frac{r_3 + \mu}{\beta_{43}\beta_3} , \quad \bar{w}_1 = 1 - \bar{w}_2 - \bar{w}_3 - \bar{w}_4 ,$$

$$\alpha_{43} = \frac{3\mu + \frac{1}{8} - \frac{\beta}{3}}{w_4\alpha_4\beta_3} \ , \quad \gamma_{43} = \beta_{43} - \alpha_{43} \ .$$

Proof: see [10].

References

[1] Butcher, J.C.: Coefficients for the study of Runge-Kutta integration processes. J.Austral.Math.Soc. 3,185-201(1963).

[2] Frank, R., J.Schneid and C.W.Ueberhuber: Order results for implicit Runge-Kutta methods applied to stiff systems. Ber.Nr.53/82, Inst.f.Numer.Math., TU Wien (1982).

[3] Kaps, P., G.Wanner: A study of Rosenbrock-type methods of high order. Numer.Math.38, 279-298(1981).

[4] Norsett, S.P.: Multiple Padé approximations to the exponential function. Rep. Mathematics and Computation No 4/74, Dept.of Math., University of Trondheim (1974).

[5] Norsett, S.P.: Semi explicit Runge-Kutta methods. Rep. Mathematics and Computation No 6/74, Dept. of Math., University of Trondheim (1974).

[6] Norsett, S.P.: Restricted Padé approximations to the exponential function. SIAM J. Numer.Anal. 15/5, 1008-1029(1978).

[7] Prothero, A. and A.Robinson: On the stability and accuracy of one-step methods for solving stiff systems of ordinary differential equations. Math.Comp. 28, 145-162(1974).

[8] Scholz, S.: Modified Rosenbrock methods with built-in global error estimation and time-lagged Jacobian. Numerical Anal. of Selected Semilin.Differential Equations, ed. by Ch.Großmann, Th.Riedrich, H.Schönheinz, 94-104(1984).

[9] Scholz, S.: On Stetter's global error estimation in the smooth phase of stiff differential equations. Computing 36, 43-55(1986).

[10] Scholz, S.: On the existence of 4-stage ROW methods of order 3 with valid global error estimation. Preprint TU Dresden 1986.

[11] Stetter, H.J.: Local estimation of the global discretization error. SIAM J. Numer.Anal. 8, 512-523(1971).

[12] Stetter, H.J.: Analysis of Discretization Methods for Ordinary Differential Equations. Berlin-Heidelberg-New York. Springer 1973.

A POSTERIORI ERROR ESTIMATES FOR SEMILINEAR PARABOLIC EQUATIONS

F.-M. Sens [1]

Abstract

In this paper a posteriori error estimates of the space discretization error in the finite element method of lines solution of semilinear parabolic equations are analyzed. We define an asymptotically exact error estimator for the energy norm of the discretization error. The theory presented here leads to the adaptive construction of time-dependent space meshes and thereby to effective computational methods.

1. Introduction

The aim of this paper is to investigate a posteriori error estimates of the space discretization error arising from handling semilinear parabolic equations in one space variable with a special method of lines. In recent years the concept presented was introduced by Babuska, Rheinboldt and Bieterman [3,5,6] to solve linear elliptic and parabolic problems. We construct an asymptotically exact error estimator for the following type of problems.

$$u_t(t,x) - (a(x)u_x(t,x))_x = g(t,u,x) \quad x,t \in (0,1) \times (0,T)$$
$$u(0,x)=u_0(x) \;,\; u(t,0)=u(t,1)=0 \tag{1.1}$$

The discrete solution of Eqs. (1.1) will be accomplished by partitioning (0,1) according to

$$\Delta = \left\{ 0 = x_0 < \ldots < x_{N(\Delta)} = 1 \right\}$$

and determining the function $U^h(t,x)$ as the solution of a weak form of Eqs. (1.1) that is piecewise linear in x and smooth in t.

We define the local error indicators $n_j(t)$ by

$$n_j^2(t) := \frac{|x_j - x_{j-1}|^2}{12a\left(\frac{x_j + x_{j-1}}{2}\right)} \int_{x_{j-1}}^{x_j} (U_t^h(t,x) - a_x(x)U_x^h(t,x) - g(t,U^h(t,x),x))^2 dx \tag{1.2}$$

---

[1] Sektion Mathematik der Technischen Universität Dresden
8027 Dresden, Mommsenstr. 13

for $1 \leq j \leq N(\Delta)$ and each $t \in [0,T)$. Using

$$\mathcal{E}(t) := \left(\sum_{j=1}^{N(\Delta)} n_j^2(t)\right)^{1/2} \qquad t \in [0,T) \tag{1.3}$$

we get a computable error estimator for the energy norm of the space discretization error $\| u(t) - U^h(t) \|_E$. In Section 3 it will be shown that the estimator $\mathcal{E}(t)$ is asymptotically exact provided that certain conditions on the solution regularity are satisfied.

In Section 4 we discuss some kinds of the adaptive use of this error estimation and its extensions to higher dimensional problems. It will be demonstrated that the error estimator $\mathcal{E}(t)$ remains asymptotically exact for a finite number of mesh modifications and a suitable transition data choice. This way we are not only able to monitor, but also to control the discretization error.

2.A Priori Estimates

Let $I = (0,1) \subset R$, for $f,g \in C^\infty(\bar{I})$ set

$$\langle f,g \rangle := \int_I f(x)g(x)dx, \quad \| f \|_0^2 := \langle f,f \rangle.$$

The Sobolev space $H^k(I)$ ($H_0^k(I)$) is the completion of $C^\infty(\bar{I})$ ($C_0^\infty(\bar{I})$) with respect to the norm

$$\| f \|_k^2 := \sum_{i=0}^{k} \left\| \frac{d^i f}{dx^i} \right\|_0^2 \qquad k \in N .$$

W e get $H^0(I) = L_2(I)$. For arbitrary real $s > 0$, $H^s(I)$ with Hilbert space norm $\| \cdot \|_s$ is defined by interpolation as in [12]. We make use of nonisotropic Sobolev spaces $H^{2v,v}(J)$ with $J = (0,T)$. For the exact definition of these spaces we refer to [12] again.

Let $a \in C(\bar{I})$, for which

$$0 < \underline{a} \leq a(x) \leq \bar{a} \qquad \forall\, x \in \bar{I} . \tag{2.1}$$

We define the bilinear form

$$B(u,v) := \langle a u_x, v_x \rangle \qquad \text{for } u,v \in H_0^1 . \tag{2.2}$$

that will induce the differential operator $Lu(x)$ of (1.1) on $H^2 \cap H_0^1$. $\| w \|_E^2 := B(w,w)$ defines a norm equivalent to $\| \cdot \|_1$ on H_0^1 because of (2.1). The following notations are used for a partition Δ of I.

$$I_j = (x_{j-1}, x_j),\ h_j = x_j - x_{j-1} \qquad ,j=1,N(\Delta)$$
$$h(\Delta) = \max_j h_j \ ,\ \underline{h}(\Delta) = \min_j h_j \ .$$

$S(\Delta)\subset H_o^1$ is the subset of functions for each Δ whose restrictions to each I_j are linear. The operators

$$P_o : L_2 \rightarrow S(\Delta)\ ,\ P_1 : H_o^1 \rightarrow S(\Delta)$$

are defined by

$$\langle P_o u - u, f\rangle = 0 \ \forall f\in S(\Delta)\ ,\ B(P_1 u - u, f) = 0 \ \forall f\in S(\Delta)\ . \qquad (2.3)$$

The following properties of these operators are well known (cf. [1]).

<u>Lemma 1 :</u> Let P denote a family of mesh partitions of I and $a\in C^1(\bar{I})$ satisfy (2.1).

There exists a $C>0$ so that for each $\Delta\in P$

$$\|(Id-P_o)u\|_o \leq C\cdot h^b(\Delta)\|u\|_b \quad \forall u\in H^b(I)\cap H_o^1(I),\ 0\leq b\leq 2 \qquad (2.4)$$

$$\|(Id-P_1)u\|_s \leq C\cdot h^{b-s}(\Delta)\|u\|_b \quad \forall u\in H^b(I)\cap H_o^1(I),\ 0\leq s\leq 1\leq b\leq 2 \quad (2.5)$$

We shall consider the following integral form of (1.1) :

Find $u : (0,T) \longrightarrow H_o^1(I)$

$$\langle u_t(t),f\rangle + B(u(t),f) = \langle g(t,u(t)),f\rangle \quad \forall t,f\in(0,T)\times H_o^1(I)$$
$$u(0) = u_o \qquad (2.6)$$

The problem defining the discrete solution of Eqs. (1.1) is given by :

Find $U^h : (0,T) \longrightarrow S(\Delta)$

$$\langle U_t^h(t),f\rangle + B(U^h(t),f) = \langle g(t,U^h(t)),f\rangle \quad \forall t,f\in(0,T)\times S(\Delta)$$
$$U^h(0) = u_o^h\in S(\Delta) \qquad (2.7)$$

We always assume : $u_o\in H^2\cap H_o^1,\ u_o^h = P_1 u_o$.

(2.7) is equivalent to a system of nonlinear ordinary differential equations, which is uniquely solvable for reasonable data u_o, g. We assume that this system can be solved exactly. Thus we keep the analysis independent of the time discretization scheme.

Define

$$\begin{aligned}
e(t) &:= u(t)-U^h(t)\ ,\ g(e(t)) := g(t,u(t))-g(t,U^h(t))\\
p(t) &:= P_1u(t)-u(t)\ ,\ g(p(t)) := g(t,P_1u(t))-g(t,u(t)) \qquad (2.8)\\
d(t) &:= P_1u(t)-U^h(t)\ ,\ g(d(t)) := g(t,P_1u(t))-g(t,U^h(t)).
\end{aligned}$$

In order to study the a posteriori error estimator, we require estimates of the form

$$\| d(t) \|_E = \underline{\underline{O}}(h^{z_1}(\Delta)) \qquad \text{as } h(\Delta) \longrightarrow 0,$$

$$\| e(t) \|_o + \| e_t(t) \|_o = \underline{\underline{O}}(h^{z_2}(\Delta)) \qquad z_1 > 1, z_2 > 0. \tag{2.9}$$

Estimates of this type have been established by many authors. Optimal L_2- and H_o^1-error bounds for $e(t)$ are given for instance in [8,9,18].

The main result of this Section is

<u>Theorem 1</u> : Let P denote a family of partitions Δ of I,

(2.10) $a \in C^1(\bar{I})$ satisfy (2.1),

(2.11) $g_u(t,u,x), g_t(t,u,x)$ and $g(t,u,x)$ be continuous in t,x on $[0,T] \times \bar{I}$, locally Lipschitz continuous in u,

(2.12) $u \in H^{2v,v}(J) \cap H^o(J, H_o^1(I))$ for $5/2 \leq v \leq 3$.

Then a constant $C > 0$ exists such that for each $\Delta \in P$

$$\| e(t) \|_o + \| d(t) \|_E \leq C \cdot h^2(\Delta) \cdot \| u \|_{2v,v} \tag{2.13}$$

$$\| e_t(t) \|_o \leq h^{2(v-2)}(\Delta) \| u \|_{2v,v} \quad . \tag{2.14}$$

<u>Remarks :</u>

1) We always assume that a unique solution of (1.1) exists, sufficient conditions are given in [11, Theorem 5.2].

2) Under some additional conditions a priori estimates of the form (2.9) are demonstrated in [17] also for $2 \leq v < 5/2$.

Proof : We essentially follow [9,18], cf. also [19, ch.10].

Substracting Eq. (2.6) from Eq. (2.7) we get

$$\langle d_t(t), f \rangle + B(d(t), f) = \langle p_t(t), f \rangle + \langle g(e(t)), f \rangle \; \forall f \in S(\Delta)$$

Setting $f = 2d(t)$ for $0 < t < T$ we get

$$\frac{d}{dt} \| d(t) \|_o^2 + 2 \| d(t) \|_E^2 \leq c \cdot (\| p_t(t) \|_o^2 + \| p(t) \|_o^2 + \| d(t) \|_o^2) \tag{2.15}$$

and thus by application of an inequality of Gronwall type

$$\| d(t) \|_o^2 \leq c \cdot h^4(\Delta) \| u \|_{2v,v}^2 . \tag{2.16}$$

Now setting $f = 2d_t(t)$ for $0 < t < T$, we get

$$2\| d_t(t) \|_o^2 + \frac{d}{dt}\| d(t) \|_E^2 = 2\langle p_t(t) + g(e(t)), d_t(t) \rangle \ . \qquad (2.17)$$

Using d(0)=0 the integration of (2.17) on (0,t) followed by the application of the Cauchy-Schwarz inequality yields

$$\| d(t) \|_E^2 \leq 2\int_0^t \| p_t(s) + g(e(s)) \|_o^2 \, ds \ .$$

From this result and together with an extension of Lemma 1 to non-isotropic Sobolev spaces and (2.5),(2.16) we obtain

$$\| d(t) \|_E^2 \leq C \cdot h^4(\Delta) \| u \|_{2v,v}^2 \ . \qquad (2.18)$$

which completes the proof of (2.13). It remains to demonstrate (2.14). We get

$$\langle d_{tt}(t), d_t(t) \rangle + 2\| d_t(t) \|_E^2 = 2\left\{ \langle p_{tt}(t), d_t(t) \rangle + \langle \frac{d}{dt} g(e(t)), d_t(t) \rangle \right\} \qquad (2.19)$$

by differentiation of the equation given above and the choice of $f=2d_t(t)$. Taking into account well known imbedding theorems,(2.5), (2.13) and the assumption (2.12) we obtain $u_t \in C(\bar{J}, C(\bar{I}))$ and the boundedness of $U^h(t)$ in $C(\bar{I})$. This leads to

$$\langle \frac{d}{dt} g(e(t)), d_t(t) \rangle \leq C \cdot (\| e(t) \|_o^2 + \| p_t(t) \|_o^2 + \| d_t(t) \|_o^2)$$

and by integration of Eqs. (2.19) on (0,t) we get

$$\| d_t(t) \|_o^2 \leq \| d_t(0) \|_o^2 + C \cdot (\int_0^t \| p_{tt}(t) \|_o^2 + \| e(t) \|_o^2 + \| p_t(t) \|_o^2 + \| d_t(t) \|_o^2 dt) \qquad (2.20)$$

The identity $d_t(0) = P_1 u_t(0) - P_o u_t(0) - P_o g(p(0))$ leads with (2.2o) to (2.14) completing the proof of theorem 1.

3. A Posteriori Error Estimation

In this Section we extend the analysis of a posteriori estimates for the space discretization error presented by Babuska, Rheinboldt and Bieterman in [3,5,6] to the situation of semilinear parabolic equations.

Let

$$g : \Delta \longrightarrow g^h \in C(\mathfrak{I}, L_2(I))$$
$$f : \Delta \longrightarrow f^h \in C(\mathfrak{I}, S(\Delta)) .$$

We define the local error indicators

$$\eta_j^2(t,g^h,f^h) := \frac{h_j}{(12a_j)^{1/2}} {}_{I_j}\| Lf^h(t) - g^h(t) \|_0 \tag{3.1}$$

for $0 \leq t < T$, $1 \leq j \leq N(\Delta)$ and the error estimator

$$\mathcal{E}(t,g^h,f^h) := (\sum_{j=1}^{N(\Delta)} \eta_j^2(t,g^h,f^h))^{1/2} \quad , t \in [0,T) . \tag{3.2}$$

We made use of the notation $a_j = a(\frac{x_{j-1}+x_j}{2})$. ${}_{I_j}\| \cdot \|$ denotes the restriction of the norm to I_j.

In this Section we prove, that the computable quantity $\mathcal{E}(t,g(U^h)-U_t^h,U^h)$ is an asymptotically exact estimator for the error $\|u(t)-U^h(t)\|_E$. The proof is based on interpreting (1.1) as one parameter family of linear elliptic boundary value problems.

For each $t \in [0,T)$ find $u(t) \in H^2 \cap H_0^1$ satisfying

$$Lu(t) = v(t) \quad \text{with } v(t) = g(t,u(t)) - u_t(t) \quad . \tag{3.3}$$

In the above formulation we interprete $u_t(t)$ and $g(t,u(t))$ as given data. Thus we get the situation analyzed by Babuska and Rheinboldt in [3] . The corresponding discretization error is $\| p(t) \|_E$,defined as in (2.8). The associated estimator $\mathcal{E}(t,g(u)-u_t,P_1u)$ is defined according to (3.1),(3.2). This error estimator is not computable and $\| p(t) \|_E$ is not of primary interest,however we are able to ensure that up to terms of higher order in $h(\Delta)$

$$\| p(t) \|_E \sim \| e(t) \|_E \;,\; \mathcal{E}(t,g(u)-u_t,P_1u) \sim \mathcal{E}(t,g(U^h)-U_t^h,U^h) .$$

At first we present the results of the investigation of (3.3).

We introduce a norm equivalent to $\| \cdot \|_1$ on H_0^1 with $|f|_1 := \| f_x \|_0$. The following results may be formulated conveniently with a decomposition of $H_0^1(I)$ into the direct sum of $S(\Delta)$ and the $H_h^1(I)$.Here we use the notation

$$H_h^1(I) := \left\{ v \in H_0^1(I) : v(x_j) = 0 \; \forall x_j \in \Delta \right\}.$$

<u>Lemma 2 :</u> Let $a \in C^1(\bar{I})$ satisfy (2.1).We denote $b_j = \inf_{c \in R} \sup_{x \in I_j} |a(x)-c|$, $j=1,N(\Delta)$.

Then it holds for all $z \in H_h^1(I)$ and arbitrary $t \in [0,T)$:

$$|p(t)-z|_1 \leq C\,\Big(\sum_{j=1}^{N(\Delta)} {}_{I_j}|z|_1^2 \cdot b_j^2\Big)^{1/2} + \sup_{w \in H_h^1(I),\,|w|_1=1} |B(p(t)-z,w)|\,\Big) \qquad (3.4)$$

In order to prove Lemma 2 we use simple properties of the bilinear form $B(.,.)$ on $S(\Delta)$, which are given for instance in [3].

We will now specify the arbitrary function $z \in H_h^1(I)$ from Lemma 2.

Lemma 3 : Suppose that (2.10)-(2.12) hold true. Let us denote the residual of (3.3) by $r(t,x) = LP_1u(t,x) + u_t(t,x) - g(t,u(t,x),x)$ for $x \in I_j$, $t \in [0,T)$.

Then a $z=z(t) \in H_h^1(I)$ exists for all $t \in [0,T)$ so that

(3.5) $B(p(t)-z,w) = 0 \quad \forall w \in H_h^1(I)$

(3.6) $\quad {}_{I_j}|\,z(t)\,|_1 \leq \dfrac{h_j}{\underline{a}\cdot\pi}\, {}_{I_j}\|\,r(t)\,\|_0 \quad , j=1,N(\Delta)$.

Proving the asymptotic exactness of $\mathcal{E}(t,g(u)-u_t,P_1u)$ we have to require -roughly speaking- that the solution of (3.3) (and so of (1.1)) is sufficiently smooth and not relatively flat at any time $t \in [0,T)$. Therefore we define some subclasses of $H^{2v,v}(J)$ by

$$H_q^{2v,v}(J) := \Big\{ u \in H^{2v,v}(J) : \inf_{t \in J} \|\,u_{xx}(t)\,\|_0 \geq q > 0 \Big\} .$$

In [3] we find a restriction of the form $|u_{xx}(x)| \geq q > 0$ for all $x \in \bar{I}$. Our subclasses $H_q^{2v,v}(J)$ are nearly the same as the subclasses $H_{\delta_1}^{\alpha,\beta}$ defined in [5] by (3.21). Our proof does not require additional conditions as those given in [5] by (3.25),(3.26).

We also have to restrict the possible families of partitions Δ of I. P is said to be a k(r)-regular family if constants $k \geq 1, r > 0$ exist so that

$$\underline{h}(\Delta) \geq r \cdot h^k(\Delta) \qquad \forall \Delta \in P.$$

Lemma 4 : Let

(3.7) $u \in H_q^{2v,v}(J)$ with $5/2 < v \leq 3$,

(3.8) P be a k(r)-regular family of partitions with $1 \leq k < 2, \Delta \in P$,

(3.9) $a \in C^2(\bar{I})$ satisfy (2.1), g satisfy (2.11).

Then for $h(\Delta) \rightarrow 0$ the following statements hold true :

a) $\|\,p(t)\,\|_E \leq C \cdot \mathcal{E}_e(t) = \mathcal{E}(t,g(t,u(t))-u_t(t),P_1u(t))$,i.e. $\mathcal{E}_e(t)$ is an upper estimator for $\|\,p(t)\,\|_E$.

B) $\|p(t)\|_E = \mathcal{E}_e(t)\,(1 + \underline{O}(h^b(\Delta)))$ with $b=1-k/2$, i.e. $\mathcal{E}_e(t)$ is an asymptotically exact estimator for $\|p(t)\|_E$.

Proof :

Property (a) easily follows from the results of Lemmas 2,3 using the triangle inequality.

The uniquely determined function $z = z(t) \in H^1_h(I)$ of Lemma 3 can be characterized by the following problems (we often drop the variable t):

$$\left.\begin{array}{l} Lz(x) = LP_1u(x) + u_t(x) - g(u(x),x)\ ,\ x \in I_j \\ \qquad z(x_{j-1}) = z(x_j) = 0 \end{array}\right\}\ j=1,N(\Delta) \tag{3.10}$$

Using Lemmas 2,3 and the identity

$$LP_1u(x) + u_t(x) - g(u(x),x) = -a_x(x)p_x(x) + a(x)u_{xx}(x)$$

we are able to show

$$\|p(t)\|_E = \Big(\sum_{j=1}^{N(\Delta)} {}_{I_j}\| z(t)\|^2_E\Big)^{1/2}\,(1 + \underline{O}(h(\Delta)))\ . \tag{3.11}$$

Now we construct approximate formulas for the ${}_{I_j}\| z(t)\|_E$ considering :

$$\left.\begin{array}{l} Lz^{(1)}(x) = a(x)u_{xx}(x) =: r^{(1)}(x)\ ,\ x \in I_j \\ Lz^{(2)}(x) = \dfrac{1}{h_j}\displaystyle\int_{I_j} a(x)u_{xx}(x)dx =: r^{(2)}(x) \equiv r_j^{(2)},\ x \in I_j \\ \qquad z^{(i)}(x_{j-1}) = z^{(i)}(x_j) = 0\ \ ;\ i=1,2\ ,\ j=1,N(\Delta) \end{array}\right\}$$

We define the associated error estimators

$$\mathcal{E}_{(i)}(t) := \Big(\sum_{j=1}^{N(\Delta)} \frac{h_j^2}{12a_j}\ {}_{I_j}\| r^{(i)}(t)\|^2_o\Big)^{1/2}\ . \tag{3.12}$$

A sequence of five asymptotic equalities related to the decompositions of the indicators defined by (3.12) leads to the desired result together with (3.11). Now we shall have to declare

$$\left.\begin{array}{ll} \text{a)} & \mathcal{E}_1(t) = \mathcal{E}_e(t)\cdot(1 + \underline{O}(h^{1/2}(\Delta))) \\ \text{b)} & \mathcal{E}_2(t) = \mathcal{E}_1(t)\cdot(1 + \underline{O}(h^b(\Delta))) \\ \text{c)} & \|z^{(2)}(t)\|_E = \mathcal{E}_2(t)\cdot(1 + \underline{O}(h(\Delta))) \\ \text{d)} & \|z^{(1)}(t)\|_E = \|z^{(2)}(t)\|_E\cdot(1 + \underline{O}(h^b(\Delta))) \\ \text{e)} & \|z(t)\|_E = \|z^{(1)}(t)\|_E\cdot(1 + \underline{O}(h^{1/2}(\Delta))) \end{array}\right\}\ \text{as } h(\Delta) \to 0 \tag{3.13}$$

Using the validity of

$$\mathcal{E}_e(t) \leqslant \Big(\sum_{j=1}^{N(\Delta)} \frac{h_j^2}{12a_j}\Big[(1+\delta)^2 \,{}_{I_j}\| r^{(1)}\|_o^2 + (1+\frac{1}{\delta})^2 \,{}_{I_j}\| r-r^{(1)}\|_o^2 \Big]\Big)^{1/2} \tag{3.14}$$

for arbitrary $\delta > 0$, we get with

$${}_{I_j}\| r - r^{(1)}\|_o^2 \leqslant C \cdot {}_{I_j}\| p\|_E^2$$

and the special choice of $\delta = h^{1/2}(\Delta)$

$$\mathcal{E}_e(t) \leqslant \mathcal{E}_1(t)\ (1 + \underline{\underline{O}}(h^{1/2}(\Delta))) \qquad \text{for } h(\Delta) \to 0.$$

We get so the complete result (3.13,a). An analogous decomposition leads together with

$$\left| r_j^{(2)} - r^{(1)}(x)\right| \leqslant C \cdot h_j \quad \forall x \in I_j \tag{3.15}$$

to

$$\mathcal{E}_1^2(t) \leqslant (1+\delta)^2 \mathcal{E}_2^2(t) + \sum_{j=1}^{N(\Delta)} C\ (1+\frac{1}{\delta})^2 \frac{h_j^5}{12a_j} \quad . \tag{3.16}$$

$r_j^{(2)}$ is the best approximant to $r^{(1)}(x)$ in $L_2(I_j)$. Thus it holds

$$\sum_{j=1}^{N(\Delta)} h_j^2 \,{}_{I_j}\| r^{(2)}\|_o^2 \geqslant C \cdot h^{2k}(\Delta) \quad \text{for } h(\Delta) \to 0 \ . \tag{3.17}$$

Combining (3.16),(3.17) we get with $\delta = h^b(\Delta)$ (3.13,b). Taking into account that $r^{(2)}$ is constant on each I_j it can be demonstrated that

$${}_{I_j}\| z^{(2)}\|_E^2 = \frac{h_j^2}{12a_j}\, {}_{I_j}\| r^{(2)}\|_o^2\ (1 + \underline{\underline{O}}(h(\Delta))) \quad \text{for } h(\Delta) \to 0 \ .$$

The result (3.13,c) is easily to derive by summation.

We omit the proof of the relations (3.13,d,e), for it works completely analogous to that given for the first asymptotic equalities.

Now we shall characterize some important properties of the estimators $\mathcal{E}_e(.)$ and $\mathcal{E}_p(.)$ where $\mathcal{E}_p(.) := \mathcal{E}(.,g(U^h)-U_t^h,U^h)$.

<u>Lemma 5</u> : Combining the assumptions of Theorem 1 and Lemma 4 there exists a $C > 0$ so that for arbitrary $\Delta \in P$ and $t \in [0,T)$

(3.18) $\mathcal{E}_e(t) \geqslant C\, h^k(\Delta)\| u_{xx}(t)\|_o$.

(3.19) $\left|\mathcal{E}_e(t) - \mathcal{E}_p(t)\right| \leqslant C \cdot h(\Delta)\left\{\| d(t)\|_E + \| e_t(t)\|_o + \| e(t)\|_o\right\}$.

For the proof of this estimates we refer to [17] .

Now we can reveal our main theoretical result, the asymptotic exactness of the a posteriori computable estimator $\mathcal{E}(t,g(t,U(t))-U_t^h(t),U^h(t))$ of the space discretization error $\|u(t)-U^h(t)\|_E$.

<u>Theorem 2</u> : Assume (2.11), (3.7)-(3.9).

Then $\mathcal{E}(t,g(t,U^h(t))-U_t^h(t),U^h(t))$ is an asymptotic exact estimator for $\|u(t) - U^h(t)\|_E$ on 0,T), i. e.

$$\|u(t) - U^h(t)\|_E = \mathcal{E}(t,g(t,U^h(t))-U_t^h(t),U^h(t))\cdot(1 + \underline{O}(h^b(\Delta)))$$

as $h(\Delta) \to 0$, where $b=1-k/2$.

<u>Proof</u> :

With application of (3.18) we may write

$$1 - \frac{|\mathcal{E}_p(t) - \mathcal{E}_e(t)|}{\mathcal{E}_e(t)} \leq \frac{\mathcal{E}_p(t)}{\mathcal{E}_e(t)} \leq 1 + \frac{|\mathcal{E}_p(t) - \mathcal{E}_e(t)|}{\mathcal{E}_e(t)}$$

and it becomes

$$1 - \frac{\|u\|_{2v,v}}{\|u_{xx}(t)\|_o} h^{2-k}(\Delta) \leq \frac{\mathcal{E}_p(t)}{\mathcal{E}_e(t)} \leq 1 + \frac{\|u\|_{2v,v}}{\|u_{xx}(t)\|_o} h^{2-k}(\Delta) \quad (3.20)$$

after application of (3.19) and the a priori estimates of Theorem 1. The definition of the operator P_1 leads by using the triangle inequality to

$$\|p(t)\|_E \leq \|e(t)\|_E \leq \|p(t)\|_E + \|d(t)\|_E \quad .$$

Thus we get

$$\frac{\mathcal{E}_e(t)\cdot\|p(t)\|_E}{\mathcal{E}_e(t)} \leq \|e(t)\|_E \leq \mathcal{E}_e(t)\left(\frac{\|p(t)\|_E}{\mathcal{E}_e(t)} + C\frac{\|u\|_{2v,v}}{\|u_{xx}(t)\|_o}\cdot h^{2-k}(\Delta)\right) \quad (3.21)$$

with (3.18) and (2.13). From (3.20) we conclude

$$\mathcal{E}_e(t) = \mathcal{E}_p(t)\,(1 + \underline{O}(h^b(\Delta))) \quad \text{for } h(\Delta) \to 0 \quad . \quad (3.22)$$

Now we apply (3.22) to (3.21) and get

$$\| e(t) \|_E = \mathcal{E}_p(t) \frac{\| p(t) \|_E}{\mathcal{E}_e(t)} (1 + \underline{O}(h^b(\Delta))) \qquad \text{as } h(\Delta) \to 0.$$

The application of Lemma 4 directly leads to the desired result of this theorem.

4. Discussion

1) By determination of the error estimator $\mathcal{E}(t)$ we can monitor and control the space discretization error. The aim of an adaptive procedure is to control this error by refining space meshes in areas of greater solution activity. To describe such practice we shall now modify the space mesh at various times $T_k \in (0,T)$.

We define the discrete solution as in (2.7), additionally we have to determine how to choose the transition data. Let us assume that we modify the partition Δ_{k-1} in T_k to Δ_k. At this time we know the discrete solution $U^{k-1}(T_k,.)$ on Δ_{k-1}. Then we compute the FE-solution on $S(\Delta_k)$ of the linear elliptic boundary value problem

$$Lv = g(T_k, U^{k-1}(T_k)) - U_t^{k-1}(T_k) \quad .$$

We define accurately

$$B(U^k(T_k),f) = \langle g(T_k, U^{k-1}(T_k)) - U_t^{k-1}(T_k) \, , f \rangle \; \forall f \in S(\Delta_k) \quad (4.1)$$

Let us assume that all partitions Δ_k are k(r)-regular and the number of mesh modifications is finite. Then the use of the transition data definition (4.1) leads to the validity of an a priori estimate

$$\| e_t(t) \|_o + \| e(t) \|_o + \| d(t) \|_E \leq C \cdot h^{2(v-2)}(\Delta) \| u \|_{2v,v} \qquad (4.2)$$

Now we are able to apply our method of developing a posteriori estimates presented for fixed partitions in Section 3. The asymptotic exactness of $\mathcal{E}(t, g(U^h) - U_t^h, U^h)$ holds true in this case.

The restriction to a finite number of mesh modifications is not necessary in case of linear parabolic problems, however we are not able to ensure the boundedness of some constants in (4.2), if we omit this restriction.

2) A practicable method to choose the partitions Δ_k is to construct

the partitions Δ_k so that the indicators (n_j^k) ,j=1,N(Δ_k) are equilibrated. This idea is based upon the hypotheses which led to Theorem 2, and was made in view of results shown in [2] where optimal meshes for elliptic boundary value problems were characterized by asymptotic equality of the error indicators.

An adaptive procedure using implicit Adams formulas to solve the systems of ordinary differential equations was implemented and tested (cf. [16] ,[17]).The results of these numerical experiments suggest the method's applicability in all selected examples.

3) First ideas to extend this concept to the situation of higher dimensional problems are given in [4] for linear elliptic equations. In this approach error estimators based upon local indicators are defined. To determine these indicators we have to solve elliptic problems defined on smaller domains G_j. Approximations to the solutions of some auxiliary problems like (3.13,c) can not be given here, in [4] the authors suggest to solve these problems numerically with a higher level of accuracy using for instance refinements of the G_j or FE-spaces of higher order.

4) Another concept to control the space discretization error is the method of moving finite elements. This method was first presented and applied to parabolic equations in [14] where the authors used time-dependent finite element spaces, whose base functions deform continuously in time. In [10] a priori estimates for general types of continuous and discontinuous mesh modifications are given.

5) The validity of a priori estimates like (2.9) leads to an extension of Theorem 2 to the situation of parabolic equations given by (1.1) with a = a(t,x). Estimates of this type are shown for $\| e(t) \|_0$ and $\| e_t(t) \|_0$ in [13] in the special cases g(t,u,x) = 0 or u_0 = 0 and g(t,u,x) = f(t,x).

References .

[1] Babuska,I.,Aziz,A.K. : Survey lectures on the mathematical foundations of the FEM. In: The mathematical foundations of the finite element method with applications to partial differential equations Aziz,A.K. (ed.), S.1-359. Academic Press, New York 1972.

[2] Babuska,I.,Rheinboldt,W. : Analysis of optimal finite element meshes in R^1. Math. Comp. 33(1979), 435-463

[3] Babuska,I.,Rheinboldt,W. : A posteriori error analysis of finite element solutions for one-dimensional problems. SIAM J. Numer. Anal. 18(1981)3, 565-589.

[4] Babuska,I.,Rheinboldt,W. : Error estimates for adaptive finite element computations. SIAM J. Numer. Anal. 5(1978)4, 736-754.

[5] Bieterman,M.,Babuska,I. : The FEM for parabolic equations. I. A posteriori error estimation. Numerische Mathematik 40(1982)3, 340-371.
[6] Bieterman,M.,Babuska,I. : The FEM for parabolic equations. II. Adaptive approach. Numerische Mathematik 40(1982)3, 373-406.
[7] Chong,T.H. : A variable mesh finite difference method for solving a class of parabolic differential equations in one space variable SIAM J. Numer. Anal. 19(1978)4, 835-857.
[8] Dendy,J.E. : Galerkin's method for some highly nonlinear problems SIAM J. Numer. Anal. 14(1977)2, 327-378.
[9] Douglas,J.,Dupont,T. : Galerkin methods for parabolic equations. SIAM J. Numer. Anal. 7(1970)4, 575-626.
[10] Dupont,T. : Mesh modification for evolution equations. Math. Comp. 39(1982), 85-108.
[11] Ladyzhenskaya,O.,Solonnikov,V.,Uralcova ,N. : Linear and quasi-linear equations of parabolic type. Nauka, Moskva 1967 (Russian).
[12] Lions,J.L.,Magenes,E. : Nonhomogeneous boundary value problems and applications I,II. Springer, N.Y.-Heid.-B. 1973.
[13] Luskin,M.,Rademacher,R. : On the smoothing property of the Galerkin method for parabolic equations. SIAM J. Numer. Anal. 19(1981)1
[14] Miller,K.,Miller,R. : Moving finite elements I/II. SIAM J. Numer. Anal. 18(1981)4, 1019-1057.
[15] Pereyra,V.,Sewell,E.G. : Mesh selection for discrete solution of boundary problems in ODE. Numerische Mathematik 23(1975)2, 261-268.
[16] Sens,F.-M. : A posteriori Fehlerschätzung für semilineare parabolische Randwertaufgaben und adaptive Gittersteuerung in Ortsrichtung. Preprint 07-12-85, Technical University Dresden 1985.
[17] Sens,F.-M. : A posteriori Fehlerschätzung bei nichtlinearen parabolischen Problemen in einer Ortsvariablen. Diplome work, Technical University Dresden (unpublished).
[18] Wheeler,M.F. : A-priori L_2-error estimates for Galerkin approximations to parabolic partial differential equations. SIAM J. Numer. Anal. 10(1973)4, 723-759.
[19] Thomee,V.: Galerkin finite element methods for parabolic problems, LNM 1054, Springer, Berlin-Heidelberg-New York-Tokyo 1984.

ON THE CONSTRUCTION AND PACKAGE REALIZATION OF INTERVAL METHODS FOR SOLVING ORDINARY DIFFERENTIAL EQUATIONS

Yu.I.Shokin, V.A.Novikov, M.V.Bazarov[1)]

New finite-difference interval methods for solving ordinary differential systems are suggested. Described is the program package to automatize solving of ordinary differential systems intervally. Numerical experiments are included.

Since its origination the interval analysis has been succesfully applied in solving ordinary differential equations (ODE). An interest to interval methods is due to the fact that they permit not only to find a solution with a guaranteed error estimation but are also applicable when either initial conditions or right-hand side are of interval form.

In a review paper of K.Nickel [1] an account is made of some achievements of theory and practice of interval methods for solving of ODE. In particular, the author points out, that "up to now there have been no publications on interval finite-difference methods". In the present paper a review is given of some of the authors' results on construction and realization of interval-difference methods of solving the initial value problem for ODE.

Consider the initial value problem

$$y' = f(x,y), \quad y(x_0) = y_0 \in Y_0, \quad x_0 \leq x \leq x_t, \tag{1}$$

where y_0, y, f are the real n-dimensional vector functions, Y_0 is the interval n-dimensional vector.

The main facts and notions of the interval analysis [2]-[4] are supposed to be known. Unless otherwise stated, real magnitudes will be denoted by small letters and interval ones by big letters.

Concerning the function $f(x,y)$ we assume it to be defined when $x \in \Delta_x = [x_0, x_t]$, $y \in \Delta = (\Delta_1, \Delta_2, \dots \Delta_n)$, $\Delta_i = [\alpha_i, \beta_i]$, $i = \overline{1, n}$ and possesses continuous derivatives up to the required order.

Let $\psi_s(x,y(x)) = y^{(s)}(x)$, where $\psi_s(x,y(x))$ is represented by $f(x,y)$ and its derivatives up to order $S - 1$ inclusively, $F(X,Y)$, $\Psi_s(X,Y)$ be inclusion monotonic interval extensions of the functions $f(x,y)$, $\psi_s(x,y)$. Besides, $F(X,Y)$ is assumed to satisfy the Lipschitz-type condition with the constant L, i.e.

---

1) Computing Center, Siberian Branch of the USSR Academy of Sciences, 660036, Krasnoyarsk, 36, Akademgorodok

$$w(F(X,Y)) \leq l(w(X) + w(Y)), \quad w([a,b]) = b - a \geq 0.$$

Interval solutions of problem (1) are computed at the points

$$x_i = x_0 + ih, \; i = \overline{0,m}, \quad (x_m - x_0)/m = h, \; x_m \leq x_t.$$

1. One of the approaches to constructing interval analogs of the finite-difference methods is to use quadrature formulae, with the corresponding residual term being taken into account by the interval analysis means rather than being discarded (as in usual methods). Such methods are easily presented by algorithms and realized with computers. To illustrate, we consider a method based upon the formula of rectangulars and prove two typical theorems concerning inclusion of sought for solution into intervals and accurate bounds of the approximate solutions obtained.

If Y_i, $i < m$ is already known, we calculate Y_{i+1} by the formula

$$Y_{i+1} = Y_i + hF(x_i + h/2,\; Y_i + (h/2)F([x_i, x_i + h/2],$$

$$Y_i + [0,h/2]\, F([x_i, x_i + h/2], \Delta))) + \tag{2}$$

$$+ (h^3/24)\Psi_3([x_i, x_{i+1}],\; Y_i + [0,h]F([x_i, x_{i+1}], \Delta)).$$

It is also necessary for the corresponding intervals not to go beyond the bounds of domains of F and Ψ_3.

If $Y_i + [0,h]\, F([x_i, x_{i+1}], \Delta) \subset \Delta$ is valid it is not difficult to show $Y_{i+1} \subset \Delta$ because of inclusion monotoncity and properties of F and Ψ_3.

Theorem 1. Let $y(x)$ be the solution of problem (1) and intervals be calculated by formula (2). Then $y(x_i) \in Y_i$ with all i for which Y_i are determined .

Proof. We have $y(x_0) = y_0 \in Y_0$.

We shall resort to complete induction now. Let $y(x_k) \in Y_k$ for $i = k$, $k < m$. Lets prove that $y(x_{k+1}) \in Y_{k+1}$. Indeed, representation

$$y(x_{k+1}) = y(x_k) + \int_0^h y'(x_k + t)\, dt \tag{3}$$

holds for $y(x_{k+1})$.

We present the integral in the right-hand side (3) by the rectangular formula in the form

$$\int_0^h y'(x_k + t)dt = hy'(x_k + h/2) + (h^3/24)y'''(x_k + \theta h), 0 \leq \theta \leq 1. \tag{4}$$

Expressing y' and y''' by the function f and its derivatives with $y' = f(x,y)$ subject to (4), instead of (3) we obtain

$$y(x_{k+1}) = y(x_k) + hf(x_k + h/2, y(x_k + h/2)) + \\ + h^3/24\,\psi_3(x_k + \theta h, y(x_k + \theta h)) \tag{5}$$

on the other hand

$$\begin{aligned} y'(x_k + h/2) &= f(x_k + h/2,\ y(x_k + h/2)) = \\ &= f(x_k + h/2, y(x_k + (h/2)\ f\ (x_k + \theta_1 h/2,\ y(x_k) + (\theta_1 h/2)\cdot \\ &\cdot f(x_k + \theta_2\theta_1 h/2,\ y(x_k + \theta_2\theta_1 h/2)))), \\ y(x_k + \theta h) &= y(x_k) + \theta hf(x_k + \theta_3 h, y(x_k + \theta_3 h)), \\ &(\theta_1, \theta_2, \theta_3 \in [0,1]). \end{aligned} \tag{6}$$

Using now relationships (5), (6) and monotoncity of the functions $F(X,Y)$, $\Psi_3(X,Y)$ due to the indicated assumption we get $y(x_{k+1}) \in Y_{k+1}$, i.e. the theorem is proved.

The following theorem gives an estimate for the width of intervals Y_i.

Theorem 2. Let

$$w(F(X,Y)) \le l(w(X) + w(Y))$$

and $h \le h^0$,
where $h^0 > 0$ such that

$$Y_i + [0,h^0]\ F\ (\ [x_i, x_i + h^0], \Delta) \subset \Delta$$

Then

$$w(Y_i) \le S_1 h^2 + t_1 w(Y_0), \tag{7}$$

where S_1, t_1 -are constants independent of i, h and $w(Y_0)$.

Proof. Making use of the properties of the function w: $w(A \pm B) = w(A) + w(B)$, $w(A) \le 2\,|A|$
obtain

$$\begin{aligned} w(Y_{i+1}) \le w(Y_i) &+ hl\ (w(Y_i) + (h/2)l(h/2 + w(Y_i)) + \\ &+ (h/2)w(F(\Delta_x, \Delta)) + (h/2)\ F(\Delta_x, \Delta)))) + \\ &+ (h^3/24)w(\Psi_3(\Delta_x, \Delta)) \le (1 + \lambda h)w(Y_i) + \gamma h^3, \end{aligned}$$

where

$$\lambda = 1 + h^0 l^2/2,\quad \gamma = \ l^2/4(1 + w(F(\Delta_x, \Delta)) + \ |F(\Delta_x, \Delta)|\ + \\ + w(\Psi_3(\Delta_x, \Delta))/24.$$

Hence

$$w(Y_i) \le (1 + \lambda h)^i w(Y_0) + ((1 + \lambda h)^i - 1)\gamma h^3/(1 + \lambda h - 1) \le$$

$$\leq \exp(\lambda(x_t - x_0))w(Y_0) + \gamma((\exp(\lambda(x_t - x_0)) - 1)/\lambda)h^2.$$

Thus, inequality (7) does take place with $t_1 = \exp(\lambda(x_t-x_0))$, $S_1 = \gamma(t_1 - 1)/\lambda$. The theorem is proved.

Methods based on other quadrature formulae are constructed analogously.

Rested upon other quadrature formulae with their error estimates following interval-difference methods have been elaborated:

1) the method using the quadrature trapezoidal rule;
2) the method on the basis of a generalized quadrature trapezoidal rule;
3) the method on the basis of the "three-Eigth" formula (Newton's rule);
4) the method on the basis of Simpson's rule;
5) interval methods of Runge-Kutta type of the S-th order;
6) interval methods of Adams type;
7) methods on the basis of the quadrature formula of Newton-Côtes of an open type.

For methods 1) - 7), theorems of the 1 and 2 type are valid.

Along with explicit methods of type (2) implicit interval methods [5] - [6] have been constructed. Consider one of them (for simplicity consider a scalar problem). Let the interval $Y_i \ni y(x_i)$ be known. Put

$$P_i = [\underline{p}_i, \bar{p}_i] = Y_i + (h/2)F(x_i, Y_i) - h^3/12\Psi_3([x_i, x_{i+1}],$$
$$Y_i + [0,h]F([x_i, x_{i+1}], \Delta))$$

ith the quadrature trapezoidal rule it is easy to show that

$$y_{i+1} \in \tilde{Y}_{i+1} = \{z \mid z - (h/2)f(x_{i+1}, z) = p_i,\ p_i \in P_i\}.$$

To find the interval $Y_{i+1} \supset \tilde{Y}_{i+1}$ one can apply an interval variant of Newton's method or one of its modifications. If $h>0$ is small enough, namely, if

$$1 - (h/2)f'_y(x_{i+1}, y) > 0,\quad y_{i+1} \in Y_i + [0,h]F([x_i, x_{i+1}], \Delta),$$

then

$$\tilde{Y}_{i+1} = Y_{i+1} = [\underline{y}_{i+1}, \bar{y}_{i+1}],$$

where

$$\underline{y}_{i+1} - (h/2)f(x_{i+1}, \underline{y}_{i+1}) = \underline{p}_{i+1},$$

$$\bar{y}_{i+1} - (h/2)f(x_{i+1}, \bar{y}_{i+1}) = \bar{p}_{i+1}.$$

The accuracy of the method is estimated by the inequality

$w(Y_1) \leq S_2h^2 + t_2w(Y_0)$.

But if $w(\psi_3(X,Y)) \leq l(w(X) + w(Y))$, then $w(Y_{i+1}) \leq S_3h^3 + t_2w(Y_0)$.

Interval implicit methods of higher orders are constructed similarly too , for example, interval analogs of methods of Runge--Kutta and Adams type. At that there is a possibility to combine different (explicit and implicit) methods, since if the sought for solution belongs to the intervals $Y_i^{(1)}$, $Y_i^{(2)}$ ($Y_i^{(1)}$ is the interval calculated by explicit and $Y_i^{(2)}$ by implicit), then it belongs to $Y_i^{(2)} \cap Y_i^{(1)}$ too. According to the above an interval method of the predictor-corrector type has been constructed (an analog of the Milne's fourth order method).

Interval calculations with computers, however, can produce unacceptably wide intervals in some cases. This circumstance is due to the following causes: insufficient accuracy of the method applied, boundedness of the word length grid when making computations, violation of some mathematical properties in the interval arithmetic, as distinct from the ordinary one,(sub-distributivity, absence of inverse operations for addition and multiplication etc.). Besides difficulties connected with peculiarities of interval arithmetic there also exist specific problems in solving ODE, for instance, the so-called "wrapping effect" first noted by R.Moore. To decrease the noted effect various ways are suggested in works [7] - [9]. It is obvious that the above methods should be used together with procedures decreasing the "wrapping effect".

2. We now discuss questions of representing methods by algorithms. In representing interval analysis methods by algorithms considerable difficulties arise because of unpracticalness of hardware means for the account of specific requirements of interval calculus. These requirements in particular are due to the fact that when generating concrete interval algorithms there arises a necessity in procedures of machine interval arithmetic, in computing interval functions, in checking relationships occuring in the interval analysis and in some analytical transformations.

Besides, in solving problem (1) by interval methods there arises a need of constructing interval extensions of the functions $f(x,y)$, $\psi_3(x,y)$ and computing interval-valued functions. This circumstance restricts number of interval methods users to some extent.

When the question is freeing the user of a necessity to familiarize oneself with the fundamentals of the interval analysis and the making of some preparatory work by replacing them on the "shoulders" of a computer, perspective and rational in our opinion is the concept to automatize utilizing of interval algorithms, namely, the concept of creation of system or applied program packages (APP) allowing to solve problems without preliminary acquaintance with the interval analysis.

By APP here is understood a generally accepted notion of pack package, i.e. the aggregate of programs and system means necessary for solving problems of a definite class. In our case this is a class of problems which are to be solved in the framework of the interval analysis. System means are standard ones of computer operational systems.

By modulus is understood a program drawn up as a subprogram (procedure) realizing some completed fragment of computation for solving a particular problem at a computer which consists of computing expressions both with real and interval numbers. Interval numbers will be considered as a pair of real numbers.

If a hardware-controlled realization of interval operations and computations of interval functions in microprogram form are available, which decreases the time of fulfilling interval-arithmetic operations, it is evident that writing and debugging programs realizing interval algorithms is simplified. The absence of hardware--controlled realization of interval operations produces necessity to include in package some "serving modules library", which can be divided into 3 groups conditionally. The first group include modules of machine interval arithmetic, the second modules of interval computations of elementary functions. Modules of special and logical interval operations constitute the third group of the library. According to the aims of the package the library list of the serving modules may be extended or contracted.

The input language of such a package can be constructed on the basis of an high level algorithmic language supplemented by mnemonic notations (key words), with which one works in the object field under consideration. The user should be given a possibility to formulate his problem as directed by input language, in a sufficiently simple form, he is familiar with. In particular the user must have possibility to point out a necessity of solving problem intervally and get a problem's interval formulation in a readable form.

According to the user's qualification it is expedient to distinguish several levels of the language. For instance, the first, the lower level should be supposed for the users having no preliminary acquaintence with the interval analysis. Such a user, when interacting with the package, formulates the initial information about the problem as directives of the input language and the package, analizing this information, constructs an interval model of the problem by the indication of the initial information and solves the interval problem.

The second level of the language should be supposed for the users-investigators wishing not merely to solve some problem but to construct their own solver extending (if necessary) abilities of the language.

We have created one of the possible variants of such a package called INAN-1 (interval analysis -1) for solving problems (1). Besides setting initial conditions in interval form the package suggested allows also setting of right-hand side intervally.

The package INAN-1 [10] is constructed according to a module principle. Composite parts of the package are the following blocs:

- main organizing program;
- library of serving modules;
- forming bloc of interval formulation of the problem;
- output bloc of interval formulation of the problem in symbolic form;
- monitor;
- external memory.

Included in the package program, namely, in its library of functional modules, are algorithms realizing interval methods solving initial-value problem, such as methods based on the quadrature formulae, interval methods of Adams type and others.

Initial information for the package is set in special directives that are close to natural mathematical form of setting the initial problem.

In initial data defined are:

- information about the method solving the problem;
- the number of equations in the system being solved;
- the number of segment points of integration;
- integration step;
- parameters to determine the solution of the initial-value problem.

Such parameters may be:

- initial values of the independent variable x;
- initial values of the dependent variable $y = (y_1, \dots, y_n)$ in an interval form;
- left-hand and right-hand boundaries of the regions Δ_1, Δ_2, ..., Δ_n, i.e. region of defining variables $y_1, y_2, \dots, y_n$ respectively;
- right-hand sides of system (1) in the analytical form.

Example. Let an unformalized request have the following form: it is required to find interval solutions of initial-value problem for the equations $y_1^1 = y_2/x$, $y_2^1 = -xy_1$ with the initial conditions $y_1(1) \in [0.999, 1.001]$, $y_2(1) = 0$ in one point. The integration step h is equal to 0.01. A solving method is the interval Moore method of the first order, the domain of the functions $f_1 = y_2/x$ and $f_2 = -xy$ are $\Delta_1 = [-2,1]$, $\Delta_2 = [0,2]$ respectively.

This request is written in the input language of the package as follows:

```
'METHOD = IMMSYST;  NUBEQ = 2; REG = 1; STEP = 0.01;
 VALUES = 1.0, 0.999, 1.001, 0,0; DEL = -2,1,0,2;   FUNC =
```

Note, that it is necessary for the user to describe and engage only the initial information file. All other operations are formed by the package itself.

3. The efficiency of the above methods and program package has been tested for ODE system with the number of variables up to 10 taken from different fields of science and technology.

As an example consider a solution of a mathematical model of governing the glucose and free fat acids levels in blood at sugar diabetes.

The considered model is in the following form 11

$$
\begin{aligned}
y_1' &= -a_1(y_1 - y_{1F})\,(1 + r_1 v_i + r_2 v_\theta) + a_2(y_2 - y_{2F}),\\
y_2' &= a_3(y_1 - y_{1F}) - a_4(y_2 - y_{2F})(1 + r_3 v_\theta)(1 + r_3 v_i + r_4 v_\theta),
\end{aligned} \tag{8}
$$

where y_1, y_{1F} is the current physiological glucose level in blood (Mg %);

y_2, y_{2F} is the current physiological FFA level;

v_i, v_θ is the rate of injecting insulin and special kind medicine;

a_j, r_j, $j = \overline{1,4}$ are the proportionality and dimension coefficients.

It is not difficult to show the interval extensions of the right-hand sides of system (8) being inclusion monotonic and satisfing Lipschitz type condition.

Results of biological experiment have been taken as the initial data. Coefficients have the following values.

$a_1 \in [0.05,\ 0.589]$, $r_1 \in [0.09,\ 0.3]$, $r_2 \in [0.01,\ 0.102]$,

$a_2 \in [19.5,\ 22.8]$, $a_3 \in [0.01,\ 0.124]$, $a_4 \in [0.24,\ 0.33]$,

$r_3 \in [0.119,\ 0.311]$, $r_4 \in [0.565,\ 0.794]$.

Glucose and FFA physiological levels in blood are presented in interval form $y_{1F} \in [80, 110]$, $y_{2F} \in [0,4;\ 0,7]$.

System (8) with certain interval coefficients is solved by "INAN-1" (interval predictor-corrector type method is chosen).

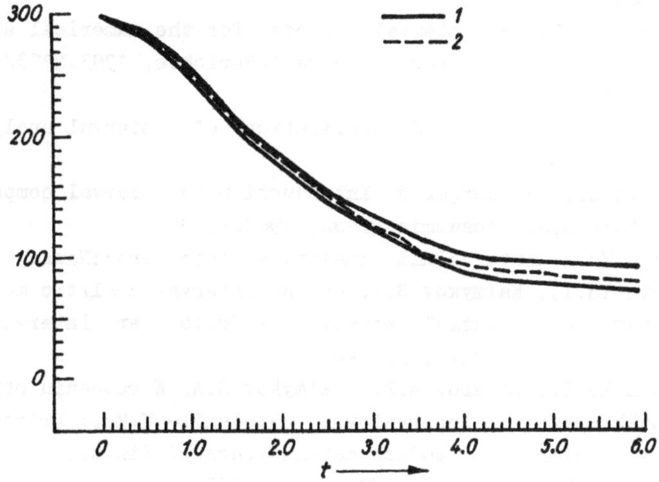

Fig.1. Comparison of model (1) and experimental (2) plots of glucose levels in blood.

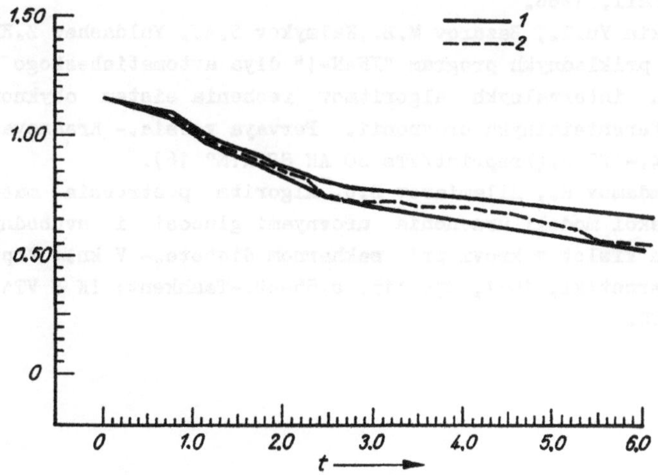

Fig.2. Comparison of model (1) and experimental (2) plots of FFA levels in blood.

The comparison of interval solutions with experimental values are presented in Figs.1,2. The computations conducted have shown a good agreement of theoretical and practical accuracy, each time double-si-

ded approximation to the exact solution being obtained.

References

1. Nickel K. Using interval methods for the numerical solution of ODE's.- Freiburger Intervall-Berichte, 1983,N°83/10, p. 13-44.
2. Moore R.E. Methods and applications of interval analysis.- - SIAM, 1979.
3. Alefeld G., Herzberger J. Introduction to interval computation. -New York e.a.: Academic Press, 1983.-333p.
4. Shokin Yu.I. Intervalnii analiz. - Novosibirsk:Nauka, 1981.
5. Shokin Yu.I., Kalmykov S.A. On the interval-analytic method for ordinary differential equations.- Freiburger Intervall-Berichte, 1982, N° 82/5, p.49-61.
6. Shokin Yu.I., Bazarov M.B., Kalmykov S.A. K resheniu obyknovennykh uravnenii intervalnymi metodami.- V kn.: Primenenie EVM v modelirovanii zadach matematicheskoi fiziki. - Krasnoyarsk, 1985, s.171-178.- VTs SO AN SSSR.
7. Harrison G.W. Compartmental models with uncertain flow rates.- Math.Biosci., 1979, v.43,p.131-139.
8. Stewart N.E. Certain equivalent requirement of approximate solutions of x = f(t,x).- SIAM J.Numer. Anal., 1970, v.7, p.256-270.
9. Moore R.E. Interval analysis.- Englewood Cliffs.-N.Y.:Prentice-Hall, 1966.
10. Shokin Yu.I., Bazarov M.B.,Kalmykov S.A., Yuldashev Z.Kh. Paket prikladnykh program "INAN-1" dlya avtomatisheskogo postroenia intervalnykh algoritmov reshenia sistem obyknovennykh differentsialnykh uravnenii. Pervaya versia.- Krasnoyarsk, 1984.- 25 s.(Preprint/VTs SO AN SSSR:N° 16).
11. Khamdamov R., Allamiarov B.U. Algoritm postroenia matematicheskoi modeli uravnenia urovnyami glucosi i svobodny zhirnykh kislot v krovi pri sakharnom diabete.- V kn.: Voptocy kibernetiki, 1981, Byp.115, s.55-62.-Tashkent: IK S VTs AN UzSSR.

MONOTONICITY AND CONTRACTIVITY PHENOMENA
IN THE NUMERICAL SOLUTION OF
STIFF DIFFERENTIAL EQUATIONS

M.N. Spijker [1)]

Abstract. In the theoretical analysis of numerical methods for solving stiff problems various stability concepts were recently introduced. This paper deals with two of these concepts, namely monotonicity and contractivity. The relevance of these two notions as well as the relations between them are discussed.

1. Introduction.

1.1. *Numerical processes for solving initial value problems*

Consider the system of ordinary differential equations

$$(1) \qquad \frac{d}{dt} U(t) = f(U(t))$$

under an initial condition $U(0) = u_0$. Here u_0 is a given vector in the n-dimensional real vectorspace $\mathbb{R}^n$ while $U(t) \in \mathbb{R}^n$ is unknown (for $t > 0$). In all of the following f is assumed to be a given continuous function from $\mathbb{R}^n$ to $\mathbb{R}^n$, $n \geq 1$.

I will deal with numerical processes which yield approximations $u_k \in \mathbb{R}^n$ to $U(k\cdot h)$, successively for $k = 1,2,3,\ldots$. Here $h > 0$ denotes the *stepsize* of the process. I assume that u_k is computed from u_{k-1} (without using the former approximations $u_{k-2}, u_{k-3}, \ldots$) .

In this paper I deal with the stability analysis of such numerical processes.

1.2. *Defining monotonicity and contractivity*

In the subsequent definitions I use so-called *pseudo-norms* $\|\cdot\|$ defined by

$$\|x\| = \sqrt{[x,x]} \qquad \text{(for all } x \in \mathbb{R}^n\text{)}.$$

Here $[\cdot,\cdot]$ stands for an arbitrary *pseudo-inner product* (i.e. $[x,y] = [y,x]$, $[\lambda x,y] = \lambda[x,y]$, $[x+y,z] = [x,z] + [y,z]$, $[x,x] \geq 0$ for all real λ and $x,y,z \in \mathbb{R}^n$)

Assume the function f and the pseudo norm $\|\cdot\|$ are such that

$$(A) \qquad \|U(t)\| \leq \|U(t-h)\| \qquad \text{(whenever } 0 < h \leq t \text{ and } U \text{ satisfies (1))}.$$

---

1) Rijksuniversiteit Leiden, Mathematisch Instituut, Niels Bohrweg 1, 2333 CA Leiden, Nederland.

In this situation it is natural to demand that the numerical process satisfies a discrete version of (A). One thus arrives at the requirement

(a) $\|u_k\| \le \|u_{k-1}\|$ (whenever u_{k-1}, u_k are related to each other by the numerical process and $h > 0$).

A numerical method is called *monotonic with respect to* f *and* $\|\cdot\|$ if requirement (a) is fulfilled.

Next assume f and $\|\cdot\|$ are such that

(B) $\|\tilde{U}(t)-U(t)\| \le \|\tilde{U}(t-h)-U(t-h)\|$ (whenever $0 < h \le t$ and $\tilde{U}, U$ satisfy (1)).

In this situation it is natural to require

(b) $\|\tilde{u}_k-u_k\| \le \|\tilde{u}_{k-1}-u_{k-1}\|$ (whenever u_{k-1}, u_k and $\tilde{u}_{k-1}, \tilde{u}_k$ are related to each other by the numerical process with the same $h > 0$).

A numerical method is defined to be *contractive with respect to* f, $\|\cdot\|$ if requirement (b) is fulfilled.

1.3. The relevance of monotonicity and contractivity

Differential equations satisfying (A) or (B) arise in many applications and comprise problems that many authors call *stiff* (cf. [2], [6], [13]). As an illustration consider the partial differential equation

$$\frac{\partial}{\partial t} v(s,t) = \alpha \cdot \frac{\partial}{\partial s} v(s,t) + \frac{\partial}{\partial s} \phi\left(\frac{\partial}{\partial s} v(s,t)\right) \qquad (0 < s < 1, t > 0)$$

under initial-boundary conditions $v(s,0) = v_0(s)$, $v(0,t) = v(1,t) = 0$. Here α is a real constant and ϕ is a given function with $\phi(0) = 0$ and continuous derivative $\phi'(\xi) \ge 0$ (for all real ξ) . Applying the method of lines (see e.g. [6]) one can arrive at a system of ordinary differential equations (1) with $y = f(x)$, for $x = (x_1, x_2, \dots, x_n)$, given by $y = (y_1, y_2, \dots, y_n)$ with

$$\begin{cases} y_i = \alpha \cdot \frac{n+1}{2} (x_{i+1}-x_{i-1}) + (n+1)\{\phi[(n+1)(x_{i+1}-x_i)]-\phi[(n+1)(x_i-x_{i-1})]\}, \\ i = 1,2,\dots,n \text{ and } x_0 = x_{n+1} = 0 . \end{cases}$$

Following well known ideas (cf. [2], [5], [6]) it can be shown that both (A) and (B) are fulfilled here with $\|x\| = \sqrt{\{(x_1)^2+(x_2)^2+\dots+(x_n)^2\}}$ (for $x = (x_1, x_2, \dots, x_n) \in \mathbb{R}^n$). A full proof is given in section 5.1.

Clearly, the more properties of the true solutions to (1) are present in the numerical approximations the better. Therefore monotonicity and contractivity are *useful demands* in the general situations (A) and (B), respectively. Further, these demands are relevant for stiff initial value problems since arbitrarily large stepsizes

$h > 0$ are included in the requirements (a), (b).

In addition to the above, contractivity is useful in that it implies a *favourable propagation of errors* in the numerical calculations. For instance let $\tilde{u}_0$ stand for a finite digit representation in a computer of the true u_0. Denoting by $\tilde{u}_k$ the approximation obtained in the presence of the rounding error $r = \tilde{u}_0 - u_0$ one sees from (b) that the pseudo norm $\|\tilde{u}_k - u_k\|$ of the propagated error $\tilde{u}_k - u_k$ never will exceed $\|r\|$.

The above concepts of monotonicity and contractivity have often been used in the recent literature (with minor modifications, cf. e.g. [1], [2], [4], [5], [6], [8], [9], [15]). However, their differences have not often been stressed. From part of the literature one might even get the impression that monotonicity and contractivity are essentially equivalent.

In the following chapters 2,3,4 I will explore the differences and relations between these two concepts for the case of Runge-Kutta methods, Rosenbrock methods and rational Runge-Kutta methods, respectively. It will appear that one should carefully distinguish between the two concepts.

The final chapter 5 contains the proofs of some statements made in the chapters 1 and 2.

2. Runge-Kutta methods.

2.1. Monotonicity implying contractivity

For given $h > 0$ and continuous $f: \mathbb{R}^n \to \mathbb{R}^n$ the general Runge-Kutta process consists in computing u_k from u_{k-1} by the formula

$$u_k = u_{k-1} + \sum_{j=1}^{m} b_j x_j \,. \tag{2.a}$$

Here $x_j \in \mathbb{R}^n$ depend on u_{k-1} in such a way that

$$x_i = h \cdot f(u_{k-1} + \sum_{j=1}^{m} a_{ij} x_j) \qquad (i=1,2,\ldots,m) \,. \tag{2.b}$$

The coefficients b_j, a_{ij} are real parameters defining the method and m is the *number of stages*. One can thus specify the method by the *coefficient scheme*

$$S = \begin{pmatrix} a_{11} & a_{12} & \cdots & a_{1m} \\ \vdots & & & \vdots \\ a_{m1} & a_{m2} & \cdots & a_{mm} \\ b_1 & b_2 & \cdots & b_m \end{pmatrix} .$$

The subsequent theorem gives an important relation between the following two properties,

(P.1) The Runge-Kutta method is *monotonic* with respect to all $f, \|\cdot\|$ satisfying (A),

(P.2) The Runge-Kutta method is *contractive* with respect to all $f, \|\cdot\|$ satisfying (B) .

Theorem 1. Property (P.1) implies (P.2).

This theorem is a variant of a very general result of Burrage and Butcher formulated in [1]. To keep the present paper sufficiently self-contained a proof of theorem 1 is given in section 5.2, by applying just the general arguments of [1] to the numerical process (2).

2.2. Does contractivity imply monotonicity?

The question arises whether property (P.2) implies (P.1).

In order to answer this question it is convenient to consider first the following (algebraic) condition on the coefficient scheme S ,

(P.3) The two $m\times m$ matrices $B = (B_{ij})$, $Q = (Q_{ij})$ with $B_{ii} = b_i$, $B_{ij} = 0$ $(i\neq j)$, $Q_{ij} = b_i a_{ij} + a_{ji} b_j - b_i b_j$ are positive semidefinite.

Theorem 2. Let S be a coefficient scheme with $(a_{i1}, a_{i2}, \ldots, a_{im}) \neq (a_{j1}, a_{j2}, \ldots, a_{jm})$ (for all $i\neq j$). Then the method specified by S has property (P.1) if and only if (P.3) holds.

For the proof of this theorem see [1] and [15].

The above question can now be settled by an analysis of the method specified by

(3) $$S = \begin{pmatrix} 1/4 & 1/4 \\ -3/4 & 5/4 \\ 1/2 & 1/2 \end{pmatrix} .$$

This coefficient scheme violates condition (P.3) and, by using theorem 2, the following result can be proved.

Theorem 3. The method specified by (3) does not have property (P.1), but it does have property (P.2).

The full proof of this theorem is given in section 5.3.

From theorem 3 one sees that, for arbitrary coefficient schemes, the answer to the above question is negative.

2.3. Equivalence of monotonicity and contractivity

The counterexample (3) considered above shows that, in general, the monotonicity property (P.1) and the contractivity property (P.2) are *not* equivalent.

The fact that the method specified by (3) has property (P.2) (although (P.3) is violated) is related to the fact that the scheme (3) is *reducible* to the coefficient scheme

(4) $$S' = \begin{pmatrix} 1/2 \\ 1 \end{pmatrix}$$

(which satisfies (P.3)). Here reducibility implies that whenever u_{k-1}, u_k satisfy (2) with coefficients (4) then these u_{k-1}, u_k also satisfy (2) with coefficients (3). For a formal definition of this concept of reducibility see [12].

Theorem 4. Let an arbitrary coefficient scheme S be given which is not reducible (in the sense of [12]). Then the three properties (P.1), (P.2), (P.3) are equivalent.

Proof. (P.1) implies (P.2) in view of theorem 1. Property (P.2) implies (P.3) by virtue of the material presented in [12]. Finally, theorem 2 shows that (P.3) implies (P.1).

This theorem thus shows that for methods specified by irreducible coefficient schemes S the monotonicity property (P.1) and the contractivity property (P.2) are equivalent.

3. A Rosenbrock method.

Consider the simple Rosenbrock method (cf. e.g. [6], [10],[13],[16],[17],[18])

(5) $$u_k = u_{k-1} + h[I-hf'(u_{k-1})]^{-1} f(u_{k-1}) .$$

Here I denotes the identity matrix and $f'(x)$ denotes the Jacobian matrix of f at x .

In the analysis of (5) I will deal with $f, \|\cdot\|$ satisfying the following condition (6).

(6a) $f: \mathbb{R}^n \to \mathbb{R}^n$; the matrix $f'(x)$ exists, is regular and depends continuously on $x \in \mathbb{R}^n$; $n \geq 1$.

(6b) $\|x\| = \sqrt{\{(x_1)^2+(x_2)^2+\ldots+(x_n)^2\}}$ (for $x = (x_1,x_2,\ldots,x_n) \in \mathbb{R}^n$) ;
$\|f'(x)^{-1}\{f'(y)-f'(x)\}\| \leq 1$ (for all $x,y \in \mathbb{R}^n$).

(6c) $f(0) = 0$; the eigenvalues of $f'(x) + [f'(x)]^T$ are ≤ 0 (for all $x \in \mathbb{R}^n$).

The inequality in (6b) refers to the spectral norm of the matrix $f'(x)^{-1}\{f'(y)-f'(x)\}$ and it can be interpreted into the requirement that the relative variation of $f'(y)$ when y varies through $\mathbb{R}^n$, does not amount to more than 100%. The requirement on the eigenvalues in (6c) implies that (B) (see section 1.2) holds, while the additional assumption $f(0) = 0$ entails property (A) (of section 1.2).

For all $h > 0$ and f satisfying (6) the process (5) is practicable since (6c) implies $[I-hf'(u_{k-1})]$ to be regular (cf. e.g. [6]).

The subsequent theorem settles the question whether the following two properties are present.

(Q.1) Method (5) is *monotonic* with respect to all $f, \|\cdot\|$ satisfying (6).

(Q.2) Method (5) is *contractive* with respect to all $f, \|\cdot\|$ satisfying (6).

Theorem 5. Method (5) has the monotonicity property (Q.1), but it does not have the contractivity property (Q.2).

This theorem reveals the surprising fact that the relation between the monotonicity property (Q.1) and the contractivity property (Q.2) is precisely opposite to the relation between the properties (P.1) and (P.2) in section 2.

Theorem 5 follows immediately from a more general theorem presented in [16].

It is worth noting that the monotonicity result formulated in theorem 5 is *optimal* in the following sense. If the inequality

$$\|f'(x)^{-1}\{f'(y)-f'(x)\}\| \le 1$$

in (6b) would be relaxed into

$$\|f'(x)^{-1}\{f'(y)-f'(x)\}\| \le \varepsilon ,$$

with any $\varepsilon > 1$, then the monotonicity property (Q.1) would no longer be valid. The latter fact follows from a counterexample presented in [11].

Interesting optimal extensions of the monotonicity property (Q.1), valid for more general methods of type

$$u_k = u_{k-1} + h[I-\theta hJ(u_{k-1})]^{-1}f(u_{k-1})$$

with $\theta \ge \frac{1}{2}$ and $J(x) \simeq f'(x)$, have recently been obtained by W.H. Hundsdorfer in [11].

4. A rational Runge-Kutta method

In this section I deal again with the inner product $[x,y] = x_1y_1 + x_2y_2 + \ldots + x_ny_n$ for $x = (x_1,x_2,\ldots,x_n)$ and $y = (y_1,y_2,\ldots,y_n)$ in $\mathbb{R}^n$.

Consider the numerical process in which u_k is calculated from

(7.a) $$u_k = u_{k-1} + h[q,q]^{-1}\{2[p,q]p-[p,p]q\} ,$$

(7.b) $$p = f(u_{k-1}) , \quad q = 2p - f(u_{k-1} + \tfrac{h}{2}p) .$$

This process was introduced by Wambecq [19] and investigated further in [8], [9], [3], [14]. It is of 2nd order and A(0)-stable (in the sense of [9]).

Following Hairer [9] I will deal with functions f of the simple linear type occurring in assumption (8).

(8) $$f(x) = Lx \quad (\text{for all } x \in \mathbb{R}^n) \quad \text{with } L = \mathrm{diag}(\lambda_1,\lambda_2,\ldots,\lambda_n) ,$$
$$\lambda_i \le 0 \quad (1 \le i \le n) ; \quad \|x\| = \sqrt{[x,x]} ; \; n \ge 1 .$$

Clearly, in the framework (8) both (A) and (B) (of section 1.2) are fulfilled.

Consider the properties

(R.1) Method (7) is *monotonic* with respect to all $f,\|\cdot\|$ satisfying (8),

(R.2) Method (7) is *contractive*, with respect to all $f,\|\cdot\|$ satisfying (8).

<u>Theorem 6.</u> Method (7) has the monotonicity property (R.1) but it does not have the contractivity property (R.2).

Property (R.1) was proved in [9], [3] while the absence of (R.2) has been proved in [16].

The above theorem is surprising in that it shows that monotonicity and contractivity fail to be equivalent even on the class of the simple linear test problems given in (8).

The absence of the contractivity property (R.2) indicates that method (7) might not be well suited for solving stiff problems. Another essential shortcoming of (7) on the class (8) was revealed by Sottas in [14].

<u>5. Proofs</u>

5.1. Proof of (A), (B) for the example of section 1.3.

Let

$$y = f(x) , \; x = (x_1,x_2,\ldots,x_n) , \; y = (y_1,y_2,\ldots,y_n)$$

as in section 1.3. Introducing the notations

$$\varepsilon = (n+1)^{-1} , \; \phi_i' = \phi'[(n+1)(x_i-x_{i-1})]$$

one has the following expressions for the partial derivatives of y_i with respect to x_j,

$$\partial y_i/\partial x_{i-1} = -\alpha(2\varepsilon)^{-1} + \varepsilon^{-2}\phi_i' \qquad (\text{for } 2 \le i \le n),$$
$$\partial y_i/\partial x_i = -\varepsilon^{-2}(\phi_{i+1}'+\phi_i') \qquad (\text{for } 1 \le i \le n),$$
$$\partial y_i/\partial x_{i+1} = \alpha(2\varepsilon)^{-1} + \varepsilon^{-2}\phi_{i+1}' \qquad (\text{for } 1 \le i \le n-1) .$$

Defining the $n\times n$ matrix $A = (a_{ij})$ by

$$a_{ij} = \tfrac{1}{2}\varepsilon^2(\partial y_i/\partial x_j+\partial y_j/\partial x_i)$$

one sees that A is symmetric and tri-diagonal with

$$a_{i,i-1} = \phi'_i \qquad \text{(for } 2 \le i \le n) ,$$
$$a_{i,i} = -(\phi'_i + \phi'_{i+1}) \qquad \text{(for } 1 \le i \le n) ,$$
$$a_{i,i+1} = \phi'_{i+1} \qquad \text{(for } 1 \le i \le n-1) .$$

By Gerschgorin's estimate there follows that all eigenvalues λ of A satisfy

$$\lambda \le 0 .$$

Consequently, the matrix $f'(x) = (\partial y_i/\partial x_j)$ has the property

$$[f'(x)v,v] \le 0 \qquad \text{(for all } v \in \mathbb{R}^n ,\ x \in \mathbb{R}^n)$$

where $[\cdot,\cdot]$ stands for the usual inner product in $\mathbb{R}^n$.

Writing $v = \tilde{U}(t) - U(t)$ (for $\tilde{U},U$ satisfying (1)) there follows

$$\frac{d}{dt}[v,v] = 2[f(\tilde{U}(t))-f(U(t)),v] = 2\int_0^1 [f'(U(t)+\tau v)\ v,v]d\tau \le 0 .$$

This proves (B) and, since $f(0) = 0$, also (A).

5.2. Proof of theorem 1.

1. Assume (P.1) and let $f,\|\cdot\|$ be such that (B) holds. Define $f^*,\|\cdot\|^*$ by

$$f^*: (x,y) \to (fx,fy) ,\ \|(x,y)\|^* = \|x-y\|$$

(for $(x,y) \in \mathbb{R}^{2n}$, $x \in \mathbb{R}^n$, $y \in \mathbb{R}^n$).

If

$$\frac{d}{dt}V(t) = f^*(V(t)) ,\ V(t) = (\tilde{U}(t),U(t))$$

then

$$\frac{d}{dt}\tilde{U}(t) = f(\tilde{U}(t)) ,\ \frac{d}{dt}U(t) = f(U(t))$$

and by (B) there follows

$$\|V(t)\|^* = \|\tilde{U}(t)-U(t)\| \le \|\tilde{U}(t-h)-U(t-h)\| = \|V(t-h)\|^* .$$

Hence $f^*,\|\cdot\|^*$ are such that (A) holds.

2. Assume

$$u_1 = u_0 + \sum_j b_j x_j \quad \text{with} \quad x_i = hf(u_0 + \sum_j a_{ij} x_j)$$

and

$$\tilde{u}_1 = \tilde{u}_0 + \sum_j b_j \tilde{x}_j \quad \text{with} \quad \tilde{x}_i = hf(\tilde{u}_0 + \sum_j a_{ij} \tilde{x}_j) \ .$$

Defining $v_k = (\tilde{u}_k, u_k)$, $y_i = (\tilde{x}_i, x_i)$ (for $k=0,1$ and $1 \le i \le m$) there follows

$$v_1 = v_0 + \sum_j b_j y_j \quad \text{with} \quad y_i = hf^*(v_0 + \sum_j a_{ij} y_j) \ .$$

In view of (P.1)

$$\|\tilde{u}_1 - u_1\| = \|v_1\|^* \le \|v_0\|^* = \|\tilde{u}_0 - u_0\| \ ,$$

which proves (P.2).

5.3. Proof of theorem 3.

Since S violates (P.3) it follows from theorem 2 that (P.1) is not fulfilled.

In order to prove (P.2) suppose $f, \|\cdot\|$ satisfy (B). From

$$\frac{d}{dt} \|\tilde{U}(t) - U(t)\|^2 = 2[f(\tilde{U}(t)) - f(U(t)), \tilde{U}(t) - U(t)]$$

one can conclude that

$$[f(\tilde{x}) - f(x), \tilde{x} - x] \le 0 \quad (\text{for all } \tilde{x}, x \in \mathbb{R}^n).$$

The relations (2.b) with S given by (3) imply

$$x_1 = hf(y_1) \ , \ x_2 = hf(y_2) \ , \ y_1 - y_2 = x_1 - x_2 \ .$$

Consequently

$$\|x_1 - x_2\|^2 = h[f(y_1) - f(y_2), y_1 - y_2] \le 0 \ ,$$

from which it follows that $x_2 = x_1 + v$ with $\|v\| = 0$. Writing $\xi = \frac{1}{2}(x_1 + x_2)$ one thus concludes from (2.b) that

$$u_k = u_{k-1} + \xi \ , \ \xi = \tfrac{1}{2}v + hf(y_1) \ , \ y_1 = u_{k-1} + \tfrac{1}{2}\xi \ .$$

Suppose also $\tilde{u}_{k-1}, \tilde{u}_k$ are related to each other by the same process. It then follows that also

$$\tilde{u}_k = \tilde{u}_{k-1} + \tilde{\xi} \ , \ \tilde{\xi} = \tfrac{1}{2}\tilde{v} + hf(\tilde{y}_1) \ , \ \tilde{y}_1 = \tilde{u}_{k-1} + \tfrac{1}{2}\tilde{\xi}$$

with $\|\tilde{v}\| = 0$.

There follows

$$\|\tilde{u}_k - u_k\|^2 - \|\tilde{u}_{k-1} - u_{k-1}\|^2 = 2[\tilde{\xi}-\xi, \tilde{u}_{k-1} - u_{k-1} + \tfrac{1}{2}(\tilde{\xi}-\xi)] =$$
$$[\tilde{v}-v, \tilde{y}_1 - y_1] + 2h[f(\tilde{y}_1) - f(y_1), \tilde{y}_1 - y_1] \leq$$
$$\leq \|\tilde{v}-v\| \cdot \|\tilde{y}_1 - y_1\| \leq (\|\tilde{v}\| + \|v\|) \cdot \|\tilde{y}_1 - y_1\| = 0 ,$$

which proves (b). The method specified by (3) thus has property (P.2).

References

[1] Burrage, K., J.C. Butcher: "Nonlinear stability of a general class of differential equations methods". BIT 20, 185-203 (1980).

[2] Butcher, J.C.: "Stability properties for a general class of methods for ordinary differential equations", SIAM J. Numer. Anal. 18, 37-44 (1981).

[3] Calvo, M., M. Mar Quemada: "On the stability of rational Runge-Kutta methods". J. CAM 8, 289-293 (1982).

[4] Cooper, G.J.: "A modification of algebraic stability for implicit Runge-Kutta methods. Submitted to JIMA, 1985.

[5] Dahlquist, G., R. Jeltsch: "Generalized disks of contractivity for explicit and implicit Runge-Kutta methods. Report TRITA-NA-7906. Dept. Comp. Sci., Roy. Inst. of Techn., Stockholm, 1979.

[6] Dekker, K., Verwer, J.G.: "Stability of Runge-Kutta methods for stiff nonlinear differential equations". Amsterdam, New York, Oxford: North Holland Publ. Comp., 1984.

[7] Frank, R., J. Schneid, C.W. Ueberhuber: "The concept of B-convergence". SIAM J. Numer. Anal. 18, 753-780 (1981).

[8] Hairer, E.: "Nonlinear stability of RAT, an explicit rational Runge-Kutta method". BIT 19, 540-542 (1979).

[9] Hairer, E.: "Unconditionally stable explicit methods for parabolic equations". Numer. Math. 35, 57-68 (1980).

[10] Hundsdorfer, W.H.: "The numerical solution of nonlinear stiff initial value problems: an analysis of one step methods". CWI Tract, Centre for Math. Comp. Science, Amsterdam, 1985.

[11] Hundsdorfer, W.H.: "A note on monotonicity of a Rosenbrock method". Report NM-R8606, Centre Math. Comp. Science, Amsterdam, 1986.

[12] Hundsdorfer, W.H., M.N. Spijker: " A note on B-stability of Runge-Kutta methods". Numer. Math. 36, 319-331 (1981).

[13] Sandberg, I.W., Shichman, H.: "Numerical integration of systems of stiff nonlinear differential equations". B.S.T.J. 47, 511-527 (1968).

[14] Sottas, G.: "Rational Runge-Kutta methods are not suitable for stiff systems of ODE's". JCAM 10, 169-174(1984).

[15] Spijker, M.N.: "Monotonicity and boundedness in implicit Runge-Kutta methods". Report Nr. 1985-15, University of Leiden, 1985, Submitted to Numer. Math..

[16] Spijker, M.N.: "A note on monotonicity and contractivity in the numerical solution of stiff differential equations". Report Nr. 1985-25, University of Leiden, 1985. Submitted to BIT.

[17] Vanselow, R.: "Stabilität und Fehleruntersuchungen bei numerischen Verfahren zur Lösung steifer nichtlinearer Anfangswertprobleme". Diplomarbeit Sektion Mathematik TU-Dresden, 1979.

[18] Verwer, J.G.: "On the practical value of the notion of BN-stability". BIT 21, 355-361 (1981).

[19] Wambecq, A.: "Rational Runge-Kutta methods for solving systems of ordinary differential equations". Computing 20, 333-342 (1978).

UNIFORMLY CONSISTENT AND CONVERGENT LINEARLY IMPLICIT RUNGE-KUTTA SCHEMES IN THE METHOD OF LINES

K. Strehmel and R. Weiner [1)]

1. Introduction

We shall be concerned with the numerical solution of initial boundary value problems for parabolic equations of the semi-linear type

$$\begin{aligned} \frac{\partial u}{\partial t} &= A_G u + Q(t,x,u) \quad \text{for} \quad x \in G, \quad 0 \leq t \leq t_e < \infty \\ u(x,t) &= \varphi(x,t), \quad x \in \partial G, \quad 0 \leq t \leq t_e \\ u(x,0) &= u_o(x), \quad x \in G \end{aligned} \tag{1.1}$$

where G is a spatial domain in R, R^2 or R^3, with boundary ∂G and A_G denotes the linear elliptic differential operator of second order

$$A_G u = \sum_{i,k=1}^{\alpha} \frac{\partial}{\partial x_i}\left(b_{ik}(x)\frac{\partial u}{\partial x_k}\right), \quad \alpha = 1,2 \text{ or } 3 .$$

The function u(x,t) may be a vector function, too. We will denote the class of problems (1.1) by F.

For the numerical solution of (1.1) we use the method of lines (MOL). In this approach the solution process is splitted in two parts, viz. the space discretization and the time integration. The discretization in space of the class F, usually performed by means of finite elements or finite differences, yields an initial value problem of ordinary differential equations

$$\begin{aligned} \frac{dU_h}{dt} &= f_h(t,U_h), \quad f_h : [0,t_e] \times R^{n_h} \to R^{n_h} \\ U_h(0) &= U_o \end{aligned} \tag{1.2}$$

where

$$f_h(t,U_h) := A_h U_h + r_h(t) + Q_h(t,U_h) . \tag{1.3}$$

Here h denotes the grid parameter which may vary over the spatial grid. $U_h(t)$ is a n_h-dimensional real vector and represents an approximation for $u_h(t)$ where $u_h(t)$ is the restriction of the exact solution u(x,t) of (1.1) to the spatial grid. The time independent matrix A_h originates from the elliptic operator A_G, the vector $r_h(t)$ arises from A_G and from the boundary function $\varphi(x,t)$ and $Q_h(t,U_h)$ from the source term Q(t,x,u). It will always be assumed that the semi-discrete parabolic problem (1.2) is consistent with (1.1), that means

$$\max_{0 \leq t \leq t_e} \|\alpha_h(t)\| \to 0 \quad \text{as} \quad \varrho(h) \to 0$$

---

1) Martin-Luther-Universität Halle-Wittenberg, Sektion Mathematik, Weinbergweg 17, 4050 Halle, GDR

where $\varphi(h)$ stands for a distance function in the grid and $\alpha_h(t)$ denotes the space truncation error defined by

$$\alpha_h(t) = f_h(t,u_h(t)) - \dot{u}_h(t) \tag{1.4}$$

(see Verwer/Sanz-Serna [16]). Furthermore, we suppose that the problem (1.1) and (1.2) always possesses an unique solution $u_h(t)$ and $U_h(t)$, respectively.

In this paper we investigate linearly implicit Runge-Kutta discretizations (LIRK-methods) of problems on F. We derive conditions for LIRK-methods to be uniformly consistent and uniformly convergent, that means that the bounds of the full local and full global error with respect to the solution $u_h(t)$ do not depend on the grid distance in space and time under the assumption that the grid distances are independent parameters. These conditions allow to construct uniformly consistent LIRK-methods of arbitrary order. The uniform order is in general smaller than the conventional order of the LIRK-method. This phenomenon is called order reduction. An interesting feature of these LIRK-methods is that for a wide subclass from F the reduction for the global error is less than for the local error, although it still may be considerable when it occurs. Finally we give some numerical results which nicely support our theoretical results.

In the MOL literature the phenomenon of order reduction was also observed by Verwer [15], Sanz-Serna/Verwer/Hundsdorfer [11]. The contribution [15] deals with diagonally implicit Runge-Kutta methods and the contribution [11] with explicit Runge-Kutta schemes.

When LIRK-methods are applied to stiff systems of ordinary differential equations, not necessarily semi-discrete partial differential equations, these methods also suffer from order reduction. This is the central theme of B-consistency and B-convergence theory which was developed by Frank/Schneid/Ueberhuber [3],[4]. In this framework Hundsdorfer[7] considers LIRK-methods for semi-linear stiff problems. He shows that under natural assumptions a LIRK-method is B-convergent of order ≥ 1. An order greater than 1 of B-convergence is proved only for the special case of linear inhomogeneous stiff systems with constant coefficients. B-consistency is not considered in [7].

2. Time discretization

For the time integration of the ODE-system (1.2) we define the s-stage linearly implicit Runge-Kutta method (LIRK-method)

$$\begin{aligned} U_{m+1}^{(1)} &:= U_m \\ U_{m+1}^{(i)} &= R_0^{(i)}(c_i\tau T)U_m + \tau\sum_{j=1}^{i-1} A_{ij}(c_i\tau T)\left[f_j - T\,U_{m+1}^{(j)}\right], \quad i=2,\dots,s+1 \\ U_{m+1} &= U_{m+1}^{(s+1)} \end{aligned} \tag{2.1}$$

where

$$f_j := f_h(t_m + c_j\tau, U_{m+1}^{(j)}).$$

Here the internal stability functions $R_0^{(i)}(z)$, $z\in\mathbb{C}$, and the $A_{ij}(z)$ are rational functions with real coefficients, the c_i are real parameters and T is a real (n_h,n_h)-matrix (usually an approximation to the Jacobian $\partial f_h/\partial U_h(t_m,U_m)$ which is constant in the time-discretization interval $[t_m,t_m+\tau]$. U_{m+1} denotes the approximation to U_h at the time level $t_m+\tau$. For the definition of LIRK-methods see also Hundsdorfer[6]. If we set T = 0 then the LIRK-method (2.1) is reduced to an explicit Runge-Kutta method. We will characterize a LIRK-method by the parameter scheme

$$\begin{array}{c|ccc} c_2 & A_{21} & & \\ \vdots & \vdots & \ddots & \\ c_s & A_{s1} & \cdots & A_{s,s-1} \\ \hline & A_{s+1,1} & \cdots & A_{s+1,s} \end{array}$$

Under the conditions

$$R_0^{(i)}(c_i\tau T) = I + \tau T \sum_{j=1}^{i-1} A_{ij}(c_i\tau T), \quad i=2,\ldots,s+1$$

(I denotes the identity matrix) the LIRK-method (2.1) is translation invariant (see Bruder/Strehmel/Weiner [1]).

Remark 2.1: The class of translation invariant LIRK-methods contains the W-methods (ROW-methods with inexact Jacobian; see Steihaug/Wolfbrandt [12]), the popular ROW-methods ($T = \partial f_h/\partial u_h(t_m,u_m)$, see e.g. Kaps/Rentrop [8], Kaps/Wanner [9]) and the adaptive Runge-Kutta methods (see Strehmel/Weiner [13]). For a W- and ROW-method the coefficients A_{ij} are defined by

$$\begin{aligned} A &= a[\Gamma - \tau\tilde{T}a]^{-1} \\ B^T &= b^T[\Gamma - \tau\tilde{T}a]^{-1} \end{aligned} \tag{2.2}$$

where

$$A := \begin{bmatrix} 0 & & & \\ A_{21} & 0 & & \\ \vdots & & \ddots & \\ A_{s1} & \cdots & A_{s,s-1} & 0 \end{bmatrix} \qquad B := (A_{s+1,1},\ldots,A_{s+1,s})^T$$

$$\tilde{T} = \operatorname{diag}(T,\ldots T)$$

and

$$a = \begin{bmatrix} 0 & & & \\ a_{21}I & 0 & & \\ \vdots & & \ddots & \\ a_{s1}I & \cdots & a_{s,s-1}I & 0 \end{bmatrix} \quad \Gamma := \begin{bmatrix} I-h\gamma T & & & \\ -h\gamma_{21}T & I-h\gamma T & & \\ \vdots & & \ddots & \\ -h\gamma_{s1}T & \cdots & -h\gamma_{s,s-1}T & I-h\gamma T \end{bmatrix}$$

$$b := (b_1,\ldots,b_s)^T \quad .$$

Here γ, γ_{ij}, a_{ij} and b_i are real parameters which define the W- and ROW-methods, respectively. By some manipulation it can be seen that with

$$\Gamma \cdot k = f$$

where $k := (k_1, \dots, k_s)^T$, $f := (f_1, \dots, f_s)^T$ we obtain from (2.1) and (2.2) with the conditions of translation invariance the W-method and ROW-method, respectively in the classical form (see [8], [9], [12]).
For an adaptive Runge-Kutta method the coefficients A_{ij} are given by

$$A_{ij} = \sum_{l=0}^{\varrho_i} R_{l+1}^{(i)}(c_i z) c_i^{l+1} \lambda_{lj}^{(i)}, \quad \lambda_{lj}^{(i)} \in R, \quad \varrho_i \leq \varrho_{i+1}$$

where the rational function $R_l^{(i)}(z)$ are recursively defined by

$$R_1^{(i)}(z) = \frac{R_0^{(i)}(z) - 1}{z}, \quad R_{l+1}^{(i)}(z) = \frac{l \cdot R_l^{(i)}(z) - 1}{z}, \quad l = 1, 2, \dots \varrho_i \qquad (2.3)$$

The coefficients of the corresponding explicit Runge-Kutta methods are given by

$$a_{ij} = \sum_{l=0}^{\varrho_i} \frac{\lambda_{lj}^{(i)} c_i^{l+1}}{l+1} .$$

For a W- and ROW-method the stability functions $R_0^{(i)}(z)$ are given by the method. For an adaptive Runge-Kutta method the stability functions can be chosen free.
In our following investigations we set $T := A_h$. Then the numerical solution of the semi-discrete parabolic problem (1.2) is given by

$$U_{m+1}^{(i)} = R_0^{(i)}(c_i \tau A_h) U_m + \tau \sum_{j=1}^{i-1} A_{ij}(c_i \tau A_h) [r_h(t_m + c_j \tau) + Q_h(t_m + c_j \tau, U_{m+1}^{(j)})] \qquad (2.4)$$

$$i = 2, \dots, s+1.$$

3. Uniformly consistent and convergent LIRK-methods

Let us assume

a1) the approximation order r_i of $R_0^{(i)}(z)$, $i = 2, \dots, s+1$, to $\exp(z)$ is sufficiently high. ($p_i \leq r_i$, where p_i denotes the classical stage order (see Dekker/Verwer [2]) of the LIRK-method)

a2) $R_0^{(i)}(z)$ has no pole for $\mathrm{Re}\, z \leq 0$ and it holds $|R_0^{(i)}(\infty)| < \infty$

a3) the functions $z A_{ij}(z)$, $i = 2, \dots, s+1$; $j = 1, \dots, i-1$, are uniformly bounded for $\mathrm{Re}\, z \leq 0$

b1) the restriction $u_h(t)$ of the exact solution $u(x,t)$ of the problem (1.1) possesses $q+1$ derivatives $u_h^{(l)}(t)$, $l = 0, 1, \dots, q+1$, and $\|u_h^{(l)}(t)\|$ can be bounded uniformly in t and h

b2) $\langle A_h w, w \rangle \leq \mu \|w\|^2$ with $\mu \leq 0$ for all $w \in R^{n_h}$ where $\langle \cdot, \cdot \rangle$ denotes some inner product and $\| \cdot \|$ the corresponding norm b3) the function $Q_h(t,U_h)$ satisfies the classical Lipschitz condition

$$\|Q_h(t,U_h)-Q_h(t,V_h)\| \leq L\,\|U_h-V_h\| \quad \text{for all } U_h,V_h \in R^{n_h}\,,\quad t\in[0,t_e]\,.$$

Remark 3.1: The assumption $\mu \leq 0$ allows to simplify the proofs of the following theorems. However, analogous results will hold for $\mu > 0$, too.

Remark 3.2: Assume that G is a domain with sides parallel to the coordinate axes, superimpose a uniformly rectilinear grid on G with grid lines parallel to the coordinate axes and let $b_{ik}(x) = b_{ki}(x)$. Then the operator A_G possesses a semi-discrete operator A_h which is nonsingular, symmetric and negative definit (see Samarskij [10], p. 165). Therefore for these problems it always holds $\mu < 0$.

3.1. Conditions for uniform consistency order

The local stage error (at $t_m + c_i\tau$) of the fully discrete solution as an approximation to $u_h(t_m+c_i\tau)$ is defined by

$$(le)^{(i)}_{m+1} := u_h(t_m + c_i\tau) - V^{(i)}_{m+1}\,, \quad i=2,\ldots,s+1$$

where $V^{(i)}_{m+1}$ is the result of a LIRK-step starting from $u_h(t_m)$.
To prove uniform consistency for a general LIRK-method we introduce the concept of uniform stage order.

Definition 3.1. A LIRK-method has the uniform stage order q_i on F if at the i-th stage for all problems in F

$$\|(le)^{(i)}_{m+1}\| \leq \delta_i(\tau^{q_i+1} + \tau \max_{0\leq t\leq t_e}\|\alpha_h(t)\|), \quad \forall\, \tau\in(0,\tau_0].$$

It is uniform consistent of order $q \geq 1$ on F if for all problems in F

$$\|(le)^{(s+1)}_{m+1}\| \leq \delta(\tau^{q+1} + \tau \max_{0\leq t\leq t_e}\|\alpha_h(t)\|)$$

where the constants δ_i, τ_0, δ are independent of t_m, τ and h (they may depend only on μ, L, M, t_e and on certain derivates of the exact solution $u_h(t)$).

Remark 3.3. $(le)^{(s+1)}_{m+1} =: (le)_{m+1}$ represents the full local error at the time level t_{m+1}.

Because of $U^{(1)}_{m+1} = u_h(t_m)$ we set $q_1 = \infty$.

For the following theorem we need the result for a rational matrix function of Hairer/Bader/Lubich [5].

Lemma 3.1. Let R(z) be a rational function and T a given matrix which satisfies $\langle Tw, w\rangle \leq \mu\,\|w\|^2$ for all $w\in R^n$. Then, for all $\tau > 0$, we have in the corresponding operator norm

$$\| R(\tau T)\| \leq \Phi_R(\tau\mu)$$

where Φ_R is the real valued function

$$\Phi_R(x) = \sup_z \{|R(z)| : \mathrm{Re}\, z \leq x\} .$$

We introduce the index-sets

$$K_i := \{ j : 1 \leq j \leq i-1,\ A_{ij}(z) \neq 0\}, \quad i=2,\dots,s+1 .$$

Now we can show the

<u>Theorem 3.1</u>: Let $q_i^{(1)} = \min_{j\in K_i} q_j$, $i=1,\dots,s$, and let

$$\sum_{j=1}^{i-1} A_{ij} c_j^{\,l} = c_i^{\,l+1} R_{l+1}^{(i)}(c_i z) \text{ for } l=0,1,\dots q_i^{(2)}. \tag{3.1}$$

Then the LIRK-method (2.4) has at the i-th stage, $i=2,\dots,s$, the uniform stage order

$$q_i = \min(q_i^{(1)}+1, q_i^{(2)}).$$

<u>Proof</u>: From (1.4) with (1.3) we have

$$\dot u_h(t) = A_h u_h(t) + r_h(t) + Q_h(t, u_h(t)) - \alpha_h(t). \tag{3.2}$$

Furthermore, it holds

$$V_{m+1}^{(i)} = R_o^{(i)}(c_i\tau A_h)u_h(t_m) + \tau\sum_{j=1}^{i-1} A_{ij}(c_i\tau A_h)\left[r_h(t_m+c_j\tau) + Q_h(t_m+c_j\tau, V_{m+1}^{(j)})\right], \quad i=2,\dots,s+1$$

and

$$\| Q_h(t_m+c_j\tau, u_h(t_m+c_j\tau)) - Q_h(t_m+c_j\tau, V_{m+1}^{(j)})\| \leq L\| u_h(t_m+c_j\tau) - V_{m+1}^{(j)}\|, \quad j\in K_i,$$

$$\leq L\,\delta_j(\tau^{q_i} + \tau \max_{0\leq t\leq t_e}\|\alpha_h(t)\|), \quad \forall \tau\in(0,\tau_o].$$

Now we get with (3.2) and Lemma 3.1

$$\| u_h(t_m+c_i\tau) - V_{m+1}^{(i)}\| \leq \| u_h(t_m+c_i\tau) - R_o^{(i)}u_h - \tau\sum_{j=1}^{i-1} A_{ij}(c_i\tau A_h)\left[\dot u_h(t_m+c_j\tau) - A_h u_h(t_m+c_j\tau)\right]\|$$

$$+ \delta_i(\tau^{q_i+1} + \tau\max_{0\leq t\leq t_e}\|\alpha_h(t)\|), \quad \forall \tau\in(0,\tau_o]$$

where

$$\delta_i = \sum_{j=1}^{i-1} \hat A_{ij}\cdot(1+L\delta_j), \quad \hat A_{ij} = \sup_z\{|A_{ij}(c_i z)| : \mathrm{Re}\, z \leq 0\} .$$

By Taylor expansion of $u_h(t_m+c_i\tau)$ and under consideration of (3.1) we obtain

$$\| u_h(t_m+c_i\tau) - V_{m+1}^{(i)}\| \leq \| \left[1 - R_o^{(i)} + c_i\tau A_h R_1^{(i)}\right]u_h + \sum_{l=1}^{q_i} \frac{1}{l!}(c_i\tau)^l \left[1 - l R_l^{(i)} + c_i\tau A_h R_{l+1}^{(i)}\right]u_h^{(l)}\|$$

$$+ \tau^{q_i+1}\sigma_{q_i} + \delta_i(\tau^{q_i+1} + \tau\max_{0\leq t\leq t_e}\|\alpha_h(t)\|)$$

where

$$\sigma_{q_i} = \frac{1}{(q_i+1)!}\| c_i^{q_i+1} u_h^{(q_i+1)}(t_m+c_i\theta_i\tau)+\sum_{j=1}^{i-1}\left[c_j\tau A_h A_{ij}-(q_i+1)A_{ij}\right]c_j^{q_i} u_h^{(q_i+1)}(t_m+c_j\theta_j\tau)\|,$$

$$0<\theta_1<1 .$$

With (2.3), the assumptions a2), a3), b1), b2) and Lemma 3.1 the statement follows immediatly.

<u>Corollary 3.1.</u> Let $q^{(1)} = \min_{j\in K_{s+1}} q_j$ and let

$$\sum_{j=1}^{s} A_{s+1,j}c_j^{\,l} = R_{l+1}^{(s+1)}(z) \quad \text{for} \quad l=0,1,\dots,q^{(2)} \tag{3.3}$$

Then the LIRK-method (2.4) has the uniform consistency order $q = \min(q^{(1)}+1, q^{(2)})$.

<u>Remark 3.4.</u> From (3.1) and (3.3) it follows that an uniformly consistent LIRK-method (2.1) is translation invariant.

3.2. Examples of LIRK-methods with different q

With the Theorem 3.1 and Corollary 3.1 we can construct LIRK-methods of uniform consistency order $q \geq 1$. In the following we give some examples.

a) <u>1-stage formula</u>: Because of $c_1 = 0$ (3.3) can only hold for $l = 0$. Therefore the resulting LIRK-method

$$\begin{array}{c|c} & \\ \hline & R_1^{(2)} \end{array} \tag{3.4}$$

is not uniformly consistent ($q = 0$). The conventional consistency order is $p = 1$. Therefore we have an order reduction.

b) <u>2-stage formula</u>: To get a method of order $q = 1$ we get the order conditions

$$A_{21} = c_2R_1^{(2)}, \quad A_{31}+A_{32} = R_1^{(3)}, \quad A_{32}c_2 = R_2^{(3)}$$

from which we obtain the LIRK-method

$$\begin{array}{c|cc} c_2 & c_2R_1^{(2)} & \\ \hline & R_1^{(3)}-\frac{1}{c_2}R_2^{(3)} & \frac{1}{c_2}R_2^{(3)} \end{array} \tag{3.5}$$

If $R_0^{(3)}$ has an approximation order ≥ 2 then the LIRK-method possesses the conventional order $p = 2$ (see Strehmel/Weiner [13]). Again an order reduction appears.

<u>Remark 3.5.</u> For a 2-stage W-method of conventional consistency order $p = 2$ it holds (see Hairer/Bader/Lubich [5])

$$a_{21} = c_2, \quad b_2c_2 = \frac{1}{2}, \quad b_1 + b_2 = 1, \quad b_2\gamma_{21} = -\gamma .$$

From (2.2) we get

$$A_{21}(c_2z) = c_2(1-\gamma z)^{-1}, \quad A_{31}(z) = \frac{b_1}{1-\gamma z} + \frac{(\frac{1}{2}-\gamma)z}{(1-\gamma z)^2}, \quad A_{32}(z) = \frac{b_2}{1-\gamma z}$$

and the conditions of translation invariance yield

$$R_0^{(2)}(c_2z) = \frac{1+(c_2-\gamma)z}{1-\gamma z}, \qquad R_0^{(3)}(z) = \frac{1+(1-2\gamma)z+(\frac{1}{2}-2\gamma+\gamma^2)z^2}{(1-\gamma z)^2} .$$

From Theorem 3.1 and Corollary 3.1 it follows

$$A_{21}(c_2z) = c_2(1-\gamma z)^{-1}, \quad A_{32}(z) = \frac{1}{c_2}\,\frac{\frac{1}{2}-\gamma^2 z}{(1-\gamma z)^2}$$

One immediately sees, that only $\gamma = 0$ or $\gamma = 1/2$ is possible. For $\gamma = 0$ the assumption a2) is not fulfilled. With $\gamma = 1/2$ we get a W-method of uniform consistency order $q = 1$:

$$\gamma_{21} = -c_2, \quad b_2 = \frac{1}{2c_2}, \quad b_1 = 1-b_2.$$

This method is A- but not strongly A-stable.

c) <u>3-stage formula</u> : For the construction of a LIRK-method of order $q = 2$ at first we remark that the values $V_{m+1}^{(j)}$ which we need to compute V_{m+1} have to be of order $q_j \geq 1$. Because $V_{m+1}^{(2)}$ is only of order $q_2 = 0$ it must hold

$$A_{42} = 0.$$

Furthermore, we get from Theorem 3.1 and Corollary 3.1 the order conditions

$$A_{21} = c_2R_1^{(2)}, \quad A_{31}+A_{32} = c_3R_1^{(3)}, \quad A_{32}c_2 = c_3{}^2R_2^{(3)}$$

$$A_{41}+A_{42}+A_{43} = R_1^{(4)}, \quad A_{42}c_2+A_{43}c_3 = R_2^{(4)}, \quad A_{42}c_2{}^2+A_{43}c_3{}^2 = R_3^{(4)}.$$

One easily sees that under the assumption a2) all conditions are fulfilled only if $c_3 = 1$ and the stability function is given by $R_0^{(4)}(z) = (1+z/2)/(1-z/2)$. The resulting LIRK-method

$$\begin{array}{c|ccc} c_2 & c_2R_1^{(2)} & & \\ 1 & R_1^{(3)} - \frac{1}{c_2}R_2^{(3)} & \frac{1}{c_2}R_2^{(3)} & \\ \hline & R_1^{(4)}-R_2^{(4)} & 0 & R_2^{(4)} \end{array} \tag{3.6}$$

possesses the uniform consistency order $q = 2$. The conventional consistency order is also $p = 2$.

If we want to obtain a LIRK-method of order $q = 2$ for an arbitrary stability function (which fulfils the assumptions a1) and a2)) we need at least four stages. From Theorem 3.1 and Corollary 3.1 only the coefficients A_{21}, A_{31}, A_{32} and A_{5j}, $j = 1(1)4$ are uniquely determined. Therefore we obtain the following class of a fourth stage LIRK-method with the uniform consistency order $q = 2$

$$\begin{array}{c|cccc}
c_2 & c_2R_1^{(2)} & & & \\
c_3 & c_3R_1^{(3)}-\frac{c_3^2}{c_2}R_2^{(3)} & \frac{c_3^2}{c_2}R_2^{(3)} & & \\
c_4 & c_4R_1^{(4)}-\frac{c_4^2}{c_3}R_2^{(4)}+(\frac{c_2}{c_3}-1)A_{42} & A_{42} & \frac{c_4^2}{c_3}R_2^{(4)}-\frac{c_2}{c_3}A_{42} & \\
\hline
 & R_1^{(5)}-\frac{(c_3+c_4)R_2^{(5)}-R_3^{(5)}}{c_3c_4} & 0 & \frac{c_4R_2^{(5)}-R_3^{(5)}}{c_3(c_4-c_3)} & \frac{R_3^{(5)}-c_3R_2^{(5)}}{c_4(c_4-c_3)}
\end{array} \tag{3.7}$$

From the order conditions in Strehmel/Weiner [13] it follows that the conventional order p is at least 3. If we set at the fourth stage

$$A_{42} = -\,{c_4^2}/{c_2}\;R_2^{(4)}$$

then for the nodes

$$c_2 = c_3 = 1/2 \text{ and } c_4 = 1$$

we get a LIRK-method of conventional order $p = 4$ which for $T = 0$ is reduced to Englands method.

3.3. Homogeneous boundary conditions

For homogeneous boundary conditions (what implies $r_h(t) = 0$) we can derive better uniform consistency results under the additional assumptions:

b4) $\left\| \frac{d^l A_h(t)}{dt^l} \right\|$ is uniformly bounded for $l=1,\ldots,q$

b5) $\left\| \frac{d^l Q_h(t,u_h)}{dt^l} \right\|$ is uniformly bounded for $l=0,\ldots,q$,

for all $t \in [0,t_e]$. It holds the

<u>Theorem 3.2.</u> Let $q_i^{(1)} = \min_{j \in K_i} q_j$ and let (3.1) hold for $l=0,\ldots,q_i^{(2)}$. Then the LIRK-method (2.4) has at the i-th stage the uniform stage order

$$q_i = \min\,(q_i^{(1)}, q_i^{(2)})+1.$$

<u>Proof</u>. Analogous to the proof of Theorem 3.1 we have

$$\|u_h(t_m+c_i\tau)-V_{m+1}^{(i)}\| \le \frac{1}{q_i!}\Big\|\Big[(c_i\tau)^{q_i}(1-q_iR_{q_i}^{(i)})+\tau^{q_i+1}\sum_{j=1}^{i-1}c_j^{q_i}A_{ij}A_h\Big]u_h^{(q_i)}\Big\|$$

$$+\tau^{q_i+1}\sigma_{q_i}+\delta_i(\tau^{q_i+1}+\tau\max_{0\le t\le t_e}\|\alpha_h(t)\|).$$

With (2.3) we obtain

$$\|u_h(t_m+c_i)-V_{m+1}^{(i)}\| \le \frac{\tau^{q_i+1}}{q_i!}\Big\|\Big[\sum_{j=1}^{i-1}c_j^{q_i}A_{ij}-c_i^{q_i+1}R_{q_i+1}^{(i)}\Big]\cdot A_hu_h^{(q_i)}\Big\| \quad . \tag{3.8}$$

$$+\tau^{q_i+1}\sigma_{q_i}+\delta_i(\tau^{q_i+1}+\tau\max_{0\le t\le t_e}\|\alpha_h(t)\|)$$

Because of

$$A_hu_h^{(l)}(t) = u_h^{(l+1)}(t)-\frac{d^l}{dt^l}Q_h(t,u_h(t))+\frac{d^l}{dt^l}\alpha_h(t)$$

and the assumptions a2), a3) and b1), b2), b4), b5) the statement follows from (3.8) with Lemma 3.1.

<u>Corollary 3.2.</u> Let $q^{(1)} = \min_{j\in K_{s+1}} q_j$ and let (3.3) hold. Then the LIRK-method is uniformly consistent of order

$$q = \min(q^{(1)},q^{(2)})+1$$

on F.

From Theorem 3.2 and Corollary 3.2 it follows that the 1-stage LIRK-method (3.4) has the uniform consistency order q = 1, the 2-stage LIRK-method (3.5) and the 3-stage LIRK-method (3.6) has the order q = 2 and the 4-stage LIRK-method (3.7) the order q = 3.

4. Uniform convergence order of LIRK-methods

The global error at the time level t_m of the fully discrete solution as an approximation to the partial differential equation solution $u_h(t_m)$ is defined by

$$\varepsilon_h(t_m) = u_h(t_m)-U_m \quad .$$

To simplify the representation it will be assumed that the time stepsize is constant.

<u>Definition 4.1.</u> A LIRK-method (2.1) is said to be uniformly convergent of order $\hat{q}$ on the problem class F if for all problems in F

$$\|\varepsilon_h(t_m)\| \le \hat{C}(\tau^{\hat{q}}+\max_{0\le t\le t_e}\|\alpha_h(t)\|) \ , \quad \forall\,\tau\in(0,\hat{\tau}_0]$$

where $\hat{C}$ and $\hat{\tau}_0$ are constants independent of h and τ.

Theorem 4.1. Let $R_0^{(s+1)}(z)$ be A-acceptable and let the LIRK-method be uniformly consistent of order q on F. Then the LIRK-method is uniformly convergent of order at least q for all problems in F.

Proof. We rewrite $\varepsilon_h(t_{m+1})$ as

$$\varepsilon_h(t_{m+1}) = V_{m+1} - U_{m+1} + (le)_{m+1} .$$

From A-stability and by applying of Lemma 3.1 we obtain

$$\|V_{m+1} - U_{m+1}\| \leq (1+C_0\tau)\|u_h(t_m) - U_m\|, \quad \forall \tau \in [0, \hat{\tau}_0] \tag{4.1}$$

where C_0 is a constant dependent on μ and the coefficients of the LIRK-method. Hence we get

$$\|\varepsilon_h(t_{m+1})\| \leq (1+C_0\tau)^{m+1}\|\varepsilon_h(0)\| + K\sum_{l=0}^{m}(1+C_0\tau)^l$$

where K (an upper bound for the full local error) is given by

$$K = \delta(\tau^{q+1} + \tau \max_{0 \leq t \leq t_e}\|\alpha_h(t)\|).$$

Let $\varepsilon_h(0)$ be the null vector. Then for $\tau \in (0, \hat{\tau}_0]$ we have

$$\|\varepsilon_h(t_{m+1})\| \leq \begin{cases} K\tau^{-1}\, t_{m+1}, & \text{if } C_0 \leq 0 \\ K\tau^{-1}\, C_0^{-1}(e^{C_0 t_{m+1}} - 1), & \text{if } C_0 > 0. \end{cases}$$

Hence, the uniform order of convergence is greater or equal to the uniform consistency order q of the method.

Remark 4.1. An integration method which satisfies the inequality (4.1) is called C-stable (see Dekker/Verwer [2]).

For problems on F where $Q(t,x,u)$ is given by

$$Q(t,x,u) = a(t)u(t,x) + b(t,x)$$

we can show that the full global error $\varepsilon_h(t_{m+1})$ often has the uniform convergence order $q+1 \leq p$. We denote this subclass of F by $\hat{F}$. For the proof of the following theorem the reader is referred to our paper [14].

Theorem 4.2. Let $R_0^{(s+1)}(z)$ be strongly A-acceptable and let

$$\sum_{j=1}^{i-1} A_{ij}c_j^{\,l} = c_i^{\,l+1} R_{l+1}^{(i)}(c_i z)\ , \quad l=0,\dots,q-1 \text{ and } i \in K_{s+1}$$

$$\sum_{j=1}^{s} A_{s+1,j}\, c_j^{\,l} = R_{l+1}^{(s+1)}(z)\ , \quad l=0,\dots,q.$$

Furthermore, let

$$[R_1^{(s+1)}(z)]^{-1}\,[R_{q+2}^{(s+1)}(z) - \sum_{j=1}^{s} c_j^{\,q+1} A_{s+1,j}(z)]$$

be uniformly bounded for Re $z \leq 0$ and let the LIRK-method be of uniform consistency order $q \geq 0$ on F. Then the LIRK-method is uniformly convergent of order q+1 for all problems in the subclass $\hat{F}$.

Examples. 1. For the 1-stage method (3.4), which has for inhomogeneous boundary values the uniform consistency order q = 0, we choose the stability function

$$R_0^{(2)}(z) = \frac{1}{1-z} \tag{4.2}$$

which has the approximation order 1 and is L-acceptable. Then

$$[R_1^{(2)}(z)]^{-1} \cdot R_2^{(2)}(z) = 1$$

holds. From Theorem 4.2 it follows that the one-stage LIRK-method (3.4) is uniformly convergent of order q = 1 on $\hat{F}$.

2. For the LIRK-method (3.5) we choose the strongly A-acceptable stability function

$$R_0^{(3)}(z) = \frac{1+(1-2\gamma)z+(\frac{1}{2}-2\gamma+\gamma^2)z^2}{(1-\gamma z)^2} \quad , \quad \gamma > \frac{1}{4} \, , \neq \frac{1}{2}$$

and for the internal stability function $R_0^{(2)}(z)$ we use (4.2). Then the method (3.5) possesses the conventional order p = 2. We recall that for inhomogeneous boundary values the uniform consistency order is 1. We obtain

$$R_1^{(3)}(z) = \frac{1+(\frac{1}{2}-2\gamma)z}{(1-\gamma z)^2} \, , \quad R_2^{(3)}(z) = \frac{\frac{1}{2}-\gamma^2 z}{(1-\gamma z)^2} \, , \quad R_3^{(3)}(z) = \frac{2\gamma(1-\gamma)-\gamma^2 z}{(1-\gamma z)^2} \, ,$$

hence it holds

$$[R_1^{(3)}(z)]^{-1}[R_3^{(3)}(z)-c_2R_2^{(3)}] = \frac{2\gamma(1-\gamma)-\frac{1}{2}\cdot c_2+\gamma^2(c_2-1)z}{1+(\frac{1}{2}-2\gamma)z} \, .$$

For $\gamma \geq 1/4$ we see that this expression is uniformly bounded for Re $z \leq 0$. From Theorem 4.2 it follows that the LIRK-method (3.5) is uniformly convergent of order 2 on $\hat{F}$. For $\gamma = 1/4$ we have

$$[R_1^{(3)}(z)]^{-1}[R_3^{(3)}(z)-c_2\, R_2^{(3)}] = \frac{3}{8} - \frac{1}{2}\, c_2 + \frac{1}{16}(c_2-1)z \, .$$

Therefore we have the uniform convergence order 2 on $\hat{F}$ only for $c_2 = 1$.

5. Some numerical examples

First we want to illustrate numerically the convergence order on two one-dimensional parabolic equations. For this purpose we consider the four stage LIRK-method

$$
\begin{array}{c|cccc}
\gamma & \gamma R_1^{(2)} & & & \\
\frac{1}{2} & \frac{1}{2}R_1^{(3)} - \frac{1}{48}R_2^{(3)} & \frac{1}{48}R_2^{(3)} & & \\
1 & R_1^{(4)} - 2R_2^{(4)} & 0 & 2R_2^{(4)} & \\
\hline
 & R_1^{(5)} - 3R_2^{(5)} + 2R_3^{(5)} & 0 & 4(R_2^{(5)} - R_3^{(5)}) & 2R_3^{(5)} - R_2^{(5)}
\end{array}
\qquad (5.1)
$$

with $\gamma = 0.4358665215084592$ which we obtain from (3.7). The stability function $R_0^{(5)}(z)$ is given by

$$R_0^{(5)}(z) = \frac{1+(1-3\gamma)z+(\frac{1}{2}-3\gamma+3\gamma^2)z^2}{(1-\gamma z)^3}$$

and the internal stability functionsare defined by

$$R_0^{(2)}(z) = \frac{1}{1-z}, \quad R_0^{(i)}(z) = \frac{1+(1-2\gamma_i)z+(\frac{1}{2}-2\gamma_i+\gamma_i^2)z^2}{(1-\gamma_i z)^2}, \quad i=3,4$$

where $\gamma_i = \gamma/c_i$. By this choice we only need one LU-decomposition per time step. The method (5.1) is L-stable and internally strongly A-stable. It holds

$$\lim_{z\to-\infty} R_0^{(2)}(z) = 0.6363\ldots \quad \text{and} \quad \lim_{z\to-\infty} R_0^{(i)}(z) = 0.9566\ldots \quad i=3,4.$$

Furthermore, the method (5.1) has the uniform consistency order $q = 2$. If A_h is a nonsingular, symmetric matrix for all $h \to 0$ then we have the uniform convergence order $\hat{q} = 3$ due to Theorem 4.2. We recall that the conventional consistency- and convergence order is $p = 3$.
All computations were performed in FORTRAN double precision with 16 decimales on an ES 1040 at the Martin-Luther University of Halle-Wittenberg.
For the discretization in space we select second order finite differences on an uniform grid. In the following two examples the initial and boundary values are taken from the exact solution. We emphasize that no space error ($\alpha_h(t) = 0$) appears in the semi-discrete parabolic problems.

<u>Example 1</u>: $\frac{\partial u}{\partial t} = \frac{\partial^2 u}{\partial x^2} + 3t^2(x+\frac{1}{2})(\frac{3}{2}-x) + 2t^3, \quad 0 < x < 1, \ 0 \leq t \leq 1$

with the exact solution

$$u(t,x) = t^3(x+\tfrac{1}{2})(\tfrac{3}{2}-x).$$

The finite difference matrix A_h in the semi-discrete problem is given by

$$A_h = \frac{1}{h^2} \cdot \begin{bmatrix} -2 & 1 & & & \\ 1 & -2 & 1 & & \\ & \ddots & \ddots & \ddots & \\ & & 1 & -2 & 1 \\ & & & 1 & -2 \end{bmatrix}, \quad h = \frac{1}{n+1}$$

The table 5.1 shows the full discretization error ε_h in the Euclidean norm at the time level $t_e = 1$ and the quantity

$$q_2(h) = \log_2 \{ \|\varepsilon_h\|_2 / \|\varepsilon_{2h}\|_2 \}$$

i.e. the order of accuracy (numerical convergence order) measured using $\tau = \frac{1}{n+1}, \frac{1}{2(n+1)}$.

Table 5.1. Order test and full global error $\|\varepsilon_h(t)\|_2$ at $t_e = 1$

$\tau = h$	$\|\varepsilon_h\|_2$	$q_2(h)$
1/10	.94512 -4	-
1/20	.13719 -4	2.784
1/40	.18900 -4	2.860
1/80	.25082 -4	2.914
1/160	.32530 -4	2.947

This table shows that the numerical order $q_2(h)$ tends to the theoretical uniform convergence order $\hat{q} = 3$ as $\tau = h \to 0$.

Example 2: We consider the quasilinear parabolic equation

$$u_t = u \cdot u_{xx} - 2t^2 \cdot \exp(-t)u + t(2-t) \cdot \exp(-t) \cdot x^2, \quad 0 \leq x \leq 1, \quad 0 \leq t \leq 1$$

with the exact solution

$$u(t,x) = 1 + t^2 \cdot \exp(-t) \cdot x^2.$$

For the time step from t_m to $t_m + \tau$ the matrix A_h in the LIRK-scheme (5.1) is given by

$$A_h = \frac{1}{h^2} \operatorname{diag}(U_{h,i}(t_m)) \cdot \begin{bmatrix} -2 & 1 & & \\ 1 & -2 & 1 & \\ & \ddots & \ddots & \ddots \\ & & 1 & -2 \end{bmatrix}$$

Table 5.2. Order test and full global error $\|\varepsilon_h(t)\|_2$ at $t_e = 1$

$\tau = h$	$\|\varepsilon_h\|_2$	$q_2(h)$
1/10	.61424 -5	-
1/20	.79684 -6	2.946
1/40	.98642 -7	3.014
1/80	.12004 -7	3.039
1/160	.14476 -8	3.052

This table shows that we have also a numerical convergence order $q_2(h) \approx 3$ for a quasilinear parabolic problem.
We remark that the same results are found when we keep h fixed and consider a finite, realistic range of τ-values.

In the second place we investigate the influence of uniform consistency- and convergence order on accuracy and computing time for implementation with stepsize control. For this purpose we consider method (3.4) (uniform consistency order q = 0, uniform convergence order $\hat{q}$ = 1 for problems from the subclass $\hat{F}$ of F) and (5.1) (uniform consistency order q = 2, uniform convergence order $\hat{q}$ = 3 for problems from the subclass $\hat{F}$ of F). The stepsize control is performed by Richardson extrapolation. The stability function $R_0^{(2)}(z)$ for method (3.4) is given by the (1,0)-Padé-approximation. Both LIRK-methods (3.4) and (5.1) were tested on a great number of semi-linear one dimensional parabolic initial boundary value problems. Here we will present only the results for the example 1 in the following Table 5.3 where we used the abbreviations:

Step : number of successful steps
r-Step: number of rejected steps
TIME : computing time in seconds
TOL : given time tolerance

The initial time stepsize was given by $\tau = 10^{-1}$.

Table 5.3. Results for example 1 with LIRK-methods (3.4) and (5.1) with time stepsize control

		LIRK-method (3.4)				LIRK-method (5.1)			
TOL	n	Step	r-Step	$\|\varepsilon_h(t_e)\|_2$	TIME	Step	r-Step	$\|\varepsilon_h(t_e)\|_2$	TIME
10^{-2}	4	26	2	.24611 -1	0.62	4	0	.36974 -3	0.48
	8	28	2	.21584 -1	1.62	4	0	.38356 -3	0.96
	16	29	2	.20260 -1	2.56	4	0	.39867 -3	1.96
	32	31	2	.18115 -1	5.38	4	0	.40917 -3	4.06
	64	31	2	.17965 -1	10.92	4	0	.41532 -3	7.84
	128	32	2	.16939 -1	22.00	4	0	.41865 -3	15.30
10^{-4}	4	347	2	.20495 -2	8.34	10	0	.14987 -4	1.00
	8	460	2	.14725 -2	19.40	10	0	.15634 -4	1.92
	16	628	2	.10231 -2	50.86	10	0	.16368 -4	3.64
	32	822	2	.74541 -3	130.28	10	0	.16886 -4	7.24
	64	944	2	.59466 -3	312.12	10	0	,17192 -4	14.28
	128	1113	2	.51617 -3	695.44	10	0	.17359 -4	29.00

For the uniformly consistent LIRK-method (5.1) the number of steps is constant for all grid parameters h and no step is repeated. The global error at the time level t_e = 1 is smaller than the given tolerance. For the LIRK-method (3.4) the number of steps is strongly increased if the grid distance decreases and the error at the time level t_e = 1 is greater than the given tolerance. Table 5.3 clearly shows the advantage of the higher order method.

References

[1] Bruder, J., K. Strehmel, R. Weiner: Partitioned adaptive Runge-Kutta methods for the solution of nonstiff and stiff systems. Submitted for Numer. Math.

[2] Dekker, K. and J.G. Verwer: Stability of Runge-Kutta methods for stiff nonlinear differential equations. North-Holland Publ. Co. Amsterdam 1984.

[3] Frank, R., J. Schneid and C.W. Ueberhuber: The concept of B-convergence. SIAM J. Numer. Anal. 18 (1981) 753-780.

[4] Frank, R., J. Schneid and C.W. Ueberhuber: B-convergence of Runge-Kutta methods. Bericht Nr. 48/81, Inst. für Numer. Math. TU-Wien (1981).

[5] Hairer, E., G. Bader and Ch. Lubich: On the stability of semi-implicit methods for ordinary differential equations. BIT 22 (1982) 211-232.

[6] Hundsdorfer, W.: The numerical solution of nonlinear stiff initial value problems. CWI-Tract 12, Amsterdam (1985).

[7] Hundsdorfer, W.: Stability and B-convergence of linearly implicit Runge-Kutta methods, Report NM-R8516, Centre for Math. and Comp.Sc., Amsterdam (1985).

[8] Kaps, P., P. Rentrop: Generalized Runge-Kutta methods of order four with stepsize control for stiff ordinary differential equations. Numer. Math. 33 (1978) 55-68.

[9] Kaps, P., G. Wanner: A study of Rosenbrock type methods of high order. Numer. Math. 38 (1981) 279-298.

[10] Samarskij,A.A.: Theorie der Differenzenverfahren. Akad. Verlagsgesellschaft Geest & Portig K.-G. Leipzig 1984.

[11] Sanz-Serna, J.M., J.G. Verwer, W. Hundsdorfer: Convergence and order reduction of Runge-Kutta schemes applied to evolutionary problems in partial differential equations. Report NW-R8525, Centre for Math. and Comp.Sc., Amsterdam (1985).

[12] Steihaug, T. and A. Wolfbrandt: An attempt to avoid exact Jacobian and nonlinear equations in the numerical solution of stiff differential equations. Math. Comp. 33 (1979) 521-534.

[13] Strehmel, K., R. Weiner: Behandlung steifer Anfangswertprobleme gewöhnlicher Differentialgleichungen mit adaptiven Runge-Kutta Methoden. Computing 29 (1982) 153-165.

[14] Strehmel, K., R. Weiner, K.D. Krannich: A study of uniform consistency and convergence of linearly implicit Runge-Kutta methods and their modification applied to parabolic equations. Submitted for SIAM J. Numer. Anal.

[15] Verwer, J.G.: Convergence and order reduction of diagonally implicit Runge-Kutta schemes in the method of lines. Report NM-R8506, Centre for Math. and Comp.Sc., Amsterdam (1985).

[16] Verwer, J.G., J.M. Sanz-Serna: Convergence of method of lines approximations to partial differential equations. Computing 33 (1984) 297-313.

MONOTONE DISCRETIZATION METHODS FOR SINGULARLY PERTURBED PROBLEMS AND ITS APPLICATION FOR ENCLOSING SOLUTIONS

Lutz Tobiska [1]

1. Introduction

Let us consider a discretization method

$$A_h u_h = f_h, \quad A_h : X_h \longrightarrow Y_h \tag{1.1}$$

for solving a boundary value problem

$$Au = f \quad , \quad A : X \longrightarrow Y \tag{1.2}$$

with the following approximation scheme

$$\begin{array}{ccc} X & \xrightarrow{\quad A \quad} & Y \\ r_h \big\downarrow \big\uparrow p_h & & \big\downarrow \tilde{r}_h \\ X_h & \xrightarrow{\quad A_h \quad} & Y_h \end{array} \tag{1.3}$$

In (1.3) X, Y, X_h, Y_h denote Banach-spaces, r_h, $\tilde{r}_h$ are restrictions to finite dimensional spaces and p_h is a prolongation.

According to the approximation scheme the error between the solutions of (1.1) and (1.2) can be measured in two different ways, in space X by $\|u - p_h u\|$ and in space X_h by $\|r_h u - u_h\|$. So we immediately obtain two kinds for enclosing u and the restriction $r_h u$, respectively. While enclosing of u is based on the fact that A is of monotone type, we shall use inverse-monotone operators A_h and derive upper and lower bounds for $r_h u$.

Instead of (1.1) we shall consider the problem

$$A_h z_h^{\pm} = f_h + \varphi_h^{\pm} \tag{1.4}$$

where $\varphi_h^{\pm}$ denote correctors determined in such a way that

---

[1] TH Otto von Guericke, Sektion Mathematik und Physik
DDR - 3032 Magdeburg, Boleslaw-Bierut-Platz 5

$$z_h^+ \geq r_h u \geq z_h^- \tag{1.5}$$

Moreover, for inverse-monotone operators A_h we have also an enclosing of the solution of (1.1)

$$z_h^+ \geq u_h \geq z_h^-$$

provided that $\varphi_h^+ \geq 0 \geq \varphi_h^-$.

Concerning the solution of (1.4) we remark that for linear operators A_h the problem corresponds to a linear system of equations with two different right sides, i.e. the expense to solve (1.4) is unessentially higher than for solving (1.1). The same is valid for nonlinear operators A_h when (1.4) is solved by Newton-Rhaphson method or Jacobi method.

2. Sufficient conditions for obtaining upper and lower bounds

Let the elements of X_h be denoted by $u_h = (u_1, \ldots, u_N)$ and the norm defined by $\|u_h\| = \max\limits_i |u_i|$. The relation $u_h \geq v_h$ means that $u_i \geq v_i$ for all i. In the same way the norm and a partial ordering in Y_h are defined. The index h characterizes the dependence on the mesh parameter h of the discretization method. An operator $T: X_h \to Y_h$ is called inverse-monotone iff the implication

$$Tu_h \geq Tv_h \Rightarrow u_h \geq v_h$$

holds.

<u>Theorem 1:</u> Suppose that the following assumptions are fulfilled:

(i) There are positive constants c_1, α such that for all $h \leq h_0$
$\|r_h u - u_h\| \leq c_1 h^\alpha$

(ii) For all $u_h, v_h \in X_h$ there are linear, inverse-monotone operators $S = S(u_h, v_h)$: $X_h \to Y_h$ such that

$$A_h u_h - A_h v_h = S(u_h, v_h)(u_h - v_h)$$

(iii) There are positive constants δ_0, δ_1 such that for all $u_h, v_h \in X_h$ there exist $e_h \in X_h$ with $e_i \geq 1$ and

$$\delta_0 \leq (S(u_h, v_h)e_h)_i \leq \delta_1 \quad \text{for any } i.$$

Then the solutions z_h^+, z_h^- of (1.4) satisfy

$$\|z_h^+ - z_h^-\| \leq \frac{1}{\delta_0} \|e_h\| \, \|\varphi_h^+ - \varphi_h^-\| \, . \tag{2.1}$$

Moreover, if the correctors φ_h^+, φ_h^- are sufficiently large, i.e.

$$(\varphi_h^+)_i, \quad (-\varphi_h^-)_i \geq \delta_1 c_1 h^\alpha,$$

the enclosing

$$z_h^- \leq r_h u \leq z_h^+ \tag{2.2}$$

is valid.

<u>Proof:</u> We start with

$$\|\varphi_h^+ - \varphi_h^-\| \geq \varphi_h^+ - \varphi_h^- = A_h z_h^+ - A_h z_h^- = S(z_h^+ - z_h^-) .$$

Since $(S \frac{e_h}{\delta_o})_i \geq 1$ the inverse-monotonicity of S yields

$$z_h^+ - z_h^- \leq \frac{e_h}{\delta_o} \|\varphi_h^+ - \varphi_h^-\|$$

and consequently (2.1).

For proving (2.2) we remark that from the definition of z_h^+ and (iii) it follows

$$\varphi_h^+ = A_h z_h^+ - A_h u_h = S\ (z_h^+ - u_h)$$

$$\delta_1 c_1 h^\alpha \geq (S\ (c_1 h^\alpha e_h))_i .$$

Using the inverse-monotonicity of S we obtain

$$z_h^+ - u_h \geq c_1 h^\alpha e_h$$

and finally

$$z_h^+ - r_h u \geq z_h^+ - u_h + u_h - r_h u \geq 0.$$

Similarly we can prove $z_h^- \leq r_h u$.

<u>Remark:</u> If A_h is continuously differentiable, we can use the representation

$$S(u_h, v_h) = \int_0^1 A_h'(t u_h + (1-t) v_h)\ dt$$

for establishing (ii), (iii).

This theorem is only of theoretical interest, because for practical use the knowledge of the value c_1 is necessary. To construct correctors the following two theorems are more appropriate.

Theorem 2: Let $A_h : X_h \to Y_h$ be inverse-monotone and

$\tau_h = A_h r_h u - \tilde{r}_h Au$ where u denotes the solution of (1.2). Then the solutions z_h^+, z_h^- of (1.4) satisfy

$$z_h^+ \geq r_h u \geq z_h^-$$

provided that

$$\varphi_h^+ \geq \tau_h \geq \varphi_h^-. \tag{2.3}$$

The proof immediately follows from the property of inverse-monotonicity.

In same cases it is advantageous to use an approximation of $r_h u$.

Theorem 3: Suppose that the assumptions (ii) and (iii) of Theorem 1 are fulfilled. Moreover, let u_h^{app} be an approximation of $r_h u$ in X_h. Then the solutions z_h^+, z_h^- of (1.4) satisfy

$$z_h^+ \geq r_h u \geq z_h^-$$

provided that

$$\begin{aligned} (\varphi_h^+)_i &\geq (A_h\, u_h^{app} - f_h)_i + \delta_i \, \| u_h^{app} - r_h u \| \\ (\varphi_h^-)_i &\leq (A_h\, u_h^{app} - f_h)_i - \delta_i \, \| u_h^{app} - r_h u \| \end{aligned} \tag{2.4}$$

Remark: For a Lipschitz mapping r_h such an approximation u_h^{app} can be obtained in the form $u_h^{app} = r_h\, u_{app}$ by means of an approximation u_{app} of u in X.

Proof: We estimate $z_h^+ - u_h^{app}$ using inverse-monotonicity of S.

$$A_h\, z_h^+ - A_h\, u_h^{app} = S\; (z_h^+ - u_h^{app}) = f_h - A_h\, u_h^{app} + \varphi_h^+$$

$$f_h - A_h\, u_h^{app} + \varphi_h^+ \geq S\; (\| u_h^{app} - r_h u \|\, e_h)$$

Consequently,

$$z_h^+ \geq u_h^{app} + \| u_h^{app} - r_h u \|\, e_h \geq r_h u$$

holds. In a similar way we can verify $z_h^- \leq r_h u$.

3. Application to a singularly perturbed two-point boundary value problem

For simplicity we restrict our considerations on two-point boundary value problems of the form

$$\begin{aligned} -\varepsilon u'' + c(x)\,u &= f(x) \quad \text{in } (0,1) \\ u(0) = \alpha,\ u(1) &= \beta \end{aligned} \tag{3.1}$$

Let the coefficients c, f be sufficiently smooth and $0 = x_1 < x_2 < \dots < x_N = 1$ denote a uniform grid with step size $h = x_{j+1} - x_j$. D_+u_j and D_-u_j are the standard notations for the divided differences.

For solving (3.1) in case $c(x) \geq c_0 > 0$ we consider the discrete problem

$$A_h\, u_h = f_h \tag{3.2}$$

where

$$(A_h u_h)_j = \begin{cases} -\varepsilon\, D_+D_-u_j + c(x_j)u_j & \text{for } j=2,\dots,N-1 \\ u_j & \text{for } j=1 \text{ or } j=N \end{cases}$$

$$(f_h)_j = \begin{cases} f(x_j) & \text{for } j=2,\dots,N-1 \\ \alpha & \text{for } j=1 \\ \beta & \text{for } j=N \end{cases} \tag{3.4}$$

Then the operator A_h fulfills the assumptions (i),(ii),(iii) of Theorem 1 with $(e_h)_j = 1$, $\delta_0 = \min(c_0, 1)$, $\delta_1 = \max(1, \max_x c(x))$, $\alpha = 2$. However, it must be mentioned that c_1 is not uniformly bounded with respect to ε.

To construct correctors we apply Theorem 2 and estimate

$$\tau_h = A_h r_h u - \tilde{r}_h A u$$

for the solution u of (3.1). Let $[u]_k$ denote the maximum of the k-th derivative of the function u. Then we have

$$\|\tau_h\| \leq \frac{\varepsilon h^2}{12}[u]_4$$

i.e. a priori estimates are necessary. The maximum principle yields

$$[u]_0 \leq K_0 = \max(\alpha, \beta, \frac{1}{c_0}[f]_0)$$

and from the differential equation it follows

$$\varepsilon\,[u]_2 \leqq K_2 = [f]_o + [c]_o\,K_o\,.$$

An estimate for the first derivative can be obtained by means of

ε \ h	1/8	1/16	1/32	1/64	1/128	1/256
1/2	2.43 - 1	6.09 - 2	1.52 - 2	3.81 - 3	9.52 - 4	2.38 - 4
1/4	2.39 - 1	6.00 - 2	1.50 - 2	3.76 - 3	9.38 - 4	2.34 - 4
1/8	2.42 - 1	6.07 - 2	1.52 - 2	3.79 - 3	9.50 - 4	2.37 - 4
1/16	2.27 - 1	6.68 - 2	1.67 - 2	4.18 - 3	1.04 - 3	2.61 - 4
1/32	3.36 - 1	8.46 - 2	2.12 - 2	5.30 - 3	1.32 - 3	3.31 - 4
1/64	4.97 - 1	1.25 - 1	3.12 - 2	7.82 - 3	1.95 - 3	4.88 - 4
1/128	8.28 - 1	2.08 - 1	5.19 - 2	1.30 - 2	3.24 - 3	8.11 - 4
1/256	1.49	3.73 - 1	9.34 - 2	2.33 - 2	5.84 - 3	1.46 - 3

table 1

interpolation inequality [3]

$$[u]_1 \leq \omega\,[u]_2 + 2/\omega\,[u]_o\,.$$

Finally, we get the following bound for τ_h

$$\|\tau_h\| \leqq \frac{h^2}{12\varepsilon}\left\{[c]_o K_2 + 2\varepsilon^{1/2}[c]_1 K_1 + \varepsilon([c]_2 K_o + [f]_2)\right\} \tag{3.5}$$

where $K_1 = 2(2K_oK_2)^{1/2}$ is a bound for $\varepsilon^{1/2}[u]_1$.

<u>Example ([1])</u>

$$-\varepsilon u'' + u = \cos^2\pi x + 2\varepsilon\pi^2\cos 2\pi x$$
$$u(0) = u(1) = 0$$

there (3.5) yields

$$\|\tau_h\| \leqq \frac{h^2}{6\varepsilon}\left\{1 + 2\varepsilon\pi^2 + \varepsilon\pi^2(1 + 4\varepsilon\pi^2)\right\} = \varkappa h^2 \tag{3.6}$$

where $\varkappa$ depends on ε.

The maximum difference between upper and lower solution $\|z_h^+ - z_h^-\|$ for the correctors $\varphi_h^+ = -\varphi_h^- = \varkappa h^2$ are given in table 1 for different values of h and ε. When for fixed h ε tends to zero $\|z_h^+ - z_h^-\|$ increases and upper and lower solution do not produce a good enclosion. This is caused by the fact that $\varkappa$ is not uniformly bounded with respect to ε.

For small ε instead of method (3.2) - (3.4) we use a modified operator $\tilde{A}_h$ producing an uniform convergent method

$$(\tilde{A}_h u_h)_j = \begin{cases} -\varepsilon\sigma(\varrho_j) D_+ D_- u_j + c(x_j) u_j & \text{for } j=2, \dots, N-1 \\ u_j & \text{for } j=1 \text{ or } j=N \end{cases} \tag{3.7}$$

where $\sigma(\varrho) = \varrho^2/(\sinh^2 \varrho)$ and $\varrho_j = \frac{h}{2}\sqrt{\frac{c(x_j)}{\varepsilon}}$ [1].

According to Theorem 3 we construct an approximation $u_h^{app} = r_h u_{as}$ using an asymptotic solution of (3.1)

$$u_{as}(x) = \frac{f(x)}{c(x)} + \left(\alpha - \frac{f(0)}{c(0)}\right)\exp\left(-x\sqrt{\frac{c(0)}{\varepsilon}}\right) + \left(\beta - \frac{f(1)}{c(1)}\right)\exp\left(-(1-x)\sqrt{\frac{c(1)}{\varepsilon}}\right)$$

It holds

$$\| r_h u_{as} - r_h u \| \leq \max_x |u_{as}(x) - u(x)| \leq c\sqrt{\varepsilon}$$

and in particular for $c(x) = 1$ we have

$$u_{as}(x) = f(x) - f(o)\exp\left(-\frac{x}{\sqrt{\varepsilon}}\right) - f(1)\exp\left(-\frac{(1-x)}{\sqrt{\varepsilon}}\right)$$

$$\| r_h u_{as} - r_h u \| \leq \max\left(\max(|f(0)|, |f(1)|)\exp\left(-\frac{1}{\sqrt{\varepsilon}}\right), \varepsilon [f]_2\right).$$

A simple calculation shows that the modified operator satisfies

$$\| \tilde{A}_h u_h^{app} - f_h \| \leq \varepsilon\sigma(\varrho)[f]_2 .$$

If ε is sufficiently small the correctors

$$\varphi_h^+ = -\varphi_h^- = \varepsilon [f]_2 (1 + \sigma(\varrho))$$

fulfil all assumptions of Theorem 3 such that the solutions of (1.4) satisfy

$$z_h^+ \geq r_h u \geq z_h^- .$$

$h = 1/8$; $\varepsilon = 2^{-j}$ $j = 7,8,\dots,13$						
4.01 - 1	1.78 - 1	8.30 - 2	4.00 - 2	1.96 - 2	9.73 - 3	4.84 - 3

table 2

For the mentioned example numerical results of the maximum difference between upper and lower solutions are given in table 2 where the correctors

$$\varphi_h^+ = -\varphi_h^- = 2\varepsilon\,[f]_2$$

were used. The same results were obtained by numerical experiment with different values of h.

4. Concluding remarks

The presented method can also be applied to other singularly perturbed boundary value problems, such as non-symmetric, nonlinear and turning point problems, provided that discretization methods with inverse-monotone operators are known. The main difficulty consists in deriving calculatable correctors and a priori bounds, respectively. If ε is small compared with the mesh size h it seems more appropriate to use asymptotic approximations which are known also in the multidimensional case [2].

References

[1] Doolan, E.P., Miller, J.J.H. and W.H.A. Schilders
Uniform numerical methods for problems with initial and boundary layers. Boole Press, Dublin 1980

[2] Goering, H., Felgenhauer, A., Lube, G., Roos, H.-G. and L. Tobiska
Singularly Perturbed Differential Equations.
Akademie Verlag, Berlin 1983

[3] Miranda, C.
Partial Differential Equations of Elliptic Type
Springer-Verlag, Berlin 1970

[4] Tobiska, L.
Diskretisierungsverfahren zur Lösung singulär gestörter Randwertprobleme
ZAMM 63(1983), 115-123

[5] Tobiska, L.
Finite difference methods for singularly perturbed two-point boundary value problems
In: Singularly perturbed Differential Equations and Applications (ed. by J. Förste)
AdW der DDR, Institut für Mathematik, Report R-MECH-03/84,Berlin1984

A MAXIMUM PRINCIPLE FOR SYSTEMS WITH DIAGONALLY DOMINANT M-MATRICES

Günther Windisch[1)]

1. Introduction

Finite-dimensional approximations of boundary or initial boundary value problems are of considerable practical interest and require us to use methods which reflect basic properties of continuous problems in discrete approximations, where maximum principles are of importance. For this purpose, several discrete maximum principles are defined and applied in the study of approximations to differential equations. In particular, discrete boundary value maximum principles, which are adapted to special types of problems, are discussed in [4], [5], [6]. In [7], [8], maximum principles are handled as matrix properties, which are independent of the original problems to be solved. However, based on the mentioned maximum principles, lower and/or upper solution bounds are deduceable only in a few exeptional cases.

Our approach discusses a maximum principle for linear equation systems with diagonally dominant M-matrices, without referring to problems solved by discretization methods, see [11]. It implies lower and/or upper solution bounds, for which sufficient conditions are given. Moreover, this maximum principle is shown to be meaningful in a number of applications.

2. Diagonally dominant M-matrices

Our major aim in this section is to list some of the results from [3], [9], [12] on diagonally dominant M-matrices, which are used in this work.

First we introduce some notations. Let $A=(a_{ij})$, etc., be real square matrices of order n, let I denote the unit matrix and let $y=(y_i)$, etc., be real n-dimensional column vectors. The superscript T in A^T and y^T denotes the transposition of matrices and vectors. Zero matrices, vectors or the number zero are denoted by the same symbol 0. Inequality relations for matrices or vectors are to be understood componentwise. Further, let $N=\{1,\dots,n\}$ and $e=(1,\dots,1)^T$.

1) Technische Hochschule Karl-Marx-Stadt, Sektion Mathematik, GDR, 9010 Karl-Marx-Stadt, Psf. 964

In our consideration, we shall use the M-matrix definition from [12], where a real matrix $A=(a_{ij})$ of order n is called an M-matrix if $a_{ij} \leq 0$ for all $i,j \in N$ with $i \neq j$, A^{-1} exists and $A^{-1} \geq 0$. As a consequence, it follows that $a_{ii} > 0$ for all $i \in N$. Here, the more general definition of M-matrices from [3] is not needed.

Taking into account only relations between the entries a_{ij} of a matrix A, the following conditions are sufficient for A to be a diagonally dominant M-matrix, see [3], [9], [12].

(i) If, for a matrix $A=(a_{ij})$,

$$a_{ii} > 0 \quad \text{for all } i \in N,$$
$$a_{ij} \leq 0 \quad \text{for all } i,j \in N \text{ but } i \neq j,$$
$$\sum_{j=1}^{n} a_{ij} = d_i > 0 \quad \text{for all } i \in N,$$

then A is a strictly row diagonally dominant M-matrix.

(ii) If, for a matrix $A=(a_{ij})$,

there does not exist a permutation matrix P such that

$$PAP^T \doteq \begin{pmatrix} A_{11} & A_{12} \\ 0 & A_{22} \end{pmatrix},$$

where A_{11} and A_{22} are square matrices,

$$a_{ii} > 0 \quad \text{for all } i \in N,$$
$$a_{ij} \leq 0 \quad \text{for all } i,j \in N \text{ but } i \neq j,$$
$$\sum_{j=1}^{n} a_{ij} = g_i \geq 0 \quad \text{for all } i \in N, \text{ where the strict inequality holds for at least one } i,$$

then A is an irreducible weakly row diagonally dominant M-matrix.

The two assertions mentioned are true, when the columns of a matrix A rather than its rows are considered.

3. The maximum principle

Suppose that we are given a linear equation system

$$A\,y = f\,, \tag{1}$$

where A is an M-matrix of order n. Then, the solution of the system (1) is for $f \geq 0$ nonnegative, that is, $y = A^{-1}f \geq 0$.

Further, we assume that the right-hand side vector f is represented by

$$f = B\,v + \tau\,C\,w\,, \tag{2}$$

where B, C are real $n \times n$ - matrices, $v=(v_1,\dots,v_n)^T$, $w=(w_1,\dots,w_n)^T$

and τ is a positive constant. In general, the right-hand side representation (2) is not unique and it is not necessary for B and C to be nonsingular.

Next, following some ideas from [5], we define the maximum principle according to [11].

Definition

We say that the linear equation system (1) with M-matrix A and the right-hand side vector (2) satisfies the maximum principle if it implies, for the solution $y=(y_1,\ldots,y_n)^T$, the estimates

$$\min_{j\in N} v_j + \tau \min_{j\in N} w_j \leq y_i \leq \max_{j\in N} v_j + \tau \max_{j\in N} w_j$$

for all $i\in N$.

In deriving sufficient conditions for the maximum principle, we formulate a theorem.

Theorem 1

Let A be a strictly row diagonally dominant M-matrix. Further, let $B=(b_{ij})\geq 0$, $C=(c_{ij})\geq 0$, where we assume that

$$A\,e = B\,e = C\,e = d = (d_1,\ldots,d_n)^T > 0 \tag{3}$$

and let $\tau > 0$. Then the linear equation system

$$A\,y = B\,v + \tau\, C\,w \tag{4}$$

satisfies the maximum principle.

Proof:

Let $y=(y_1,\ldots,y_n)^T$ be the unique solution of the system (4). Assuming $y_{i_o} = \min_{i\in N} y_i$, we obtain from the i_oth equation of the system (4)

$$a_{i_o i_o} y_{i_o} = \sum_{j\neq i_o} (-a_{i_o j}) y_j + \sum_{j=1}^{n} b_{i_o j} v_j + \tau \sum_{j=1}^{n} c_{i_o j} w_j$$

the estimate

$$a_{i_o i_o} y_{i_o} \geq y_{i_o} \sum_{j\neq i_o} (-a_{i_o j}) + \min_{j\in N} v_j \sum_{j=1}^{n} b_{i_o j} + \tau \min_{j\in N} w_j \sum_{j=1}^{n} c_{i_o j}.$$

Using the assumptions (3), we have

$$d_{i_o} y_{i_o} \geq d_{i_o} \left(\min_{j\in N} v_j + \tau \min_{j\in N} w_j \right),$$

which shows that

$$\min_{j\in N} v_j + \tau \min_{j\in N} w_j \leq y_i \quad \text{for all } i\in N.$$

The proof for the upper bound is analogous.

■

If a matrix A satisfies the assumptions of Theorem 1, we can apply the maximum principle to the system $Ay=f$, putting $B=Id$, $v=(Id)^{-1}f$ and $w=0$. In this case, the estimate of the solution components y_i is in general sharper than the bounds which we get from $\|y\|_\infty \le \|A^{-1}\|_\infty \|f\|_\infty \le (\min_{i\in N} d_i)^{-1}\|f\|_\infty$, see [10].

Putting $\underline{y}=(\min_{j\in N} v_j + \tau \min_{j\in N} w_j)e$ and $\overline{y}=(\max_{j\in N} v_j + \tau \max_{j\in N} w_j)e$, we have

$$A\,\underline{y} \le B\,v + \tau\,C\,w \le A\,\overline{y}\,,$$

where $\underline{y}$ and $\overline{y}$ may be starting vectors for an iterative improvement of the enclosure of the solution y of the system (4), see [1], [2].

A weakening of the previous result is possible if we change the condition (3). Then, we can deduce only one-sided bounds for the solution components of the system (4).

Theorem 2

Let A be as in Theorem 1 and let $B=(b_{ij})\geqslant 0$, $C=(c_{ij})\geqslant 0$, where it is assumed that

$$0 < d = A\,e \le c_{min}e \quad \text{with} \quad c_{min} = \min_{k\in N}\left\{(Be)_k,(Ce)_k\right\}.$$

Further, let $v\geqslant 0$, $w\geqslant 0$, $\tau > 0$ in the right-hand side representation (2) of the system (4). Then, the solution $y=(y_1,\dots,y_n)^T$ of the linear equation system (4) satisfies

$$\min_{j\in N} v_j + \tau \min_{j\in N} w_j \le y_i \quad \text{for all } i\in N.$$

Proof:

It follows from the assumptions that $f=Bv+\tau Cw\geqslant 0$, hence, $y=A^{-1}f\geqslant 0$. Let $y_{i_o} = \min_{i\in N} y_i$. Considering the i_oth equation of the system (4), we get

$$a_{i_oi_o}y_{i_o} \geqslant y_{i_o}\sum_{j\neq i_o}(-a_{i_oj}) + c_{min}\left(\min_{j\in N} v_j + \tau\min_{j\in N} w_j\right).$$

Thus, we obtain the assertion from the inequalities

$$d_{i_o}y_{i_o} \geqslant c_{min}\left(\min_{j\in N} v_j + \tau\min_{j\in N} w_j\right) \quad \text{and} \quad c_{min} \geqslant d_{i_o} > 0,$$

which proves the theorem. ■

Theorem 3

Let A be as in Theorem 1, let $B=(b_{ij})\geqslant 0$, $C=(c_{ij})\geqslant 0$, where we assume that

$$A\,e = d \geqslant c_{max}e > 0 \quad \text{with} \quad c_{max} = \max_{k\in N}\left\{(Be)_k,(Ce)_k\right\}.$$

Further, let v, w and τ be as in Theorem 2. Then, the solution $y = (y_1,\dots,y_n)^T$ of the linear equation (4) satisfies

$$y_i \le \max_{j\in N} v_j + \tau\max_{j\in N} w_j \quad \text{for all } i\in N.$$

Proof:

As in the proof of Theorem 2, it follows from the assumptions that $y=A^{-1}f \geq 0$. Let $y_{i_o} = \max_{i\in N} y_i$. From the i_oth equation of the system (4) we get

$$a_{i_o i_o} y_{i_o} \leq y_{i_o} \sum_{j \neq i_o} (-a_{i_o j}) + c_{max} (\max_{j\in N} v_j + \tau \max_{j\in N} w_j).$$

Thus,

$$d_{i_o} y_{i_o} \leq c_{max} (\max_{j\in N} v_j + \tau \max_{j\in N} w_j)$$

and the assertion follows with $0 < c_{max} \leq d_{i_o}$.

■

If A is a weakly row diagonally dominant M-matrix, the proofs of the Theorems 1 to 3 fail. But it is possible to prove the following theorem.

Theorem 4

Let A be an irreducible weakly row diagonally dominant M-matrix. Further, let $B=(b_{ij}) \geq 0$, $C=(c_{ij}) \geq 0$, where

$$0 < q = (q_1,\dots,q_n)^T = B\,e = C\,e\ , \tag{5}$$

and let $v \geq 0$, $w \geq 0$, $\tau > 0$ in the right-hand side representation (2) of the system (4). If there exists a diagonal matrix $D \geq 0$ such that

$$q = (A + D)e\ , \tag{6}$$

then the linear equation system (4) implies

$$\min_{j\in N} v_j + \tau \min_{j\in N} w_j \leq y_i \quad \text{for all } i \in N.$$

Proof:

Instead of the system (4) we consider the perturbed system

$$(A + D)\,z = B\,v + \tau\,C\,w\ , \tag{7}$$

where $f=Bv + \tau Cw \geq 0$ by the assumptions made above. According to (5) and (6), the matrix A+D is a strictly row diagonally dominant M-matrix, thus, the system (7) has a unique sulution $z=(z_1,\dots,z_n)^T \geq 0$. Applying Theorem 1 to the system (7), we have

$$\min_{j\in N} v_j + \tau \min_{j\in N} w_j \leq z_i \leq \max_{j\in N} v_j + \tau \max_{j\in N} w_j \tag{8}$$

for all $i \in N$.

For M-matrices A and diagonal matrices $D \geq 0$, the matrix A+D is an M-matrix which satisfies

$$(A + D)^{-1} \leq A^{-1}, \tag{9}$$

see [3], [12]. Multiplying the inequality (9) by the right-hand side vector $f=Bv + \tau Cw \geq 0$ of the systems (4) and (7) from the right, we deduce $z \leq y$. This estimate combined with (8) completes the proof.

■

4. Applications of the maximum principle

In this section, we shall present a number of examples which show the efficiency of the introduced maximum principle.

Example 1

Consider, from [4] for constant $\varepsilon > 0$, the two-point boundary value problem

$$\begin{aligned} -\varepsilon u''(x) + u(x) &= 0, \quad 0 < x < 1 \\ u(0) = u(1) &= a \end{aligned} \tag{10}$$

which has the unique solution

$$u(x) = a(1 + \exp(-\tfrac{1}{\sqrt{\varepsilon}}))^{-1}(\exp(-\tfrac{x}{\sqrt{\varepsilon}}) + \exp(-\tfrac{1-x}{\sqrt{\varepsilon}})).$$

If $a > 0$, then the problem (10) satisfies the maximum principle

$0 \leq u(x) \leq a$ for $0 \leq x \leq 1$ (for $a < 0$, conversely, $a \leq u(x) \leq 0$).

For small ε, the solution $u(x)$ has boundary layers near $x=0$ and $x=1$. Suppose that the problem (10) is approximated by the following finite difference method over the grid of the step size $h=1/n$:

$$\begin{aligned} -\varepsilon D_+D_-y_i + y_i &= 0, \quad i=1,\dots,n-1 \\ y_0 = y_n &= a, \end{aligned} \tag{11}$$

where $D_+y_i = (y_{i+1} - y_i)/h$, $D_-y_i = (y_i - y_{i-1})/h$. The approximation (11) is of the matrix form $Ay = f = Bv$, where

$$y = \begin{pmatrix} y_0 \\ \vdots \\ y_n \end{pmatrix}, \quad A = \begin{pmatrix} 1 & 0 & & & \\ -\varepsilon & 2\varepsilon+h^2 & -\varepsilon & & \\ & \ddots & \ddots & \ddots & \\ & & -\varepsilon & 2\varepsilon+h^2 & -\varepsilon \\ & & & 0 & 1 \end{pmatrix}, \quad f = Bv = \begin{pmatrix} 1 & & & \\ & h^2 & & \\ & & \ddots & \\ & & & h^2 \\ & & & & 1 \end{pmatrix}\begin{pmatrix} a \\ 0 \\ \vdots \\ 0 \\ a \end{pmatrix},$$

putting $w=0$ for $C=B$ and an arbitrary $\tau > 0$. The requirements of Theorem 1 are fulfilled, thus,

$$\begin{aligned} &\text{if } a > 0, \quad \text{then} \quad 0 \leq y_i \leq a \quad \text{for } i=0,\dots,n, \\ &\text{if } a < 0, \quad \text{then} \quad a \leq y_i \leq 0 \quad \text{for } i=0,\dots,n. \end{aligned}$$

This shows that our maximum principle is a discrete analogue to the maximum principle of problem (10). More complicated problems and approximations, for which the maximum principle works, are available, for instance from [4] .

Example 2

For a constant $T > 0$, we consider the following initial boundary value problem on the region $(0,1)\times(0,T)$ in the (x,t)-plane:

$$\begin{aligned} \frac{\partial u}{\partial t} &= a\frac{\partial^2 u}{\partial x^2} + b\frac{\partial u}{\partial x} + g, \quad (0,1)\times(0,T) \\ u(x,0) &= \varphi(x), \quad 0 \leq x \leq 1, \\ u(0,t) &= g_0(t), \quad u(1,t) = g_n(t), \quad 0 < t \leq T. \end{aligned} \tag{12}$$

We assume that $\varphi(x)$, $g(x,t)$, $g_0(t)$ and $g_n(t)$ are sufficiently smooth functions.

Suppose for simplicity that a and b are constants, where $a > 0$. Using for the discretization a space grid of step size $h=1/n$, a time grid of step size $\tau=T/m$ and denoting the approximations by $y_i^k \approx u(x_i,t_k)$, we apply a finite difference upwind scheme, see [6]:

$$(y_i^k - y_i^{k-1})/\tau = aD_+D_-y_i^k + b^+D_+y_i^k + b^-D_-y_i^k + g_i^k, \quad i=1,..,n-1$$
$$y_o^k = g_1^k, \quad y_n^k = g_n^k, \tag{13}$$

where $b^+ = (b + |b|)/2$, $b^- = (b - |b|)/2$.

The matrix representation of the discrete problem (13) is as follows

$$A\,y = B\,v + \tau\,C\,w,$$

where

$$y=(y_o^k,..,y_n^k)^T,\ v=(g_o^k,y_1^{k-1},..,y_{n-1}^{k-1},g_n^k)^T,\ w=(0,g_1^k,..,g_{n-1}^k,0)^T,\ B=C=I,$$

$$A = \begin{pmatrix} 1 & 0 & & & \\ -\alpha & 1+2\alpha & -\alpha & & \\ & \ddots & \ddots & \ddots & \\ & & -\alpha & 1+2\alpha & -\alpha \\ & & & 0 & 1 \end{pmatrix} + \frac{\tau}{h}\begin{pmatrix} 0 & 0 & & & \\ b^- & |b| & -b^+ & & \\ & \ddots & \ddots & \ddots & \\ & & b^- & |b| & -b^+ \\ & & & 0 & 0 \end{pmatrix}, \quad \alpha = \frac{a\tau}{h^2}.$$

We see that A is a strictly row diagonally dominant tridiagonal M-matrix, and from Theorem 1, it follows that

$$\min\left\{g_o^k, g_n^k, \min_{1\le j\le n-1} y_j^{k-1}\right\} + \tau \min\left\{0, \min_{1\le j\le n-1} g_j^k\right\} \le y_i \le$$
$$\le \max\left\{g_o^k, g_n^k, \max_{1\le j\le n-1} y_j^{k-1}\right\} + \tau \max\left\{0, \max_{1\le j\le n-1} g_j^k\right\} \quad \text{for } i=0,..,n.$$

This inequality is a discrete analogue to the maximum principle of problem (12), see [5].

Example 3

The linear equation systems $Ay=f$, which arise in the discrete approximations to first kind elliptic or parabolic boundary value problems are of the following special structure, see [8]:

$$A = \begin{pmatrix} I & 0 \\ A_{21} & A_{22} \end{pmatrix}, \quad f = \begin{pmatrix} \bar{f} \\ 0 \end{pmatrix}, \qquad \bar{f}=(\bar{f}_1,..,\bar{f}_m)^T, \quad 1 < m < n,$$

I unit matrix of order m .

It is assumed that the grid has m boundary grid points at which the boundary values $\bar{f}_j$, $j=1,..,m$ are given. Furthermore, let A be a strictly row diagonally dominant M-matrix with row sums

$$0 < d_i = (Ae)_i \quad \text{for all } m+1 \le i \le n.$$

Then, we rewrite the right-hand side vector f to get the form required by Theorem 1. Introducing the matrix $B=\mathrm{diag}(1,..,1,d_{m+1},..,d_n)$, we have $f=Bv$, where $v=f$, putting $w=0$. Applying now Theorem 1, it follows that the solution components y_i satisfy

$$\min\left\{0, \min_{1\le j\le m} \bar{f}_j\right\} \le y_i \le \max\left\{0, \max_{1\le j\le m} \bar{f}_j\right\} \quad \text{for all } i \in N.$$

Example 4

We consider the linear equation system Ay=f, where it is assumed that $f>0$ and A=tridiag(-1,2,-1). In this case, the matrix A is an irreducible weakly row diagonally dominant M-matrix, which follows, for instance, from the finite difference approximation of $-u''(x)$.
We can apply Theorem 4, putting B=C=I, w=0 and D=diag(0,1,...,1,0). Then, we have

$$0 < \min_{j\in N} f_j \leq y_i \quad \text{for all } i\in N ,$$

which coincides with the statement for an irreducible M-matrix that there holds $A^{-1}>0$, see [9].

5. Remarks

An example is given in [11], where Theorem 3 is applied. The maximum principle is meaningful in the application of the finite element method, see for instance [5]. Under certain conditions, it is applicable to iteration methods of the type $Qy^{k+1} = Ry^k + f$ to solve linear equation systems Ay=f, and to nonlinear problems A(y)y=f, where A(y) is a diagonally dominant M-matrix, for any vector y.

References

[1] Adams,E.,Ames,W.F., On contracting interval iteration for nonlinear problems in R^n, Part I: theory, Nonlinear Analysis TMA 3, 1979, 773-794

[2] Adams,E.,Ames,W.F., On contracting interval iteration for nonlinear problems in R^n, Part II: applications, Nonlinear Analysis TMA 5, 1981, 525-542

[3] Berman,A.,Plemmons,R.J., Nonnegative matrices in the mathematical sciences, Academic Press, New York, 1979

[4] Doolan,E.P.,Miller,J.J.H.,Schilders,W.H.A., Uniform numerical methods for problems with initial and boundary layers, Boole Press, Dublin, 1980

[5] Ikeda,T., Maximum principle in finite element models for convection-diffusion phenomena, North-Holland Publishing company, Amsterdam, New York, Oxford, 1983

[6] Samarski,A.A., The theory of difference schemes, Nauka, Moscow, 1977 (russ.)

[7] Stoyan,G., On a maximum principle for matrices and on conservation of monotonicity with applications to discretization methods, ZAMM, 62, 1982, 375-381

[8] Stoyan,G., On maximum principles for monotone matrices, Linear Algebra Appl., in print

[9] Varga,R.S., Matrix iterative analysis, Prentice Hall, Englewood Cliffs, 1962

[10] Varga,R.S., On diagonal dominance arguments for bounding $\|A^{-1}\|_\infty$, Linear Algebra Appl., 14, 1976, 211-217

[11] Windisch,G., A discrete maximum principle, Technische Hochschule Karl-Marx-Stadt, Preprint Nr. 5, 1986

[12] Young,D., Iterative solution of large linear systems, Academic Press, New York, 1971

THE DEVELOPMENT OF COMPUTER ARITHMETIC

Ulrich Kulisch *)

The first electronic computers were developed in the middle of this century. Before that time, highly sophisticated electromechanic computing devices were used. Several very interesting techniques provided the four basic operations addition, subtraction, multiplication and division. Many of these calculators were able to perform an additional operation which could be called "accumulating addition/subtraction" or continued summation. The machine was equipped with an input register of about 10 to 13 digits. Compared to that the result register was much broader and had perhaps 30 digits. It was situated on a sled which could be shifted back and forth relatively to the input register. This allowed an accumulation of a large number of summands into different positions of the result register. There was no rounding executed after each addition. As long as no overflow occurred this accumulating addition/ subtraction was error free. Addition was associative, the result was independent of the order in which the summands were added.

This accumulating addition without intermediate roundings was never built for electronic computers. Only recently several /370 compatible systems appeared which simulate this process on general purpose machines by making use of the cache memory. This allows the elimination of a large number of roundings and contributes essentially to the stability of the computational process.

The first electronic computers executed their calculations in fixed-point arithmetic. Fixed-point addition and subtraction is error free. Even very long sums can be accumulated with only one final rounding in fixed-point arithmetic, if a carry counter is provided which gathers all intermediate positive or negative overflows or carries. At the very end of the summation a normalization and rounding is executed. Thus accumulation of fixed point numbers is associative again. The result is correct to one unit in the last place and it is independent of the order in which the summands are added. Fixed-point arithmetic, however, imposed a scaling requirement. Problems are to be pre-processed by the user so that they can be accomodated by the fixed-point number representation.

*) Institut für Angewandte Mathematik, Universität Karlsruhe, Kaiserstraße 12, D-7500 Karlsruhe.

With increasing speed of computers, problems that could be solved, became larger and larger. The necessary preprocessing soon became an enormous burden.

The introduction of floating-point representation in computation largely eliminated this burden. A scaling factor is appended to each number in floating-point representation. The arithmetic itself takes care of the scaling. Multiplication and division require an addition respectively subtraction of the exponents which may result in a big change in the value of the exponent. But multiplication and division are relatively stable operations in floating-point arithmetic. Addition and subtraction, in contrary, are troublesome in floating-point.

As an example let us consider the two floating-point vectors

$$x = \begin{bmatrix} 10^{20} \\ 1223 \\ 10^{24} \\ 10^{18} \\ 3 \\ -10^{21} \end{bmatrix} \qquad y = \begin{bmatrix} 10^{30} \\ 2 \\ -10^{26} \\ 10^{22} \\ 2111 \\ 10^{19} \end{bmatrix}$$

A computation of the inner or scalar product of these two vectors delivers

$$x \bullet y = 10^{50} + 2446 - 10^{50} + 10^{40} + 6333 - 10^{40} = 8779$$

Most digital computers will return zero as the answer although the exponents of the data vary only within 5 % or less of the exponent range of large systems. This error comes about because the floating-point arithmetic in these computers is unable to cope with the large digit range required for this calculation.

Floating-point representation and arithmetic in computers was introduced in the middle of this century. Computers then were relatively slow and able to execute about 100 floating-point operations per second. The fastest computers today are able to execute billions of floating-point operations per second. This is a gigantic gain in speed by a factor of 10^7 over the electronic computers of the early fifties. Of course, problems that can be dealt with, became larger and larger. The question is, whether floating-point representation and arithmetic which fails already in simple calculations, as illustrated above, are still adequate to be used in computers of such gigantic speed for huge problems.

We think that the set of floating-point operations should be extended by a fifth operation, the "accumulating addition/subtraction" without intermediate rounding, an operation which was already available on many electromechanic calculators. This additional operation can be executed with extreme speed. In the simplest case it can be realized by adding the floating-point summands into a fixed-point number over the full floating point range. Thus "accumulating addition/subtraction" becomes error free. Even very long chains of additions/subtractions can be executed with only one single rounding at the very end of the summation. The result of "accumulating addition/subtraction" is correct to one unit in the last place and it is independent of the order in which the summands are added.

The fifth operation "accumulating addition/subtraction" combines the advantages of fixed-point arithmetic - error free addition and subtraction even for very long sums - with the advantages of floating-point arithmetic - no scaling requirements.

With the "accumulating addition/subtraction", in particular scalar products can be precisely accumulated. The scalar or inner product turns out to be a key operation for obtaining highly accurate results in scientific computation.

As the next step then it can be shown that with the optimal scalar product and different roundings (rounding to the nearest floating-point number, downwardly and upwardly directed roundings,etc.) all operations in the most commonly used linear spaces can be performed correctly to one unit in the last place on digital computers. These spaces are: the real and complex numbers, the real and complex vectors and matrices, as well as their interval correspondents: the real and complex intervals and the real and complex interval vectors and interval matrices. The operations in these spaces enlarge the set of basic computer operations considerably. This enlarged set of arithmetic operations is sometimes called "advanced computer arithmetic".

Advanced computer arithmetic changes the interplay between numerical analysis and computer arithmetic in a qualitative way. Very general methods and concepts have been developed which solve practically all standard problems of numerical analysis with high accuracy and provide upper and lower bounds for the solution in general with full machine accuracy. Such methods exist, for instance, for: linear systems of equations, inversion of matrices, eigenvalue and eigenvector problems, problems with sparse matrices, polynomial evaluation, zeros of polynomials, evaluation of rational arithmetic expressions, optimization problems, nonlinear systems of equations, classes of initial and boundary value problems of ordinary differential equations and so on.

These methods do more than provide highly accurate bounds for the solution. They also permit the computer to verify simultaneously the existence and uniqueness of the solution within the computed bounds. If there is no unique solution, for instance in the case of singular matrix, the computer delivers this information to the user. These methods which allow the computer to verify or validate the correctness of computed results open a new dimension for scientific computation.

Whenever a standard problem of numerical analysis can be solved with high accuracy and automatic verification of the correctness of the computed result it can be interpreted as a higher order arithmetic operation. The user does not need to execute an error analysis for this part of a problem. He can use this routine as a higher order basic operation in his computation.

There is still another, still more important aspect which comes about with the possibility of automatic verification of the correctness of computed results by the computer itself. Similar to the meaning of the conditional statement for an automatic program execution is the automatic result verification or validation if a numerical expert system is going to be built up.

If a computer user or an engineer has a certain problem to solve he is, in general, only interested in the solution and not in the method which solves his problem. A clear mathematical description of the problem and the interpretation of the solution is all he is primarily interested in. If numerical analysts teach him numerical methods they - by the way - put the responsibility and the burden to prove the correctness of the computed result on his shoulders.

For many problems it is certainly possible to store the information which algorithm is most favourably used in the computer itself. If the computer then gets a certain problem to solve, it can choose among the various criteria the particular method which seems to be best suited to obtain the solution. The computer should now be able to detect by itself whether the computed solution is correct or not. This requires methods for an automatic verification of the correctness of the computed result. If the answer delivered by such a method is "okay", the problem is solved. If the method failed to find the solution the computer may choose a more elaborate algorithm or start a conversation how to proceed with the user.

Methods which finally verify the correctness of computed results or so called self-validating methods are certainly necessary for the development of what is called automatic problem solving environments.

We conclude our brief sketch of the development of computer arithmetic with a survey of the actual implementations of the methodology which are already in existence. We remark first that a more detailed introduction into the ideas sketched in this article as well as references to the corresponding literature can be found in: U.W. Kulisch and W.L. Miranker: "The Arithmetic of the Digital Computer: A New Approach", SIAM Review, Vol. 28, No. 1, March 1986.

Basic, advanced and higher order arithmetic routines embedded in an extended PASCAL environment, called PASCAL-SC, are available for several micros, like the ZILOG Z-80, the MOTOROLA 68 000 and the INTEL 8086 processor series. Several hardware units have been developed which support all advanced computer arithmetic routines in PASCAL-SC environments.

IBM has adopted these ideas into its product line. Basic arithmetic, including correct continued addition/subtraction, is available in all IBM System/370 computers; on several such systems as for instance the 4361 or the 9377 by hardware. Based on this, many parts of advanced and of higher order computer arithmetic routines are available in the form of subroutines and program packages. A commercially available such package is called ACRITH. Whenever such a package delivers a result it is verified to be correct by the computer. ACRITH can also be used on any system /370 compatible computer, as for instance on Hitachi or NAS systems. Meanwhile, Siemens has followed IBM's lead with its own development of an ACRITH-like capability called ARITHMOS. It is expected that these features will shortly find their way into the super computing environment.

Bei Fragen zur Produktsicherheit wenden Sie sich bitte an:
If you have any questions regarding product safety,
please contact:

Walter de Gruyter GmbH
Genthiner Straße 13
10785 Berlin
productsafety@degruyterbrill.com